Basics for Builders:

PLAN READING & MATERIAL TAKEOFF

Wayne J. DelPico

RSMeans

Basics for Builders:

PLAN READING & MATERIAL TAKEOFF

- *Step by step instructions— from site work through electrical*
- *Takeoff procedures based on a full set of working drawings*
- *Includes Light Commercial*

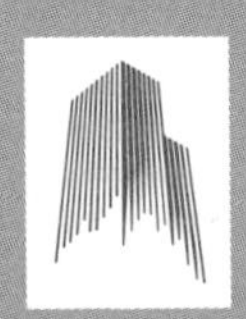

Wayne J. DelPico

R.S. MEANS COMPANY, INC.
CONSTRUCTION PUBLISHERS & CONSULTANTS

100 Construction Plaza
P.O. Box 800
Kingston, MA 02364-0800
(617) 585-7880

The editors for this book were Howard Chandler, Kevin Foley, and Suzanne Morris; the managing editor was Mary Greene; the production coordinator was Marion Schofield. Composition was supervised by Karen O'Brien. The book and jacket were designed by Norman R. Forgit.

Printed in the United States of America

10 9 8 7 6 5 4

Library of Congress Catalog Number 95-113562

ISBN 0-87629-348-8

Dedication

To my wife Kris, for support and encouragement.
To my daughters Maria-Laina and Kristina, for missed bedtime stories.
To my parents Arman and Gloria, for inspiring a thirst for knowledge.

TABLE OF CONTENTS

FOREWORD

Somewhere along the line, nearly everyone who works in construction has to convey or gather information from a set of drawings. This book explains and demonstrates the fundamentals of plan reading and takeoff. While there can be no substitute for experience, the information and techniques presented here will give the beginner a tremendous advantage. It is also very helpful for established contractors who wish to better organize their operations based on an accepted, efficient standard. In today's fast-paced industry, decisions must be made quickly and with confidence, and an error in interpreting the drawings can cost thousands of dollars. A solid knowledge of the practices conveyed in this book will allow you to solve the unknowns and to move things along.

In my years in the construction industry, I have learned that there is rarely only one right way to do something. Construction drawings and specifications convey what the owner wants. The experienced plan reader is able to interpret the drawings in relation to the specifications and perform whatever function is required, whether it is taking off material quantities, providing an estimate, or building the project. The contractor or estimator must decide how to approach the drawings in order to accomplish whichever of these tasks is at hand.

Plan reading is, of course, directly related to estimating. Depending on the purpose of the estimate, the plan reader must determine how much detail will be required and in which format the information will be recorded. To create a unit price estimate, the contractor must have a basic understanding of all trades and the work they perform. The unit price estimate must be carefully considered and directly related to the plans, as it will serve as the basis for ordering materials and setting the project schedule.

Perhaps the most important benefit of being able to read and interpret plans is being able to communicate effectively about the building process. As you become more and more familiar with plans and specs, you can develop your own style of plan reading and takeoff. You will discover that you have the ability to communicate through drawings as well, and will consequently have better control over the project.

It is important to remember that plan reading is a skill that, like any other, requires practice and constant application if one is to excel. Starting with this text you will have at least one standard, correct method from which you can build your knowledge and skills. No matter how advanced construction methods become, you will continue to depend on your fundamental knowledge of plan reading to get the job done.

Howard M. Chandler

Howard Chandler, a contractor for over 15 years, has managed residential, commercial and industrial projects. He was Associate Professor at Wentworth Institute of Technology College of Design & Construction, and Director of Education for the Massachusetts/Rhode Island Associated Builders and Contractors, where he organized and taught construction courses, including Plan Reading. He is currently a National Board Member of the Commercial Builders Council of the National Association of Home Builders, and a member of the Associated Schools of Construction. He is a former editor of Means Residential Cost Data *and* Light Commercial Cost Data.

ACKNOWLEDGMENTS

The author would like to express sincere thanks to Kevin Foley, Howard Chandler, and Suzanne Morris of the R. S. Means Company for their individual contributions at various stages of the book's development.

A special debt of gratitude to Mary Greene, Managing Editor of Means Reference Books, whose patience, guidance, and overall professionalism never wavered from start to finish.

I would also like to acknowledge the contribution of Home Planners, Inc.®, who provided the full set of plans for the sample building project and several of the detail drawings. Home Planners is the leading provider of home plans through nationally distributed books and magazines.

INTRODUCTION

This book was created as a tool for contractors who wish to learn organized, efficient, standard methods for plan reading and takeoff, or who want to improve their skills. For contractors or estimators who are new to the profession, *Plan Reading and Material Takeoff* is a basic resource that conveys the fundamentals of tasks that are an essential part of their work. Experienced contractors will find the book useful for reviewing and enhancing their knowledge of plan reading and takeoff for their own and other various trades. Because the book outlines accepted practices, along with tips from experienced professionals, it also offers a standard that can be implemented to ensure that all employees are working in a uniform, organized way on any given project. Subcontractors will benefit from using the information in this book as a guide to where they fit into the overall construction process.

Much of the book is organized according to the Construction Specifications Institute's MasterFormat, the most widely recognized and used system of organizing construction information. Following the fundamentals of plan reading, instructions for calculating area and volume, and a review of the project specifications in Chapters 1–3, each MasterFormat "Division" has a chapter of its own. In this way, all categories of construction materials are covered, from site work through electrical. Each chapter describes how a material or construction item is likely to appear on the plan, and how that information can be translated into a reliable quantity takeoff for the estimate that will follow. Each chapter includes not only the basic rules and principles for plan reading and takeoff, but also offers guidance based on the "real-life" problems and considerations contractors encounter.

The book includes a complete set of residential building plans, courtesy of Home Planners, Inc.®, as part of a sample project to illustrate plan reading and material takeoff methods. Individual drawings are shown in the appropriate chapters, and filled-in quantity takeoff sheets are presented for all elements included in the sample project. Since this book includes both residential and light commercial construction, it would be difficult, if not impossible, to select one sample project that would adequately illustrate all facets of plan reading and takeoff. For this reason, the book also includes many additional illustrations that are not part of the sample project, to support the text that describes plan reading and takeoff methods for all common construction items. (This is especially true in divisions that do not typically relate to residential construction.) Consequently, not all of the illustrations shown are part of the sample project and therefore are not recorded on the quantity takeoff sheets at the end of the chapters.

The Appendix contains a comprehensive list of symbols and abbreviations that are found on construction drawings. This section is referred to often throughout the chapters. The Appendix also contains the full set of actual house plans that are used to demonstrate the takeoff of materials in Chapters 5 through 19.

Chapter One

READING AND UNDERSTANDING PLANS

Chapter One

Reading and Understanding Plans

Before construction begins, a process must occur that translates the owner's wishes and dreams into reality. That process is the design of the project. At the risk of oversimplifying a complex and highly specialized profession, this chapter will present an outline of how the *working drawings* come to be.

General Information

The process usually begins with the need or desire for more or different space. Collecting the information required to design that space is the responsibility of the architect or design team. The initial meetings between owner and designer require answers to some very basic questions:

- What is the intended function of the structure?
- Who will occupy it?
- What is the budget?

Answers to these questions enable the designer to establish project parameters.

Stages of Drawings

Initial drawings are used to show a basic scheme or "how the building will work." These are referred to as *schematic drawings* and are of a conceptual nature. The schematic drawings are the first graphic representation of the owner's needs, spatial requirements, and how the building will ultimately function.

The drawings are presented to the owner for review. Resulting changes are incorporated into the next phase: *preliminary drawings*. The preliminary drawings provide a graphic view of the project, more refined detail of how the proposed project will look and work, often showing elevations and the key design theme of the building.

Preparing *working drawings* represents the final step in the design process. The completed drawings become a "set" that incorporates all the adjustments, changes, and refinements made by the designer as he or she turns the schematic drawings into working drawings. Working drawings should be in compliance with all regulatory agency requirements. They include all the detail that the contractor will need to prepare a detailed estimate. (See Appendix B for a set of working drawings that will be used for the sample takeoff project throughout this book.) The working drawings are organized so that each particular scope of work is represented:

- *Architectural drawings* show the layout of the project: floor plans, elevations and details.
- *Structural drawings* depict how the various load-carrying systems will be built.
- *Mechanical/electrical drawings* show the physical plant of the structure such as lighting, power, plumbing, fire protection and HVAC.
- *Site drawings* show the relationship of the structure to the property it will occupy, including various site improvements such as sanitary system, utilities, etc.

The intentions of the design team are presented in several ways on the drawings—in *plan views, sections, elevations, details,* and *schedules.* Written instructions called *specifications* are prepared and issued by the architect as part of the project. For projects of a more limited scope, the specifications may be printed on the drawings. The specifications will be explained in detail in Chapter 3.

The Cover Sheet

The cover sheet is one of the most important pages in a set of drawings. It contains much of the information that the estimator, and ultimately the person responsible for construction, will need. The cover sheet provides the contractor with information essential to understanding the drawings and the project as a whole. It typically lists such basic information as the name of the project; the location; and the names of the architects, engineers, owners, and other consultants involved in the design.

The cover sheet lists the drawings that comprise the set in the order they will appear. The drawing list is organized by the number of the drawing and the title of the page on which it appears.

The cover sheet also lists the specific requirements of the building code having jurisdiction over the design of the project. Information required includes the total square foot area of the structure, the use group the structure will fall under, and the type of construction.

An important element in the cover sheet is the listing of the abbreviations or graphic symbols that are used in the set. In addition, there is often a section that contains general notes for the contractor, such as "All dimensions shall be verified in the field," or "All dimensions are to face of masonry."

In the absence of a separate set of bound specifications, the cover sheet may list the general technical specifications that will govern the quality of materials used in the work. Optional information, such as a Locus Plan that locates the project with respect to local landmarks or roadways or an architectural rendering of the structure, may be part of the cover sheet.

Revisions

Often after the set of working drawings has been completed, recommendations are made for correction or clarification of a particular detail, plan, or elevation. Major changes may require the re-drafting of the entire sheet. Smaller changes are shown as a revision of the original. All changes must be clearly recognizable. Changes should be indicated with a revision marker, and circled with a scalloped line that resembles a "cloud."

The revision marker is a triangle that encloses the number of the revision. Revisions are noted in the title block, or close to it, by date and revision number (see Figure 1.1).

Conventional Nomenclature

Certain conventions have been adopted and practiced to provide standardization of drawings from one design firm to another. While most of the following conventions are widely accepted and practiced, there will always be minor deviations based on local practices. This is most apparent in the use of abbreviations and symbols. In most cases, any unfamiliar symbols and abbreviations will become clear as the contractor studies the drawings.

Title Block

The title block is located in the lower right-hand corner of the drawing, although more frequently design firms are using customized sheets that extend the title block from the lower right to the upper right hand side of the paper. Regardless of the style, the title block should include the following information: the prefixed number of the sheet (so that it may be identifiable as to the group and order in which it belongs); the name of the drawing (e.g., "First Floor Plan"); the date of the drawing; the initials of the draftsperson; and any revisions to the final set of drawings. The date and scope of the revisions should be noted within the title block; if insufficient space is available within the title block, the revisions should be noted close to it. The title block should specify whether the entire drawing

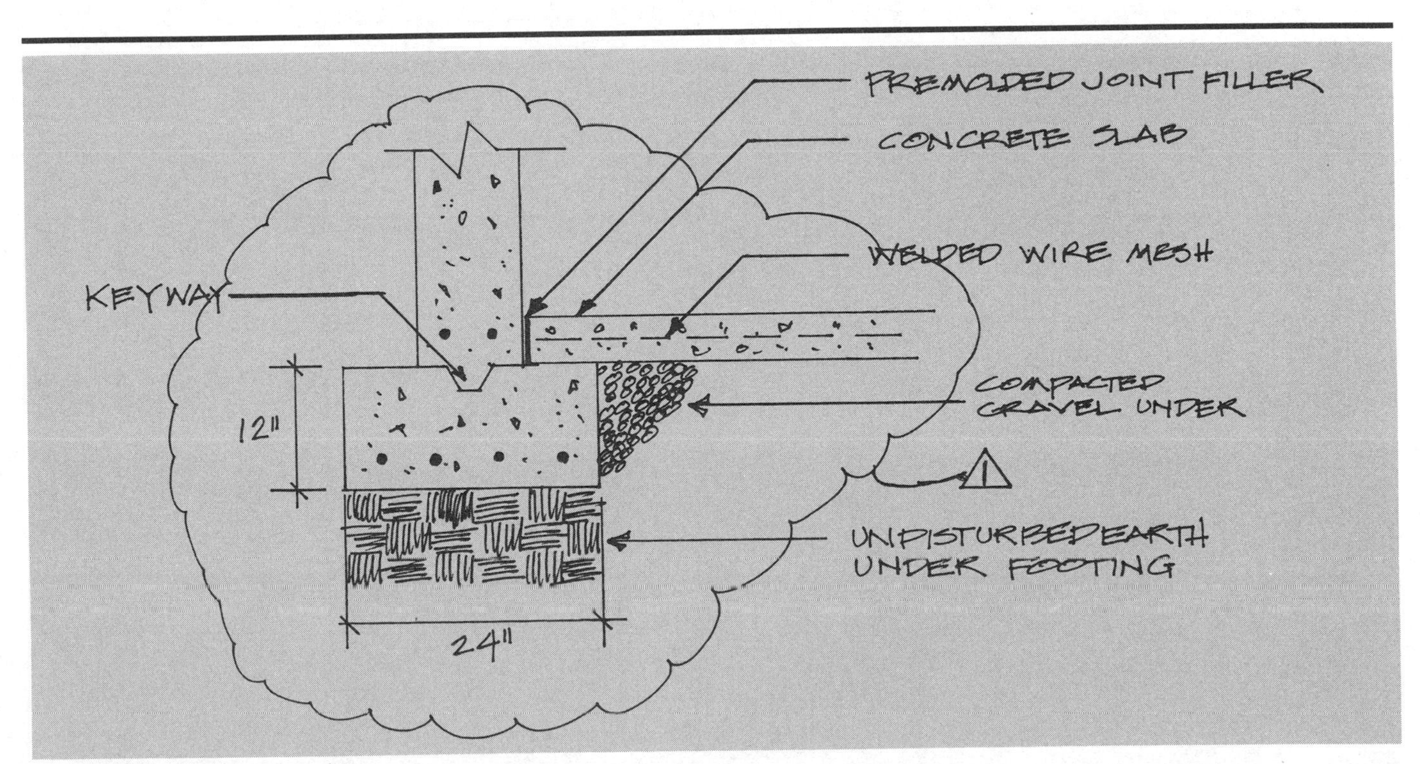

Figure 1.1 Revision Marker

is one scale—sometimes, as in the case of a sheet of details, the scale may vary per detail. See Figure 1.2 for an example of a title block.

Sets of drawings for commercial projects require the stamp of the design professional. The stamp contains the architect's or engineer's name and registration number. The signature of the individual is usually required over the stamp. See Figure 1.3 for an example.

Lines

Since drawings must indicate a great deal of information in a relatively small space, the use of words would be impractical. Other means of communicating information include the use of *lines*. The most commonly encountered lines are discussed below.

The *main object line* is the most important, as it defines the outline of the structure or object. It is a heavy, unbroken line that shows the main outlines of the wall, floor, elevation, detail, or section.

Dimension lines show the measurements of the main object lines. A dimension line is a light line with arrowheads at each end. The arrowheads fall between extension lines that extend from the main object

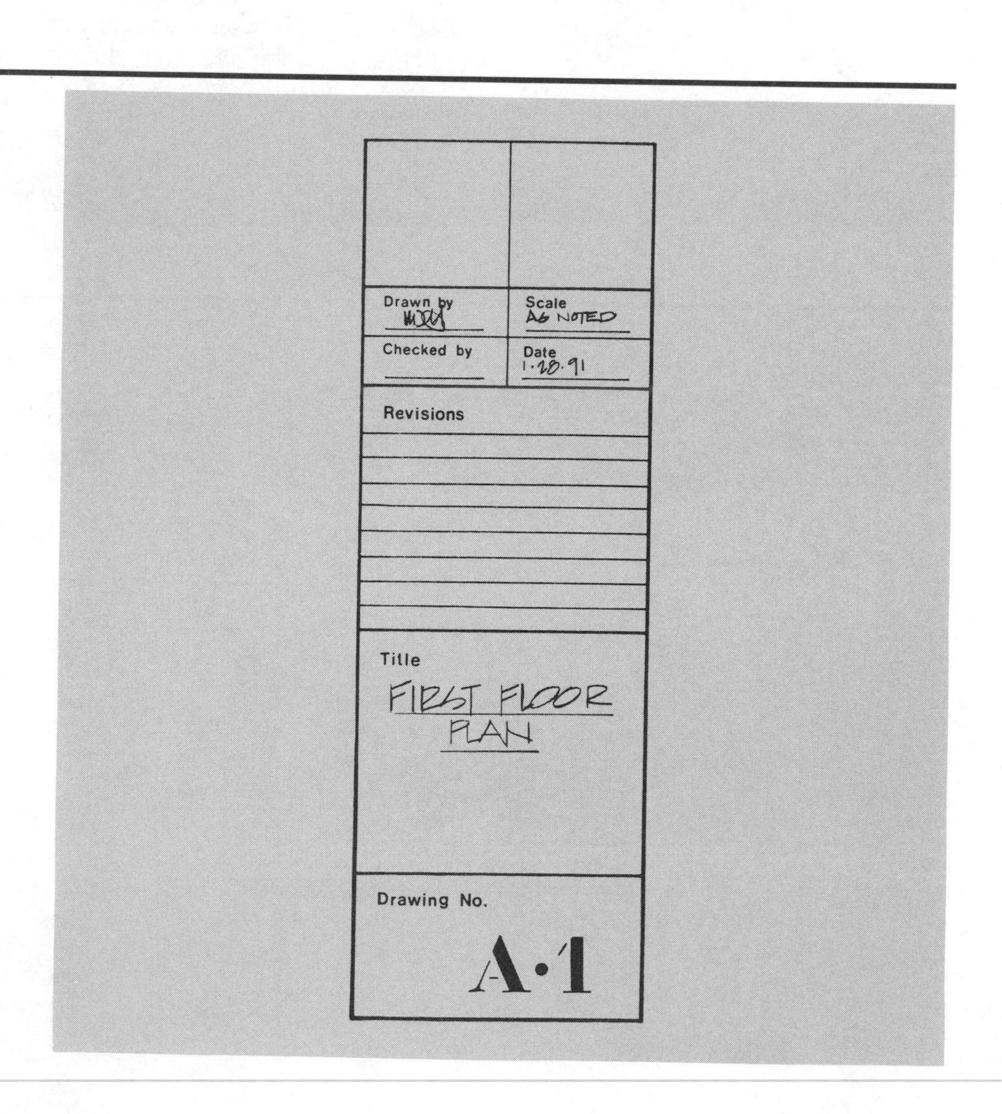

Figure 1.2 Title Block

lines to show the limits of the item drawn. The number that appears in the break in the dimension line indicates the measurement of the item indicated beween the extension lines.

Extension lines are used together with dimension lines, as mentioned above. The extension line is a light line that extends from the edge or end of the main object line; the arrowheads touch this line.

Hidden or invisible lines are light dashed lines that indicate the outlines of an object hidden from view, under or behind some other part of the structure. The dashes are typically of equal length.

Center lines are light lines of alternating long and short dashes. They indicate the center of a particular object, and are frequently labeled with the symbol of the letter C superimposed over the letter L.

Materials Indication Symbols

Another means of conveying information without words is to use Materials Indication Symbols. They allow the designer to define the composition of the object in view. Because of the different views used on drawings, various materials must be recognizable from plan to section to elevation. The Materials Indication Symbols Table in Appendix A lists some of the most widely used symbols for detailing the material composition of an item shown on the plans.

Graphic Symbols

Graphic symbols provide the reader with a standardized form of recognizing the information shown. They are indispensable tools used by the designer to depict repetitive information on the drawings. They are listed and defined in a table in Appendix A.

Trade-Specific Symbols

Like graphic symbols, trade-specific symbols depict items that are common to the various trades. Refer to Appendix A for a table listing the symbols that relate to the work of the mechanical and electrical trades.

Figure 1.3 Architect Stamp

Abbreviations

Abbreviations are used to save the designer's time as well as space on the drawings. The most common abbreviations used on working drawings are shown in Appendix A.

Types of Drawings

Scale

Since there are various physical limitations to drawing a building actual size on a piece of paper, drawings must be smaller than the actual size of the building they represent. The drawings retain their relationship to the actual size of the building using a ratio. This practice of using an accepted ratio between full size and what is seen on the drawings is called *scale*. There are two major types of scales used in reading plans.

Architect's Scale

The *architect's scale* may be either flat or three-sided. The three-sided architect's scale has ten separate scales: ⅛" and ¼", 1" and ½", ¾" and ⅜", 3⁄16" and 3⁄32", and 1½" and 3". The one remaining side is in inches similar to a ruler. For example, look at a ¼" scale. When used on a floor plan that is ¼ scale, each ¼" delineation represents one foot. The same rules apply for ⅛ scale, in that each ⅛" segment on the drawing represents 1′-0″ of actual size. The same approach applies to each of the other scales.

There is no strict convention for which scale is used on what drawings. Most floor plans and elevations are in ¼ or ⅛ scale. In some circumstances in which the building is very large, smaller scales are used. Sections can be drawn to ¼", ½", or ¾" scales, depending on the size required for clarification. Details can vary from ½" scale to 3" scale, and even full size scale for certain millwork details. Often, to conserve space, the designer will use breaks in dimension lines to describe a detail. Refer to "Break in a Continuous Line" in the Graphic Symbols, Appendix A.

Engineer's Scale

The *engineer's scale* is physically very similar to the architect's scale. The difference is the size of the increments on the sides of the scale. The engineer's scale has six scales: 10, 20, 30, 40, 50, and 60. Other specialty scales are divided into even smaller increments, such as 100. The 10 scale refers to 10 feet per inch; the 20 scale is 20 feet per inch, and so on. The engineer's scale is used to measure distance on site plans.

Occasionally, the architect includes a detail strictly for visual clarification. These details are labeled NTS, meaning "not to scale". This label tells the plan reader that these details are not for extracting quantities and measurements, but for illustration.

Civil Drawings

Plans for light commercial projects typically include a site plan because the structure is designed in accordance with the selected site. The site plan is often omitted from residential plans because the actual site may not have been determined. The site plan illustrates the relationship of the proposed structure to the building lot as well as the various improvements to the lot required to accommodate the new building. The grouping of different types of site drawings, such as utility and drainage plans, grading plans, site improvement plans, and landscaping plans, are known under the general classification of *civil drawings*. Civil drawings encompass all the work that pertains to projects other than the structure itself.

The most obvious difference between the civil drawings and the architectural drawings is the use of the engineer's scale. As mentioned earlier, smaller scales are used on site drawings to indicate the much larger areas covered. It is important to make careful note of the scale in order to avoid errors in measuring.

To avoid confusion, it is best to use the title block to clarify the type of drawing and scale.

Civil Drawing Nomenclature and Symbols

Like architectural drawings, civil drawings have symbols and graphics that convey intent with a minimum of words. Some of the symbols and graphics are unique to site plans, and others are very similar to those used on the architectural plans. Some civil drawings offer a legend to decipher the symbols and graphics on a specific set of drawings. The following is a discussion of the more common terms and symbols associated with civil drawings.

Site Plan

The main purpose of the basic site plan is to locate the structure within the confines of the building lot. Even the most basic of site plans clearly establish the building's dimensions, usually by the foundation's size and the distances to the respective property lines. The latter, called the *setback dimensions*, are shown in feet and hundredths of a foot, versus feet and inches on the architectural drawings. For example, the architectural dimension of 22′-6″ would be 22.50′ on a site plan.

To obtain the most information about the site in preparation for the site design, a *site survey* is performed by a registered land surveyor. The land surveyor records special conditions present on the building lot. This includes locating existing natural features such as trees or water, as well as man-made improvements such as walks, paving, fences, or other structures. The new site plan shows how the existing features are kept, modified, or removed to accommodate the new design.

Another chief purpose of the site plan is to show the unique surface conditions, or *topography*, of the lot. The topography of a particular lot may be shown right on the site plan. For projects in which the topography must be shown separately for clarity, a grading plan is used. The topographical information includes changes in the elevation of the lot such as slopes, hills, valleys, and other variations in the surface. These changes in the surface conditions are shown on a site plan by means of a *contour*, which is a line connecting points of equal elevation. An elevation is a distance above or below a known point of reference, called a *datum*. This datum could be sea level, or an arbitrary plane of reference established for the particular building.

The existing contour is shown as a dashed line, with the new or proposed contour shown as a solid line. Both are labeled with the elevation of the contour in the form of a whole number. The spacing between the contour lines is at a constant vertical increment, or interval. The typical interval is five feet, but intervals of one foot are not uncommon for site plans requiring greater detail (or where the change in elevation is more dramatic).

Two important characteristics of the contour need to be observed when reading a site plan.

- Contours are continuous, and frequently enclose large areas in comparison to the size of the building lot. For this reason, contours are often drawn from one edge to the other edge of the site plan.
- Contours do not intersect or merge together. The only exception to this rule is in the case of a vertical wall or plane. For example, a retaining wall shown in plan view would show two contours touching, and a cliff that overhangs would be the intersection of two contours.

A known elevation on the site for use as a reference point during construction is called a *benchmark*. The benchmark is established in reference to the datum and is commonly noted on the site print with a physical description and its elevation relative to the datum. For example: "Northeast corner of catch basin rim—Elev. 102.34'" might be a typical benchmark found on a site plan. When individual elevations, or *grades*, are required for other site features, they are noted with a "+" and the grade. Grades vary from contours in that a grade has accuracy to two decimal places, whereas a contour is expressed as a whole number.

Some site plans include a small map, called a *locus*, showing the general location of the property in respect to local highways, routes, and roads. Sophisticated site plans showing utilities and drainage services often require a legend. The legend is similar to that on the architectural drawings, listing the different symbols and abbreviations found in the particular group of site plans.

The "north arrow" clearly shows the direction of magnetic north as a reference for naming particular sides or areas of the project. In addition, the surveyor labels the property lines in accordance with the directions normally found on a compass. This reference, in the form of an angle and its corresponding distance, is called the *bearing* of a line. The bearings of the encompassing property lines are often the legal description of the building lot. Figure 1.4 is an example of a simple site plan.

Drainage and Utility Plans

Larger projects have several site plans showing different scopes of related or similar work. One such plan is a *drainage and utility plan.* Utility drawings show locations of the water, gas, sanitary sewer, and electric utilities that will service the building. Drainage plans detail how surface water will be collected, channeled, and dispersed on or off site. Drainage and utility plans illustrate in plan view the size and type of pipes, their length, and the special connections or terminations of the various piping. The elevation of a particular pipe below the surface is given with respect to its *invert.* The invert of a pipe is the bottom of the pipe trough through which the liquid flows. This is typically noted with the abbreviation for invert and an elevation; for example, "INV. 12.34'." The inverts are shown at the intersections of pipes or other changes in the continuous run of piping, such as a manhole, catch basin, sewer manholes, and so on. It should be noted that inverts are rarely provided for piping that does not pitch or have a gravity flow. With the use of the benchmark, contours, or spot elevations, the reader can quickly calculate the distance of the piping below the surface and the direction of the flow. Figure 1.5 shows a partial drainage plan between two drain manholes.

If one subtracts the elevation of INV.#2 of 12.34' from the elevation of the rim of the drain manhole DMH #2 of 18.34', the resulting difference of

6.00′ is the distance of the invert of the pipe from the rim of the manhole. If the reader performs the same calculation for DMH #1 and INV.#1, the direction of flow can be determined.

In addition to the plan view, certain site plans require clarification in the form of a detail, similar to the architectural detail. Classic examples of site details are sections through paving, precast structures, pipe trenches, and curbing. Details are not limited to scaled drafting, but occasionally appear in the form of a perspective drawing. Again, perspectives are not drawn to scale, and are used as a means of clarification only.

Landscaping Plan

Other site plans involving more specialized types of work are required for complex projects. *Landscaping plans* show the location of various species of plantings, lawns, and garden areas. The plantings are noted with an abbreviation, typically three letters, along with the quantity of the particular species. This designation corresponds to a planting schedule that furnishes a complete list of the plantings by common name, Latin or species name, and the quantity and size of each planting. Certain notes describing planting procedures or handling specifications accompany the planting schedule.

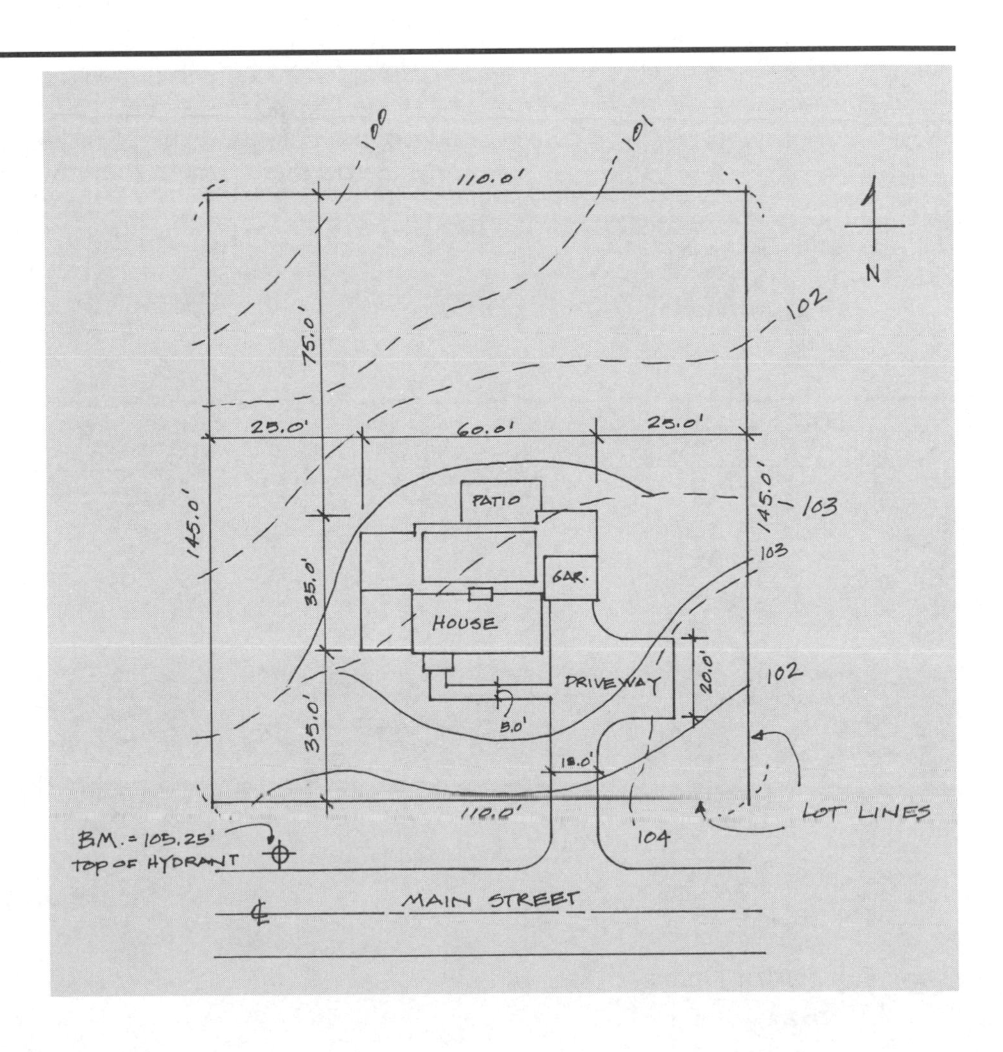

Figure 1.4 Site Plan

Figure 1.6 is an example of a typical landscaping plan and its accompanying planting schedule. Landscaping plans have additional graphics and symbols unique to the profession, some of which are listed in Appendix A.

Site Improvement Drawings

With projects of a more sophisticated nature, separate drawings showing various site improvements may be needed for clarification. These *site improvements* include such items as curbing, walks, retaining walls, pavings, fences, steps, benches, and flagpoles.

Paving and curbing plans show the various types of bituminous, concrete, and brick paving and curbing, and the limits of each. This information allows for the calculation of areas and measurements of the paving and curbing. Again, it is important to review the legend symbols in order to clearly delineate where one material ends and another begins. No assumptions should be made by the plan reader. Details showing sections through the surface are used to differentiate between thickness, and the substrate below.

Architectural Drawings

The core drawings are the *architectural drawings.* Architectural drawings are typically numbered sequentially with the prefix "A" for architectural and, in order, basement or ground-floor plans, upper-level floor plans, exterior elevations, sections, interior elevations, details, and window, door, and room finish schedules.

Plan View

One of the architect's tools is the *plan view.* The most common type of plan view is the *floor plan.* The function of the architectural floor plan is to show the use of space. Floor plans identify the locations of rooms, stairs, means of egress, and where and how rooms will be accessed. Floor plans

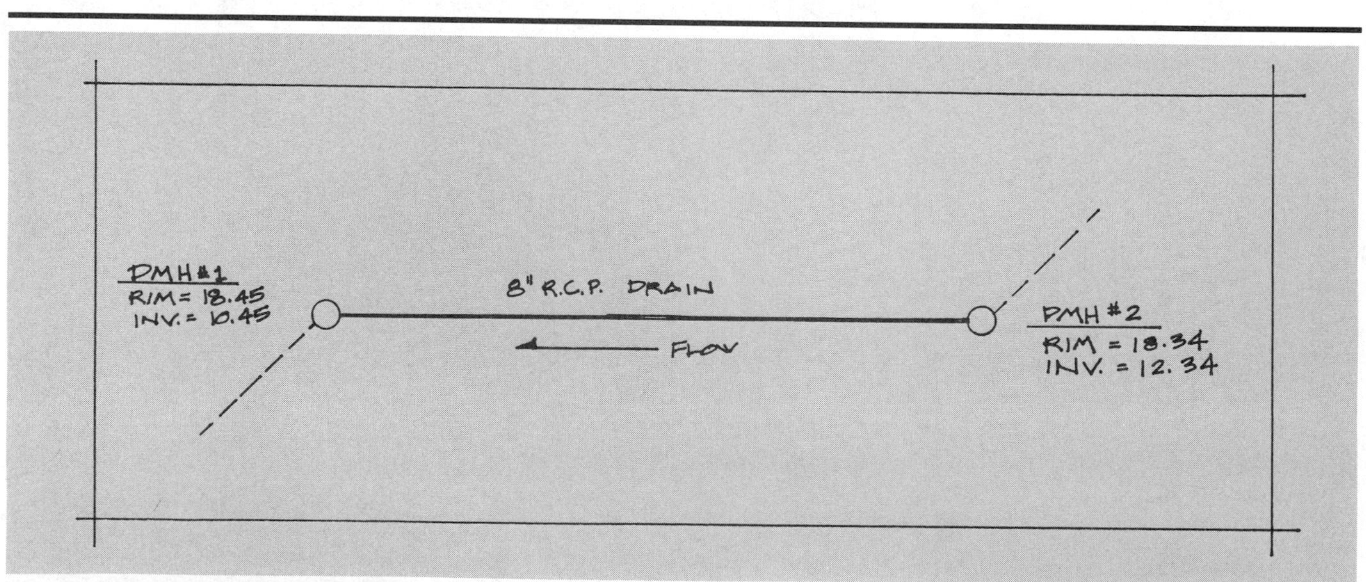

Figure 1.5 Partial Drainage Plan

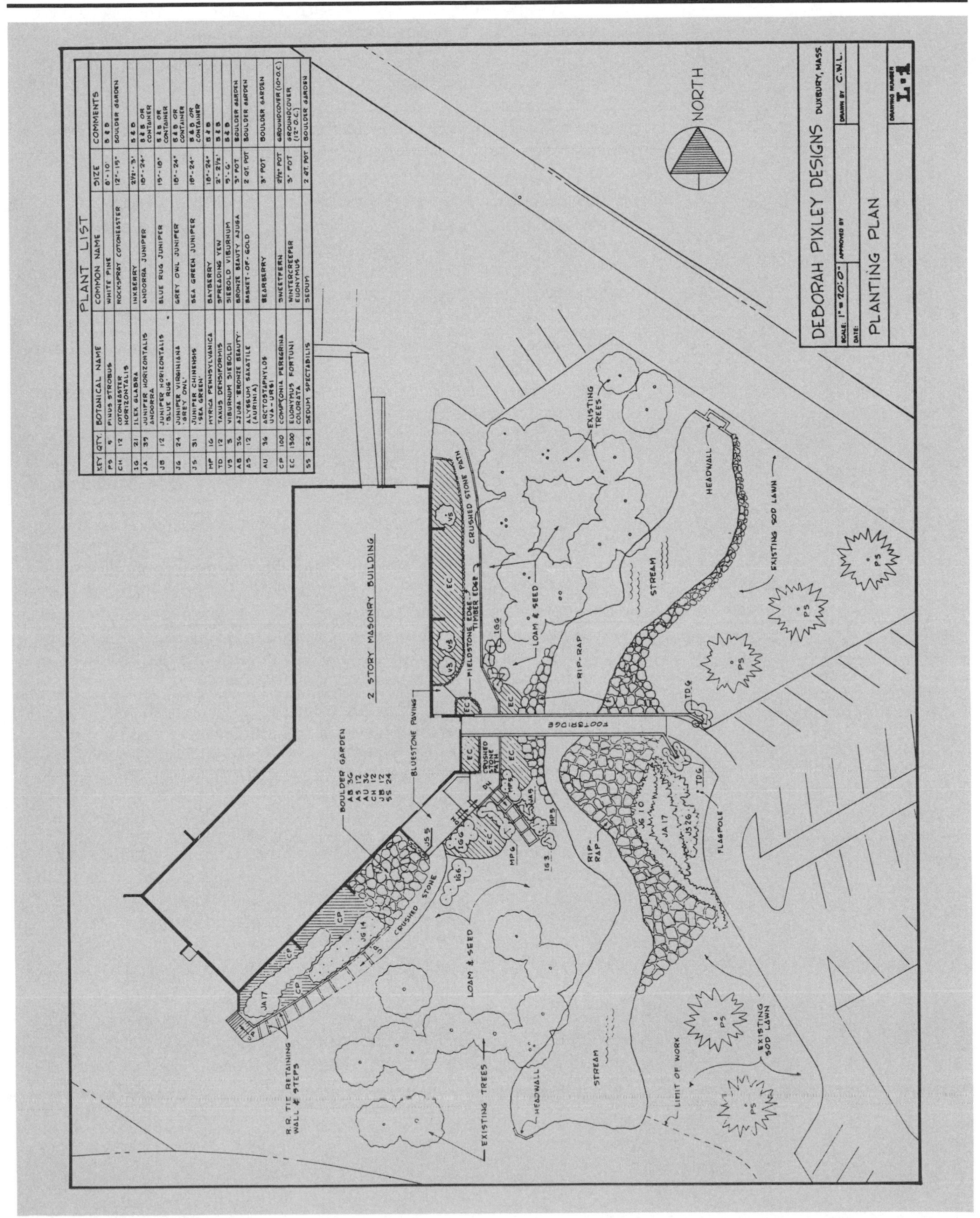

Figure 1.6 Landscaping Plan

show the locations of major features such as windows, partitions, interior doorways, and built-ins such as cabinetry and bookcases.

Architectural plans should be dimensioned to show actual length and width, thereby allowing the reader to calculate areas. Dimensions should be accurate, clear, and complete, showing both exterior and interior measurements of the space. The floor plan in Figure 1.7 shows the exterior dimensions as "outside of frame to outside of frame," while the windows are located to the centerline.

Architectural plans include notes that further define and provide information about a particular segment of work, or reference another drawing. In residential and light commercial construction, a separate set of specifications is not always issued, depending on the size of the project and financial constraints. In many cases, it is assumed that the notes will suffice.

The plan view approach is not limited to floor plans, but can be used in other plans such as roof or demolition plans to provide the same perspective. *Demolition plans* show proposed changes to the existing floor plan. *Partial floor plans* show an enlarged view of smaller areas such as bathrooms, bedrooms, kitchens, and stairs. A *reflected ceiling plan* gives a view of the ceiling from the floor, indicating the location of lights, sprinkler heads, and smoke detectors, among other items that may be located on the ceiling (see Figure 1.8).

Exterior Elevations

An important part of the architectural drawings are the *exterior elevations*. They provide a pictorial view of the exterior walls of the structure. An elevation provides a view like that seen in a photograph of an exterior wall taken perpendicular to both the vertical and horizontal planes. Elevations are not in perspective view, which makes it difficult to determine depths or changes in direction without looking at all the elevations.

Exterior elevations may be titled based on their location with respect to the headings of a compass (North Elevation, South Elevation, East Elevation, and West Elevation), or may be titled Front Elevation, Rear Elevation, Right and Left Side Elevations. The scale of the elevation is noted either in the title block or under the title of the elevation.

The function of the elevation is to provide a clear depiction of exterior doors, windows, and the facade of the building, often using numbers or letters in circles to show types that correspond to information provided in the door and window schedule. In addition, elevations show the surface materials of the exterior wall, and any changes in the surface materials within the plane of the elevation.

While the floor plan shows measurements in a horizontal plane, elevations provide measurements in a vertical plane with respect to a horizontal plane. These dimensions provide a vertical location of floor-to-floor heights, window sill or head heights, floor-to-plate heights, roof heights, or a variety of dimensions from a fixed horizontal surface. By using these measurements, the reader can calculate quantities of materials needed. Sometimes the elevation dimensions are given as decimals (10.5′ as opposed

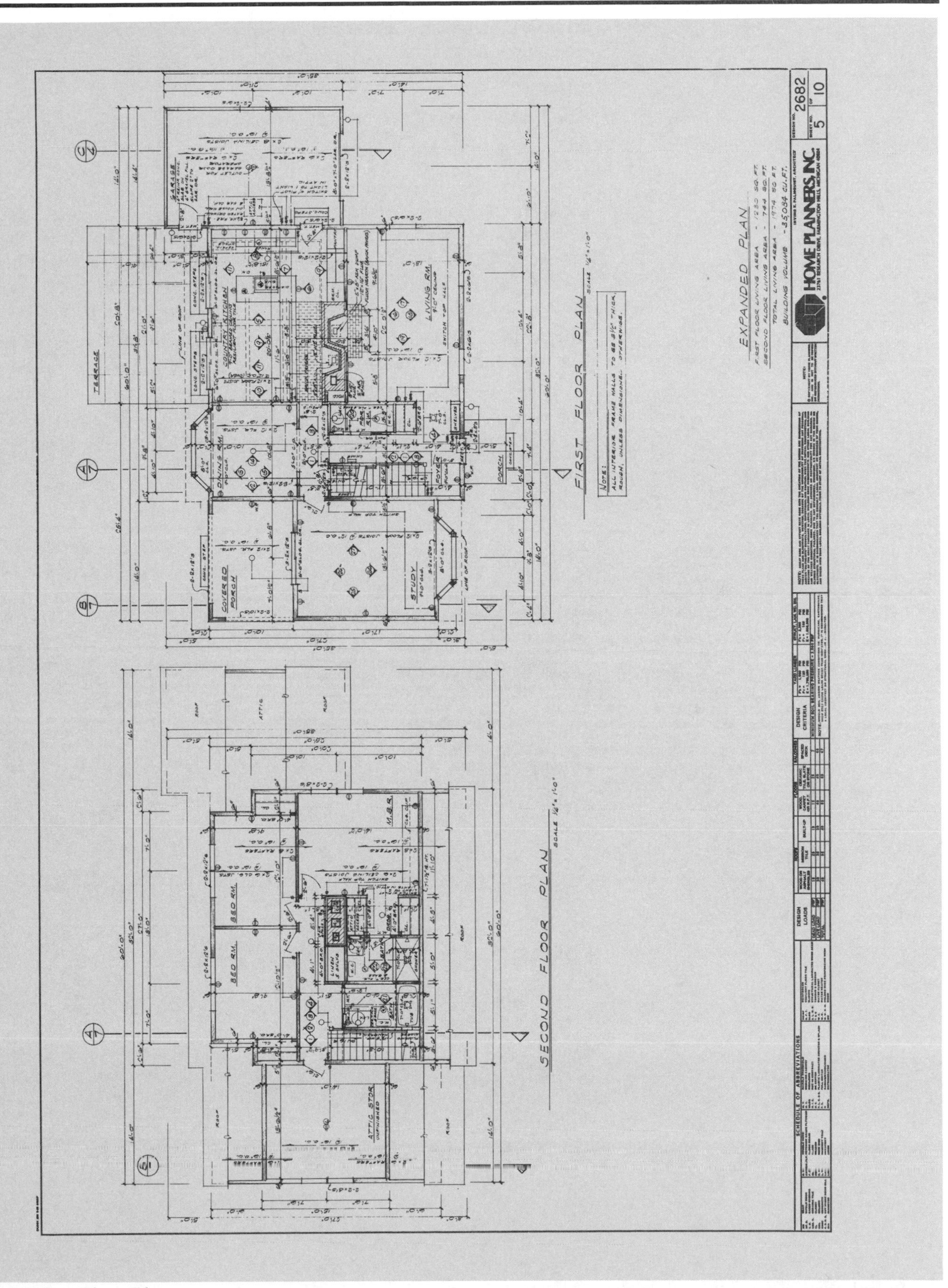

Figure 1.7 First Floor Plan

to 10′-6″). Along with the dimensions on the elevations, notes are included to help clarify what can be seen in a floor plan. Figure 1.9 is an example of an exterior elevation.

Building Section

The building section, commonly referred to as the *section,* is a "vertical slice" or cut through a particular part of the building. It offers a view through a part of the structure not found on another drawing. Several different sections may be incorporated into the drawings. Sections taken from a plan view are called *cross-sections;* those taken from an elevation are referred to as *longitudinal sections,* or simply *wall sections.* Wall sections offer the reader an exposed view of the components and the arrangement within the wall itself. Using sections in conjunction with floor plans and elevations, the reader begins to get a "feel" for how the building goes together.

The scale of the section may be different than previously encountered in the plan and elevations. A general rule is that the scale increases as the section in question becomes smaller.

The wall section in Figure 1.10 clearly shows the components that comprise the wall and how one material is connected to another to form the exterior

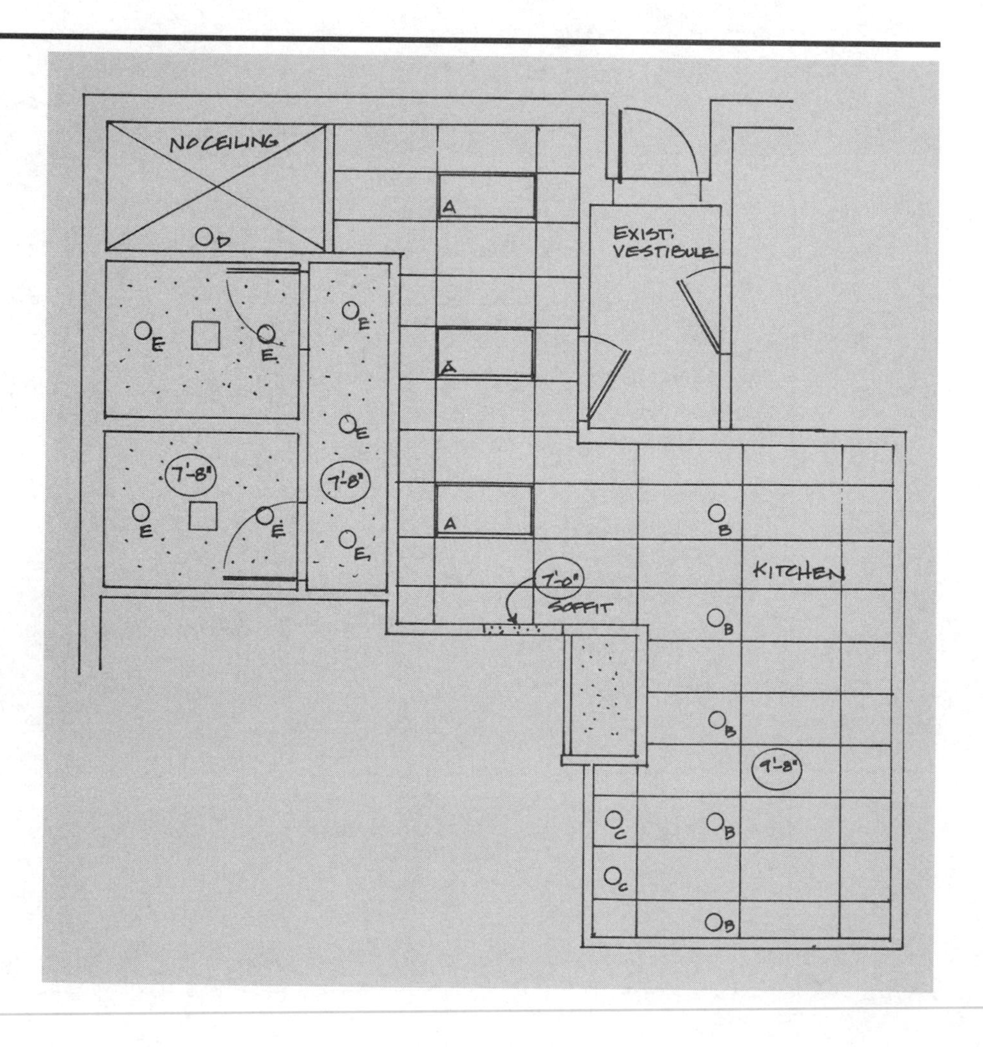

Figure 1.8 Reflected Ceiling Plan

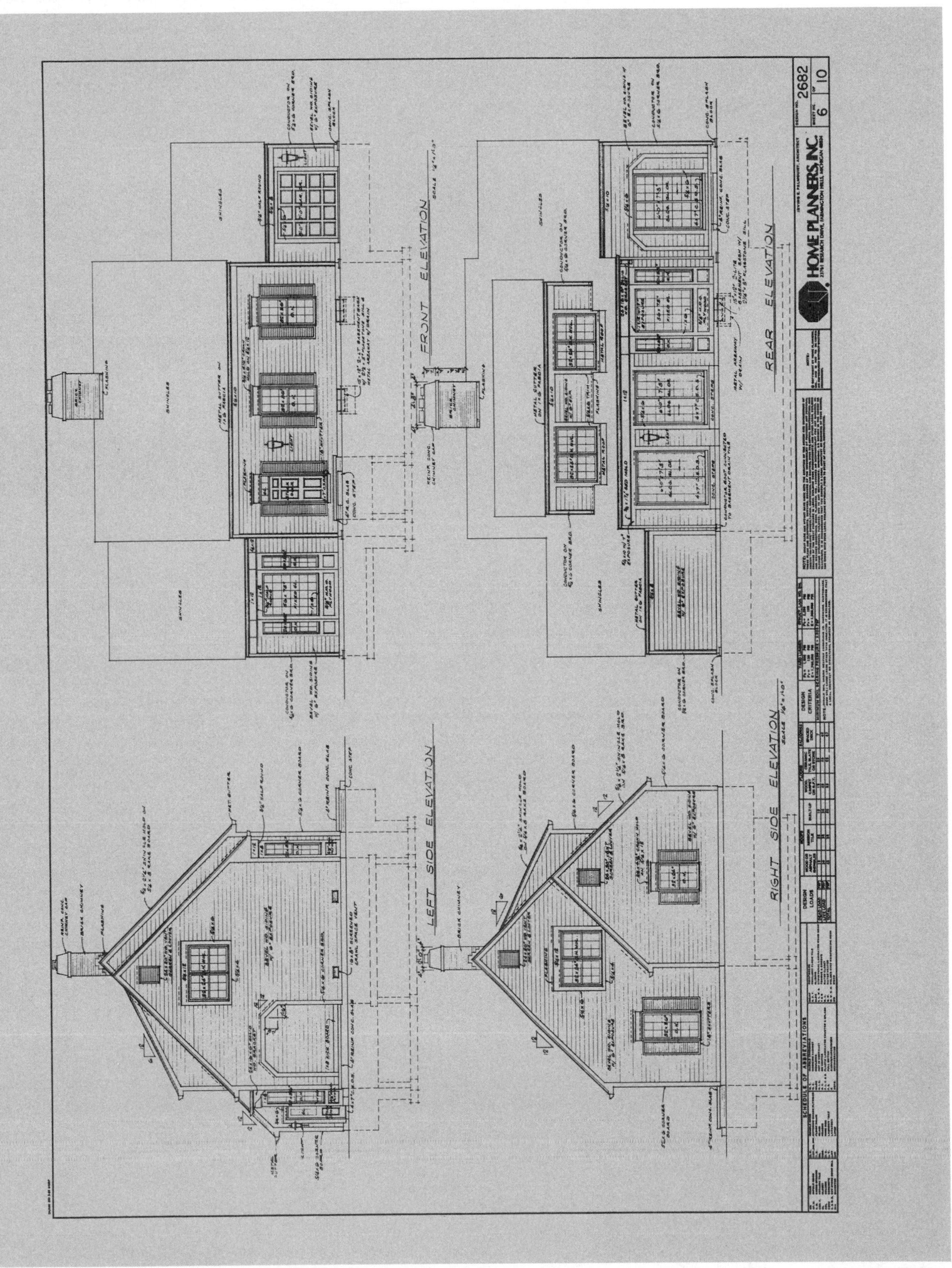

Figure 1.9 Exterior Elevations

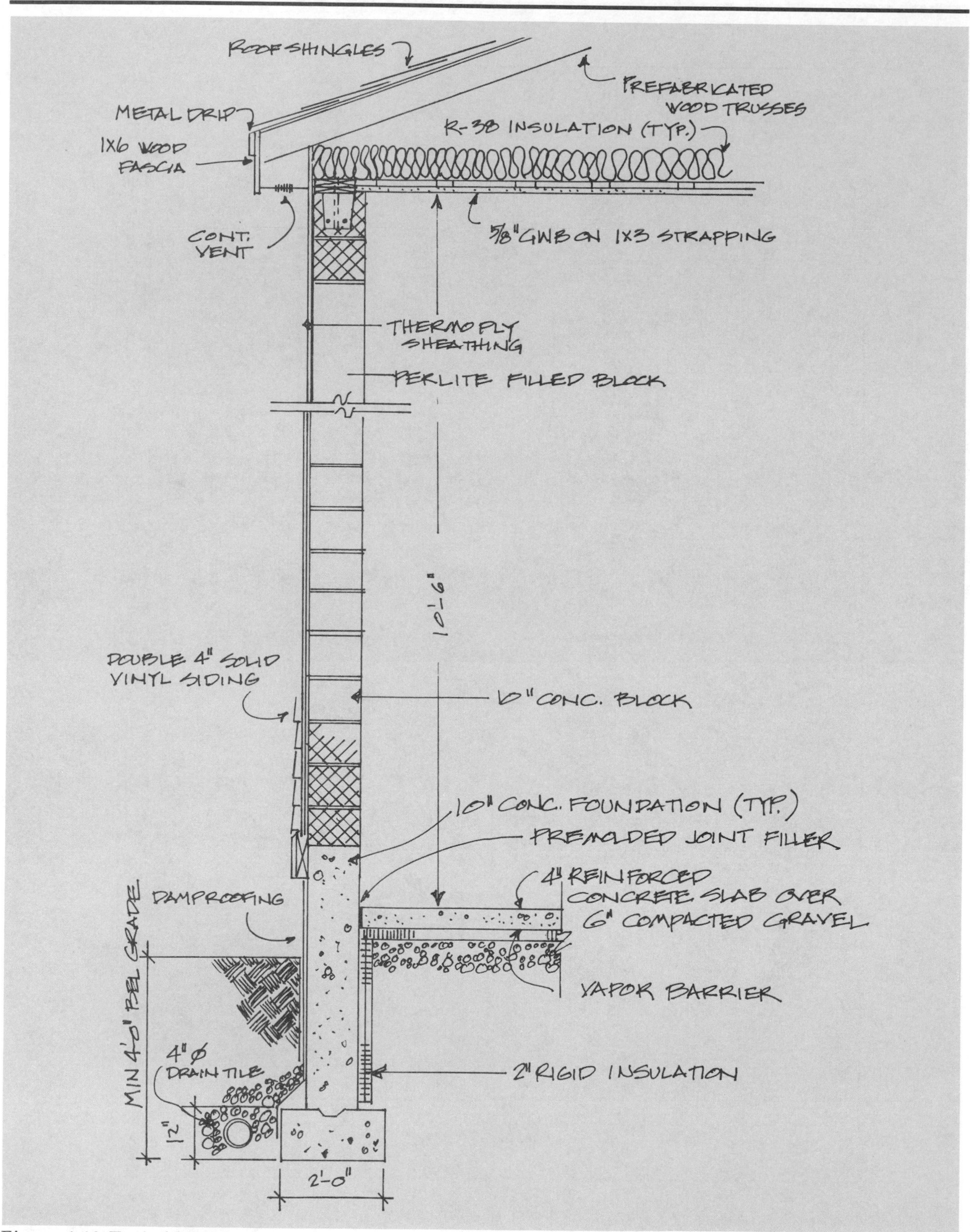

Figure 1.10 Typical Wall Section

wall of the structure. A typical wall section should provide dimensions vertically, and any relative dimensions horizontally, such as wall thickness, setbacks, overhangs, or similar changes in the vertical plane.

Details

For greater clarification and understanding, often certain areas of a floor plan, elevation, or a particular part of the drawing may need to be enlarged. Enlargements are drawn to a larger scale and are referred to as *details*. Details can be found either on the sheet where they are first referenced, or grouped together on a separate detail sheet included in the set of drawings. Details are one of the most important sources of information available to the contractor. Again, the detail is shown in larger scale to provide additional space for recording dimensions and notes. Details are not limited to architectural drawings but can be used in structural and site plans and, to a lesser degree, in mechanical or electrical plans.

Figure 1.11 shows a detail of the top portion of the wall section from Figure 1.10.

Schedules

In an effort to keep the drawings from becoming cluttered with too much printed information or details, architects have devised a system to

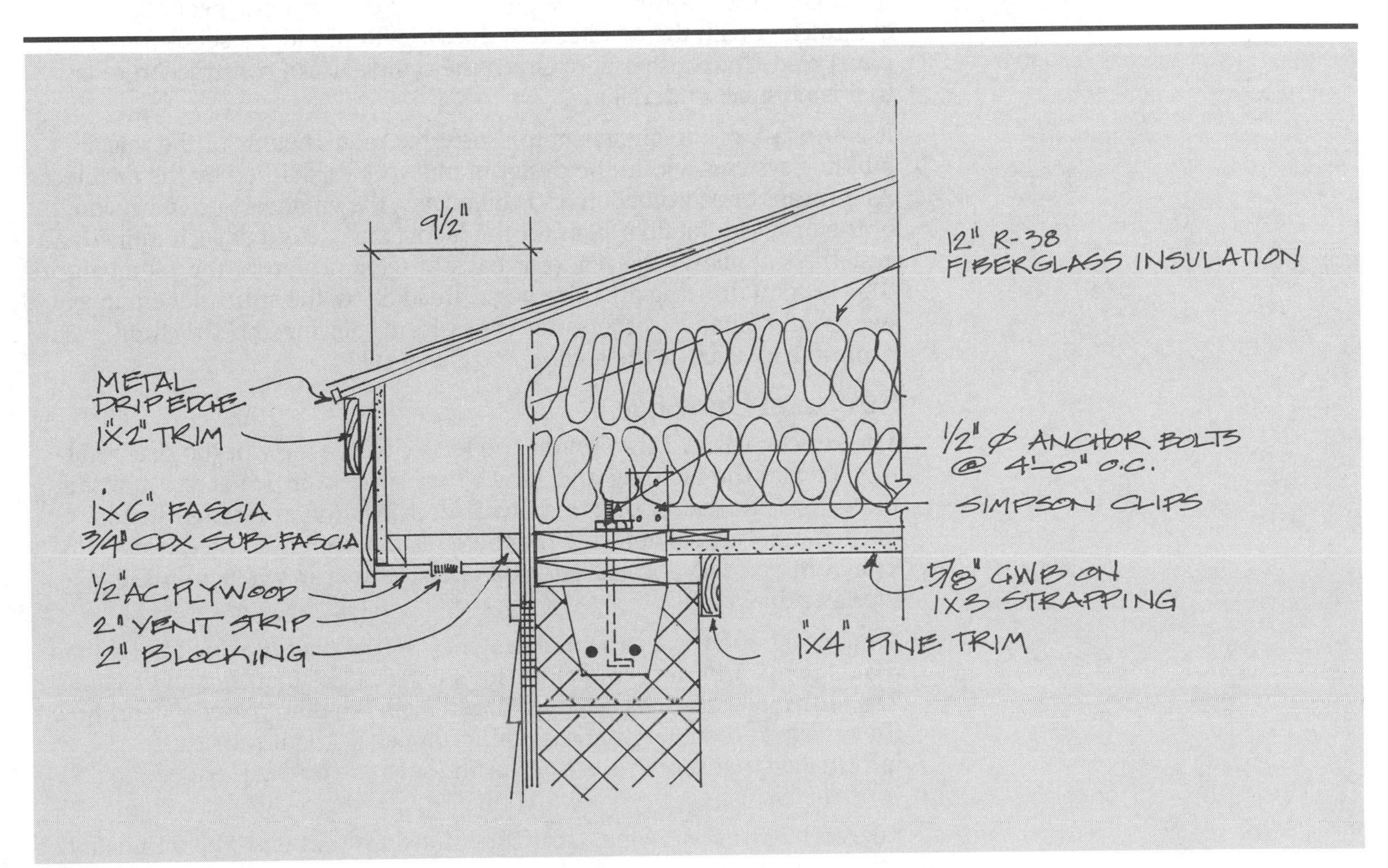

Figure 1.11 Detail

incorporate information pertaining to a similar group of items. For example, listings of doors, windows, room finishes, columns, trusses, and light or plumbing fixtures can be set in an easy-to-read table. These tables are called *schedules.*

Figure 1.12 is a typical door schedule. Note that each door is listed by number and provides information such as size and type, thickness, frame material, and hardware. In addition, the door schedule often states any specific instructions for an individual door, such as fire ratings, undercutting, weather stripping, or vision panels. The "Remarks" portion of the schedule offers the architect the opportunity to list any nonstandard requirements for the door.

Figures 1.13 is a typical window schedule. Figure 1.14 is a room finish schedule. Both schedules are for a light commercial project. They provide the information in a table format.

In the previous pages we have seen the various views that comprise the architectural drawings. It is easy to see how the architectural drawings have become the focal point of a set of working drawings. But the architectural drawings alone do not provide sufficient information to construct a building. There are many other systems that are part of a completed structure that the architectural drawings do not address. These include the structural components of the building, the plumbing, heating, and electrical systems, and the site plan for the structure. Aside from the fact that all of this information could not possibly fit on the architectural drawings, the actual design of these systems requires the talents of engineers specializing in each discipline. Normally it is the responsibility of the architect to coordinate the efforts of the different engineers in conjunction with the architectural drawings to produce a set of completed plans ready for construction. Given the complexity of construction today, that is no small undertaking.

It is the task of the engineers to satisfy the requirements of the major building systems within the design framework established by the architect. As a means of coordination and uniformity, the engineers use a version of the architectural floor plan for the layout and design of each individual system. This allows the reader a basis of reference, from the plumbing to the electrical drawing and vice versa. In addition, the structural engineer will employ many of the same views for displaying the structural components, such as the section and the detail.

Structural Drawings

The structural drawings provide the reader with a view of the structural members of the building and how it will support its loads and transmit those loads to the ground. Structural drawings (often referred to as "structurals") are sequentially numbered and are prefixed by the letter "S." They are normally located after the architectural drawings in a set of working drawings.

The benefit of having structural drawings is that they provide information that is useful and can stand alone for subtrades like framers and erectors. The structural drawings clearly indicate main building members and how they relate to the interior and exterior finishes, without providing information that is not necessary for the subtrade or this stage of construction.

Like architectural drawings, structural drawings start with the foundation plans, ground- or first-floor plan, upper-floor plans, and the roof plan. The

DOOR SCHEDULE								
NO.	SIZE	TYPE	MATL.	FRAME	TRSH.	CLOSER	HARDWARE	REMARKS
101	$3^0 \times 7^0 \times 1\frac{3}{4}''$	A	WD/GL	WD.	ALUM.	✓	BRASS PUSH BAR/PULL LOCKSET	MORGAN M-5911
102	$3^0 \times 7^0 \times 1\frac{3}{4}''$	A	"	WD	"	✓	"	"
103	$3^0 \times 7^0 \times 1\frac{3}{4}''$	A	"	WD.		✓	"	"
104	$3^0 \times 6^8 \times 1\frac{3}{8}''$	B	WD	WD	MARBLE		PRIVACY SET	MORGAN 5 CROSS PANEL
105	$3^0 \times 6^8 \times 1\frac{3}{8}''$	B	WD.	WD	"		"	"
106	BY WALK-IN MANUFACTURER							2'6" WIDE MAX.
107	$3^0 \times 6^8 \times 1\frac{3}{8}''$	D	WD	WD			SPRING HINGE	BY OWNER
108	$3^0 \times 7^0 \times 1\frac{3}{4}''$	C	H.M.	P.M.		✓	EXISTING	EXIST H.M. DOOR
109	$3^0 \times 7^0 \times 1\frac{3}{4}''$	C	H.M.	P.M.		✓	"	"

Figure 1.12 Door Schedule

WINDOW SCHEDULE						
MARK	MANUF/MODEL	TYPE	ROUGH OPEN.	GLASS	JAMB	REMARKS
A	ANDER/C24	CASEMENT	4'-0 1/2" × 4'-0 1/2"	HIGH PERFORM.	4 9/16"	GRILLES, SCREENS
B	ANDER/C34	CASEMENT	6'-0 1/2" × 4'-0 1/2"	"	"	" "
C	ANDER/CW14	CASEMENT	2'-4 7/8" × 4'-0 1/2"	"	"	" "
D	ANDER/CW25	CASEMENT	4'-9" × 5'-0 3/8"	"	"	" "
E	ANDER/24210	DOUBLE HUNG	2'-6 1/8" × 3'-1 1/4"	"	"	" "
F	ANDER/2842	DOUBLE HUNG	2'-10 1/8" × 4'-5 1/4"	"	"	" "
G	ANDER/30-3646-18	30° BAY	7'-0" × 4'-10 3/4"	"	"	" "
H	ANDER/A330	AWNING	3'-0 1/2" × 3'-0 1/2"	"	"	" "
I	ANDER/AW31	AWNING	3'-0 1/2" × 2'-4 7/8"	"	"	" "
J	ANDER/CW13	CASEMENT	2'-4 7/8" × 3'-0 1/2"	"	"	" "
K	ANDER/C12	CASEMENT	2'-0 5/8" × 2'-0 5/8"	"	"	" "

Figure 1.13 Window Schedule

significant difference is that only information essential to the structural systems is shown. For example, a second-floor structural plan would show the wood or steel framing and the configuration and spacing of load-bearing members, but would not show doors, nonload-bearing partitions, and so on. Following the plan views are the sections and details, in the same basic format as the architectural drawings. Schedules are used to record such information as footings, columns, and trusses. Figure 1.15 is a typical foundation plan.

In order to see what takes place within the foundation wall and footing, a section through the wall becomes necessary. Figure 1.16 is a typical foundation section.

In cases where further clarification is required, the engineer provides a detail such as that shown in Figure 1.17.

Mechanical Drawings

After the structural drawings are the mechanical drawings, which comprise the different mechanical systems of the building. Specifically, they include the plumbing drawings, sequentially numbered and prefixed by the letter "P"; the heating, ventilating, and air-conditioning drawings, commonly referred to as the *HVAC drawings,* prefixed by the letter "H"; and the fire protection drawings, typically prefixed by the letters "FP". Most of the work shown on these three types of drawings is in plan view form. Because of the diagrammatic nature of mechanical drawings, the plan view offers the best illustration of the location and configuration of the work. Figure 1.18 is an example of a partial plumbing plan.

Because of the large amount of information required for mechanical work and the close proximity of piping, valves, and connections, the engineer uses a variety of symbols to convey the intent. These symbols and abbreviations are incorporated in a chart or table called the *legend.* The legend explains symbols and abbreviations shown on the drawings. (See Figure 1.19 for an example that corresponds with the plumbing plan in Figure 1.18.)

ROOM FINISH SCHEDULE

NO.	NAME	FLOOR	BASE	WALLS NORTH	EAST	SOUTH	WEST	CEILING MATERIAL	HEIGHT	REMARKS
101	VESTIBULE	QUARRY TILE	WOOD	PLASTER PAINTED	PLASTER PAINTED	PLASTER PAINTED	PLASTER PAINTED	"V" GROOVE WOOD	VARIES	PROVIDE NON SKID FL. MAT
102	LOUNGE	Q.T.	WOOD	"	" WAINSCOT	" WAINSCOT	"	"	"	
103	BAR	Q.T.	4" VINYL	"	—	—	"	"	9'-8"	
104	DINING ROOM	SLATE	WOOD	" WAINSCOT	P. PTD/WAINS.	P. PTD/WAINS	P. PTD/WAINS	PLASTER PTD.	9'-8"	
105	SERVICE	Q.T.	WD/VINYL	P. PTD.	P. PTD	—	P. PTD.	"	9'-8"	
106	CORRIDOR	Q.T.	WOOD	"	"	—	"	"	7'-8"	
107	WOMEN	C.T.	C.T.	"	"	P. PTD.	"	"	7'-8"	
108	MEN	C.T.	C.T.	"	"	"	"	"	7'-8"	
109	KITCHEN	CONC. PTD	VINYL	ST. STL./C.T.	C.T./PTD GWB	C.T./PTD GWB	PTD. GWB	2x4 VINYL A.C.T	9'-8"	
110	PREP./STORAGE	CONC. PTD	VINYL	PTD. GWB	PTD. GWB	—	PTD. GWB	"	9'-8"	NO CLG ABOVE WALK-IN

Figure 1.14 Room Finish Schedule

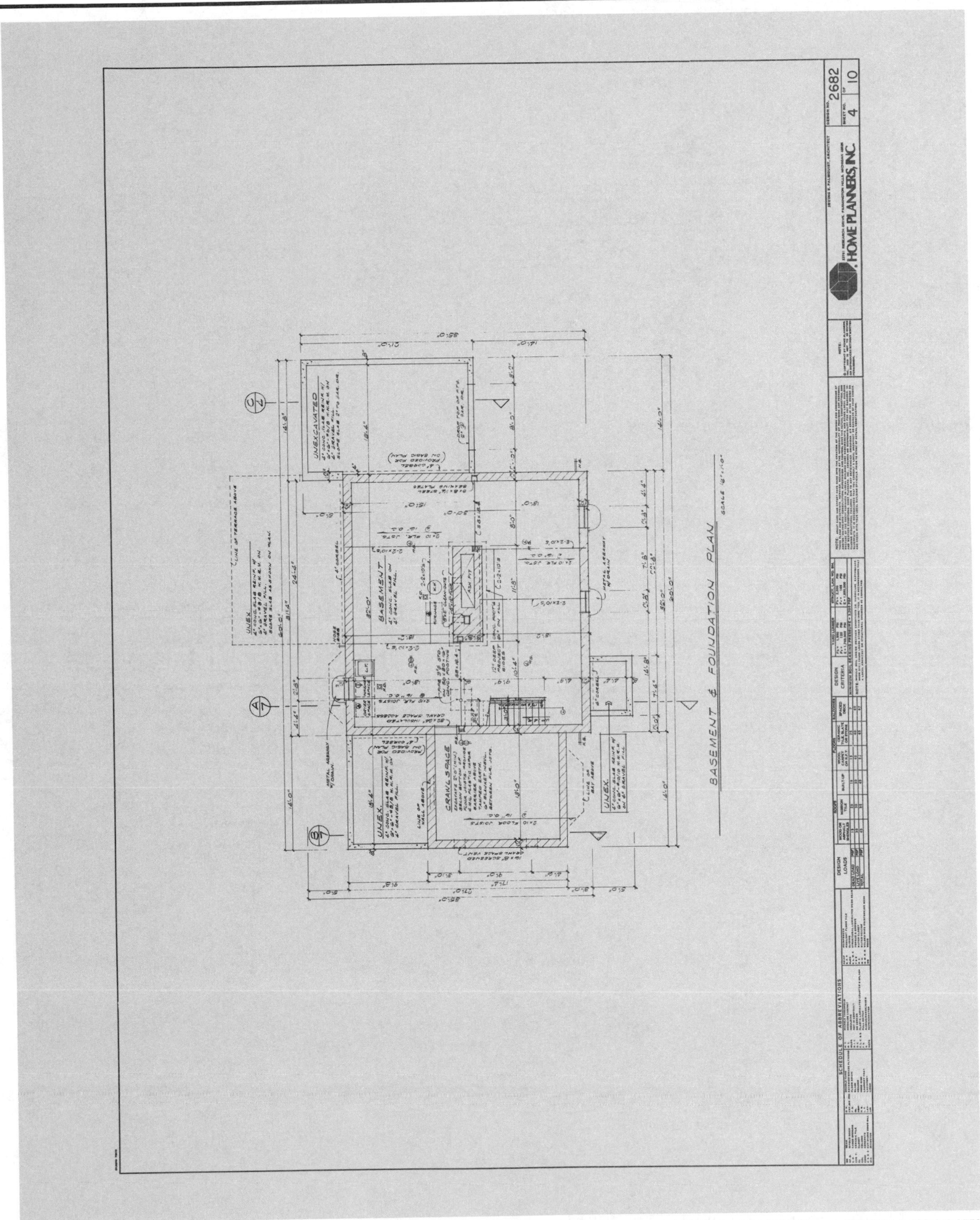

Figure 1.15 Foundation Plan

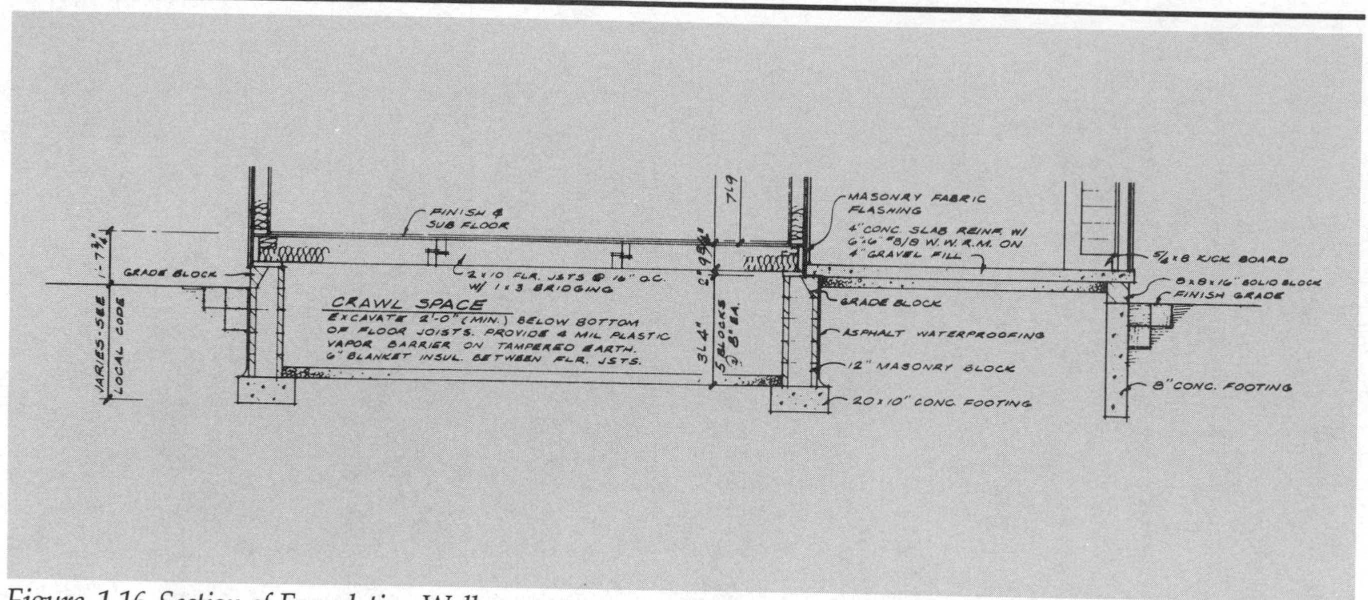

Figure 1.16 Section of Foundation Wall

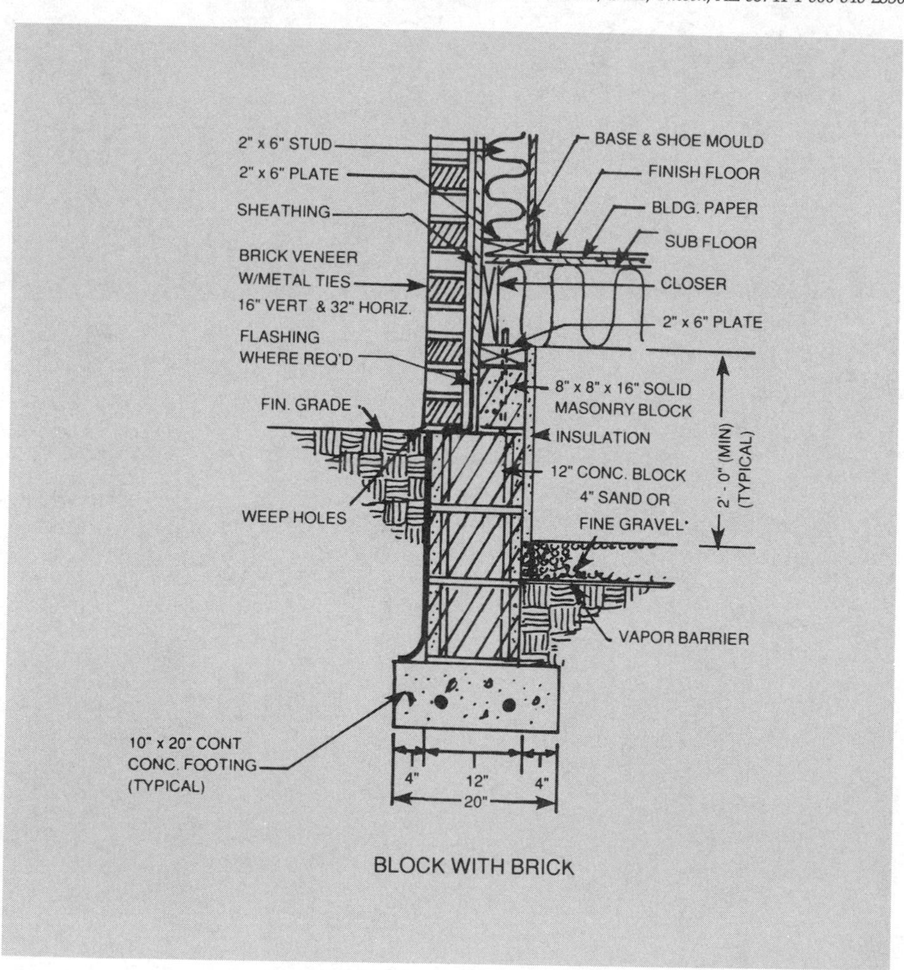

Figure 1.17 Detail at Top of Wall

In addition, the engineer makes use of the schedule format discussed in the preceding pages to list such items as the plumbing fixtures, HVAC diffusers, and gas-burning appliances. Figure 1.20 is a typical plumbing fixture schedule.

A less frequent option on mechanical drawings is the use of detail drawings. These drawings differ considerably from their architectural counterpart in that they are rarely drawn to scale, and they are illustrated

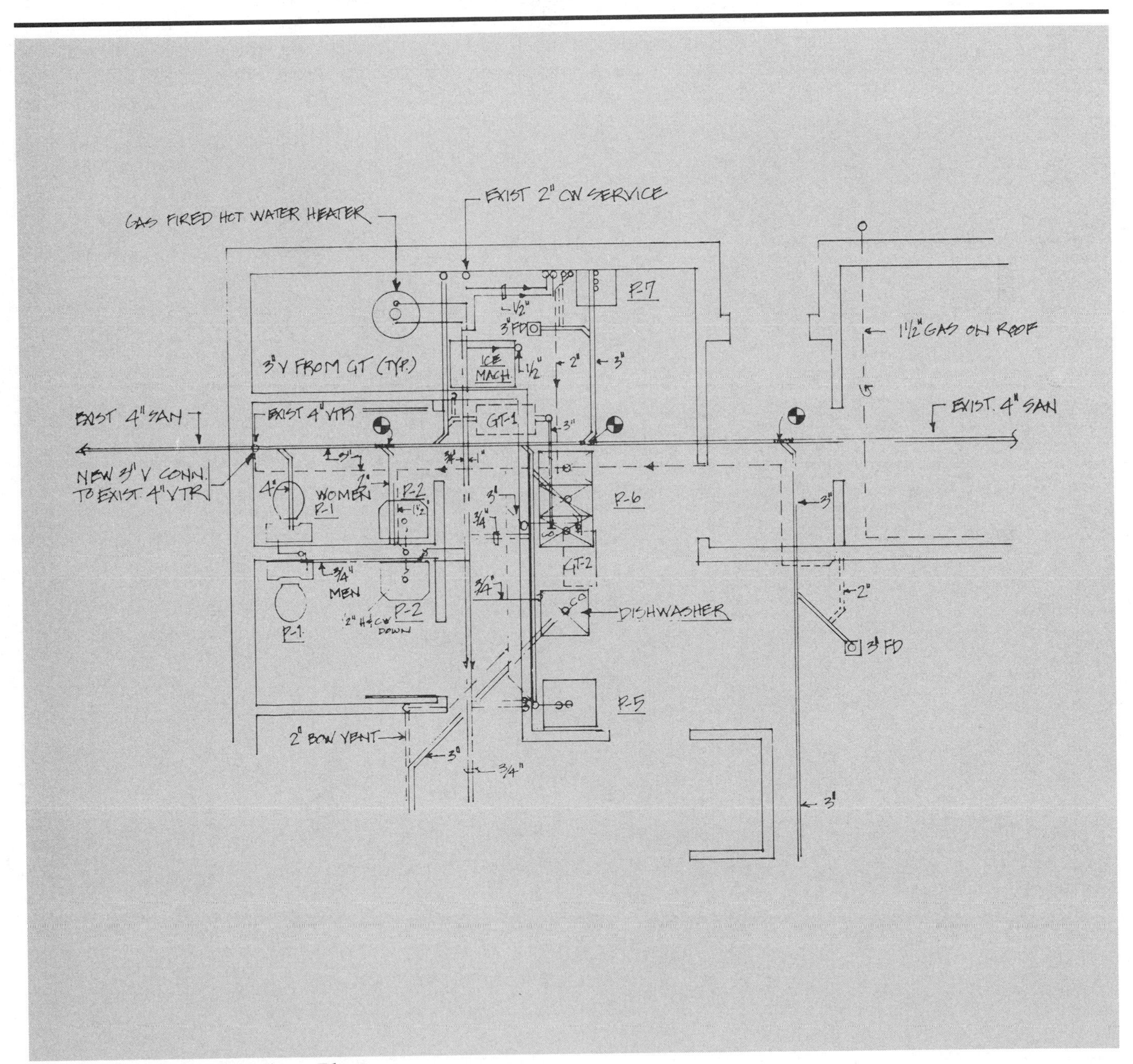

Figure 1.18 Partial Plumbing Plan

as an elevation or in perspective view. They are used to embellish the configuration of valves or appurtenances on a particular mechanical item. Figure 1.21 is a detail of a water heater. Without the detail, a clear understanding of the piping and valves needed for the water heater would be unavailable.

Finally, the mechanical engineer employs *riser diagrams* to illustrate what is often a complicated configuration of vertical piping. Riser diagrams are not drawn to scale, but offer a perspective view strictly for enhancement.

Electrical Drawings

The last part of the working drawings are usually the electrical drawings, which show the various electrical and communication systems of the building. The electrical drawings are numbered sequentially, using the letter "E" as a prefix. They include electrical power and lighting plans, telecommunications, and any specialized wiring systems such as fire or security alarms. The electrical drawings are similar in format to the mechanical drawings in their use of plan view drawings for layout and the

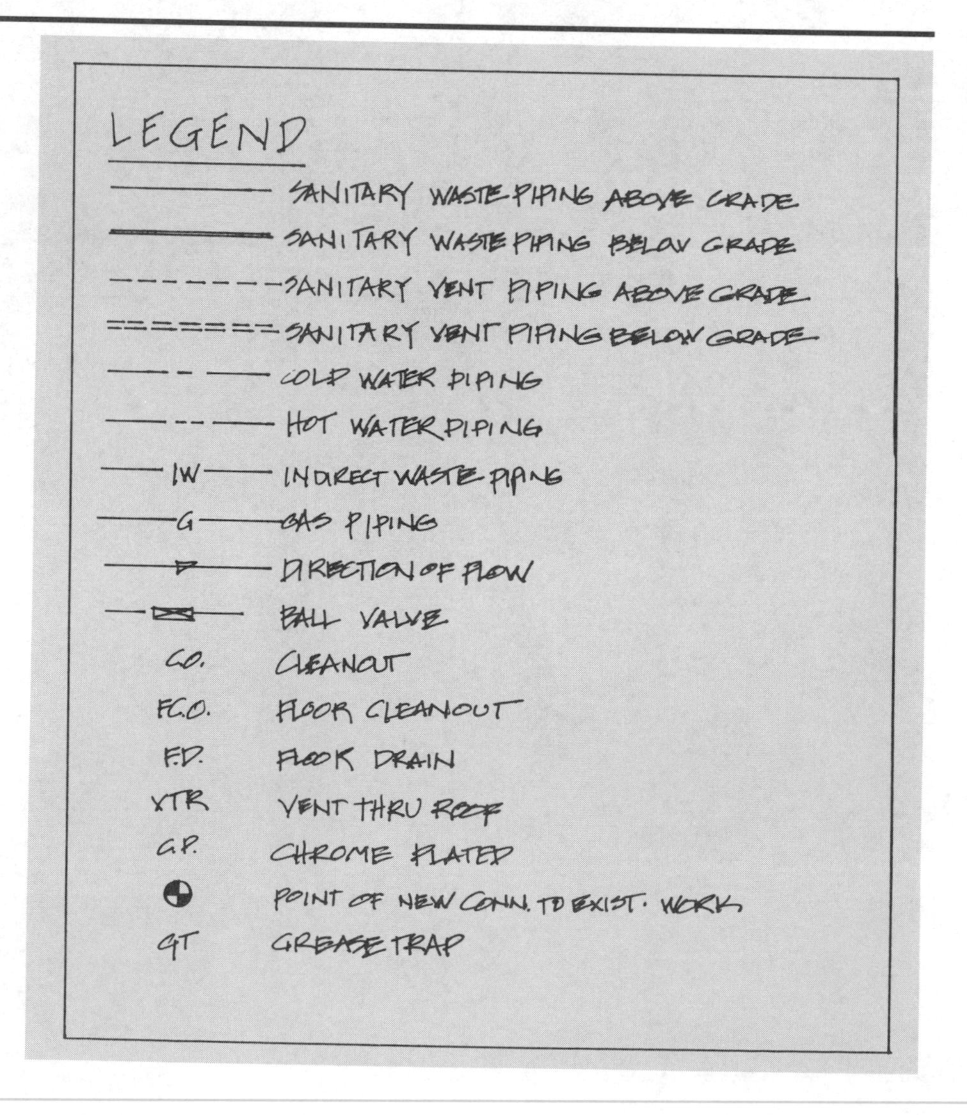

Figure 1.19 Sample Legend

MARK	FIXTURE	PIPE SIZE					REMARKS
		TRAP	WASTE	VENT	HW	CW	
P-1	WATER CLOSET	INT.	4"	2"	—	1/2"	FLOOR MOUNTED TANK TYPE HANDICAPPED
P-2	LAVATORY	1 1/2"	1 1/2"	1 1/2"	1/2"	1/2"	WALL HUNG HANDICAPPED
P-3	HAND SINK	1 1/2"	1 1/2"	1 1/2"	1/2"	1/2"	FURNISHED BY OTHERS
P-4	SINK/WASHER	↓	↓	↓	↓	↓	↓
P-5	RINSE SINK	↓	↓	↓	↓	↓	↓
P-6	POT SINK (3 COMPART.)	3@2"	3@2"				↓
P-7	JANITORS MOP SINK	3"	3"	2"	1/2"	1/2"	↓
P-8	BAR SINK	1 1/2"	1 1/2"	1 1/2"	1/2"	1/2"	↓

NOTES:
1. THE CONTRACTOR SHALL FURNISH AND INSTALL ALL C.P. SUPPLIES WITH SHUT OFFS, C.P. TRAPS & TAIL PIECES, C.P. FAUCETS ETC. FOR A COMPLETE PLUMBING SYSTEM.

Figure 1.20 Plumbing Fixture Schedule

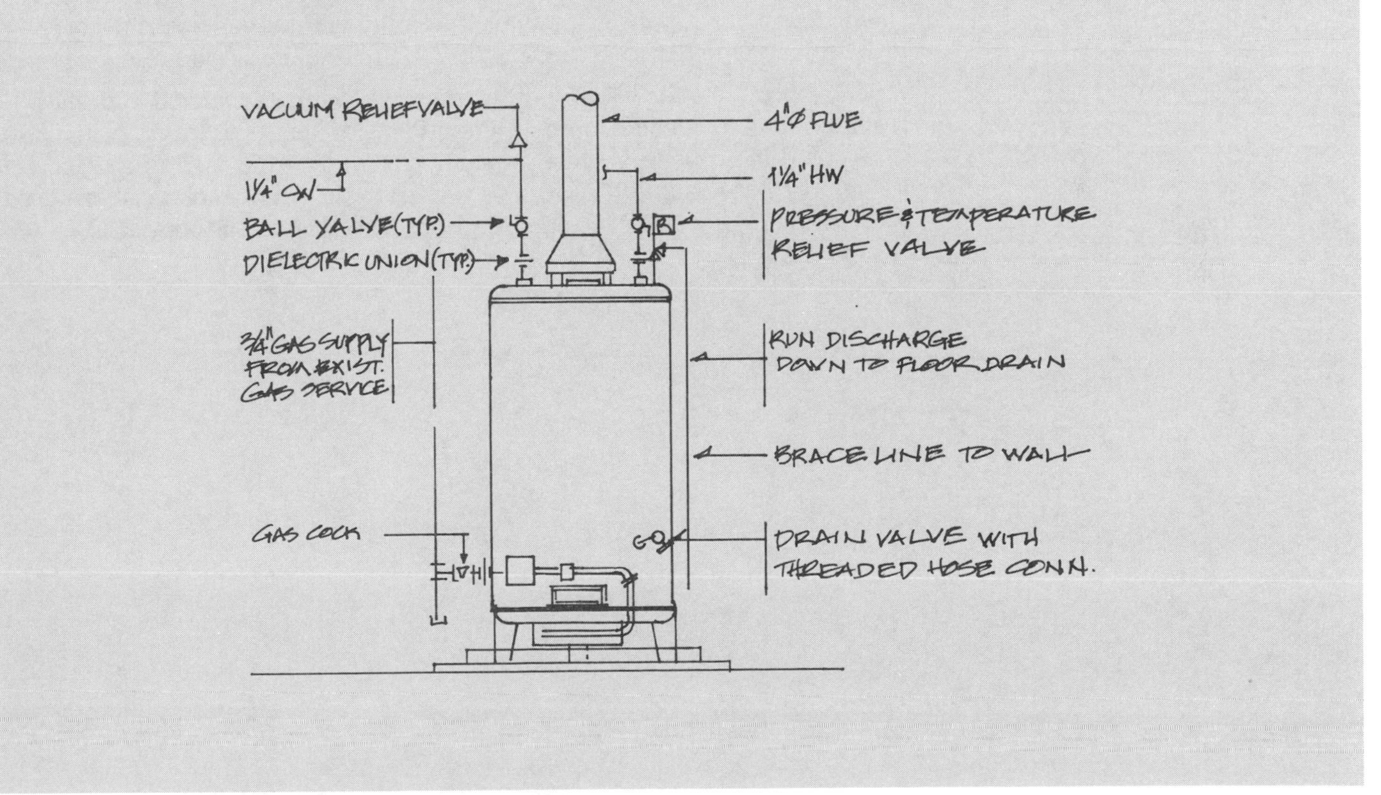

Figure 1.21 Detail—Gas Fired Water Heater

use of details and schedules for clarifications. Often power, lighting, and telecommunications layouts are shown on one drawing. For more complex structures, the systems are separated. See Figure 1.22 for a sample plan.

The power plan illustrates the power requirements of the structure, locating panels, receptacles, and the circuitry of power-utilizing equipment. Included on the power plan are the panel schedules that list the circuits and power requirements for a project with multiple panels. This is similar to the gas equipment schedule found on the mechanical drawings. Panel schedules list the circuits in the panel, plus the individual power required for each panel and the total power requirement for the electrical system. The schedules total the power requirements to help the electrical contractor size the panel; then the power company sizes the service requirements. Figure 1.23 is a sample panel schedule.

The lighting plan locates the various lighting fixtures in the building and is complemented by the use of a fixture schedule that lists the types of light fixtures to be used. The fixture schedule lists, by number or letter, the manufacturer and model, the wattage of the lamps, voltage, and any special remarks concerning the fixture. The lighting plan locates such devices as switches, smoke and fire detection equipment, emergency lighting, cable TV, and telephone outlets.

Summary

This chapter has introduced the different types of plans and the drawing elements that together comprise a full set of working drawings. It is essential that the contractor become familiar with the drawings prior to the site inspection and quantity takeoff. A thorough review of the drawings will reveal discrepancies or omissions, and will help the contractor determine whether to proceed with the next step in bidding the job.

See Appendix A for a listing of the most widely used material indication symbols, which detail material compositions of items shown on plans. Also included in Appendix A are window and door symbols, and specific symbols that relate to the work of various trades. Many abbreviations used on working drawings and in discussions throughout this book follow the symbols.

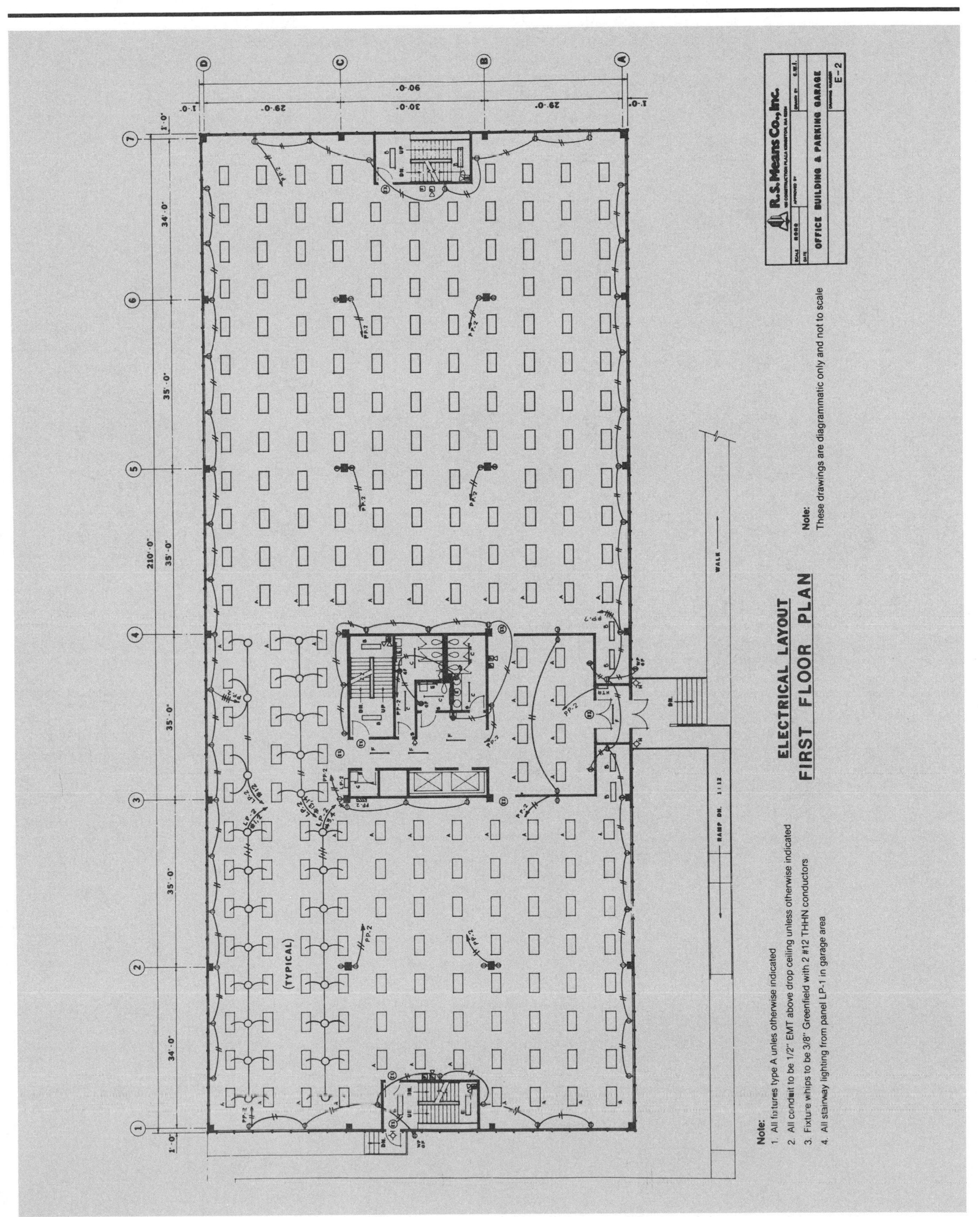

Figure 1.22 Electrical Plan

PANELBOARD "A" SCHEDULE

SERVICE: 120/208Y - 400A - 3PH - 4W - 10KAIC
CABINET: SURFACE - 42CKT

NO.	LOAD	KVA	CB	NO.	LOAD	KVA	CB
A1	EXIT/EGRESS LTG.	0.6	20A-1P	A2	EXIST. 10 TON A/C UNIT	20.0	60A-3P
A3	RECEPTACLES		20A-1P	A4	↓		
A5	↓		20A-1P	A6	↓		
A7	WALKIN COOLER COMPRESSOR	2.1	20A-2P	A8	ICE MACHINE	1.5	20A-1P
A9	↓			A10	WATER BOOSTER HEATER	12.0	40A-3P
A11	WALKIN COOLER	0.4	20A-1P	A12	↓		
A13	DISHWASHER	8.6	30A-2P	A14	↓		
A15	↓			A16	SLICER	1.6	20A-1P
A17	↓			A18	LOBOY REFRIG	1.2	20A-1P
A19	SALAD UNIT	0.9	20A-1P	A20	GLASS DR. REFER	1.5	20A-1P
A21	CONVECTION OVEN	1.5	20A-1P	A22	ENTRANCE HEATER	4.0	30A-1P
A23	RANGES	0.5	20A-1P	A24	↓		
A25	COOKING HOOD MAKEUP & EXHAUST FANS	2@1HP (2.0)	20A-3P	A26	LIGHTING	1.4	20A-1P
A27	↓			A28	↓	1.5	↓
A29	↓			A30	↓	0.9	↓
A31	LIGHTING	1.0	20A-1P	A32	↓	1.1	↓
A33	SPARE		20A-1P	A34	↓	1.0	↓
A35	↓		↓	A36	SPARE		↓
A37	↓		↓	A38	↓		↓
A39	↓		↓	A40	↓		↓
A41	↓		↓	A42	↓		↓

Figure 1.23 Sample Panel Schedule

Chapter Two

CALCULATING AREA AND VOLUME

Chapter Two

Calculating Area and Volume

To perform accurate quantity surveys or takeoffs, the contractor should be fluent in the calculations of area and volume. A basic understanding of mathematical relationships is also important. This chapter will review the basic formulas and relationships needed to perform these calculations.

Units of Measure

It is important to use correct units of measure for area and volume, and to remember to keep the units the same. For example, feet multiplied by feet result in *square feet*, yards multiplied by yards result in *square yards*, and so on. Multiplying a dimension in feet by a dimension in inches leads to an erroneous value. Dimensions given in differing units should be changed to the same units to yield a usable result.

For example, when calculating the area of a space that is 24′-6″ × 20′-3″, the feet and inches dimensions should be changed to feet in order to arrive at a measure in square feet for the area. This means converting 24′-6″ and 20′-3″ to decimal form.

24′-6″ is equal to 24.5′
20′-3″ is equal to 20.25′

When multiplied, the result is an area of 496.13 square feet. The following are decimal equivalents of inches in feet.

1″ = 0.08 feet	7″ = 0.58 feet
2″ = 0.17 feet	8″ = 0.67 feet
3″ = 0.25 feet	9″ = 0.75 feet
4″ = 0.33 feet	10″ = 0.83 feet
5″ = 0.42 feet	11″ = 0.92 feet
6″ = 0.50 feet	12″ = 1.00 feet

With calculators as commonplace as pencils today, decimal equivalents of fractions of an inch are as easy as pushing a few buttons, literally. For example:

3-1/2″ = 3.5″
3.5″ divided by 12″/ft. = 0.29 feet

Note that the decimal equivalent of 2″ is not precisely two times the decimal equivalent of 1″. This is because two-place accuracy after the decimal point is sufficient for estimating purposes in most cases. While the actual value of 2″ in decimal form is .0166666 of a foot, this level of accuracy hardly seems necessary; so the value for 2″ is rounded up to 0.17.

Area and Square Measure

If, for a moment, we imagine the floor of a building as simply a planar surface with no depth, the sum of the sides of that floor would be called its perimeter. Remember that area calculations do not take into account the depth of that surface. Areas are expressed in square units—most commonly square feet, square yards, or square inches.

The area of a rectangle or square is defined as the product of its length and width (see Figure 2.1). The formula is:

$$A = l \times w$$

where A = area, l = length, and w = width.

Another familiar shape in building construction (e.g., roof rafters, stair stringers) is the right triangle, a shape having three sides—a base, an altitude, and a hypotenuse—and one angle that is 90 degrees. The base is the side of the triangle in the horizontal plane, the altitude is the side in the vertical plane, and the hypotenuse is the longest side, opposite the right angle. The definition of the area of a right triangle is one-half the product of its base and its altitude (see Figure 2.2). Expressed as a formula,

$$A = 1/2 \times b \times a$$

where A = area, b = base, and a = altitude.

In Figure 2.3, it is necessary to calculate the length of unknown legs of the right triangle. To do so, the *Pythagorean Theorem* or the *Right Triangle Law* is used. The Right Triangle Law is the same as the "3-4-5 triangle" that framers and contractors use to "square up" work in the field. It states:

The square of the hypotenuse of a right triangle is equal to the sum of the squares of the other two sides.

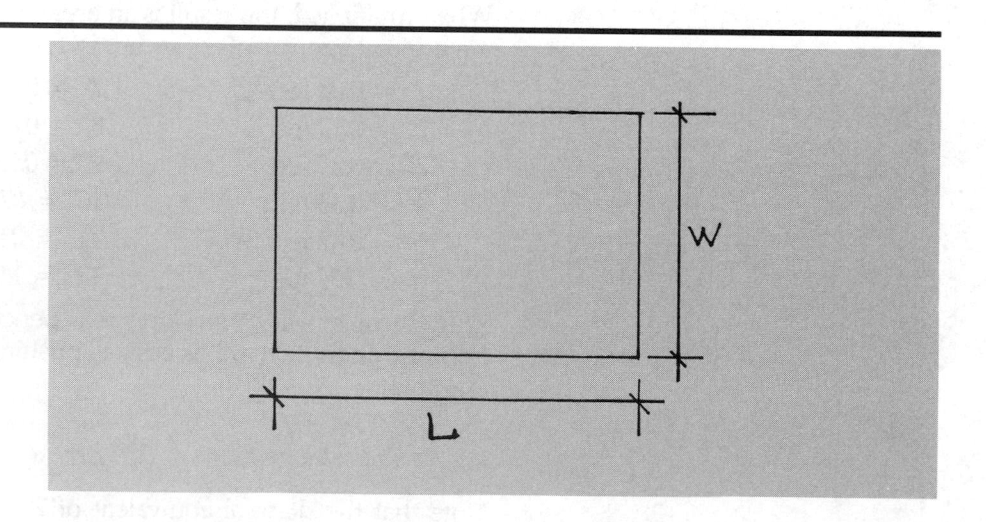

Figure 2.1 Area of a Rectangle

In Figure 2.3, therefore, the corresponding formula is:

$C^2 = A^2 + B^2$
where C = *the length of the hypotenuse,*
A = *the length of the altitude, and* B = *the length of the base.*

It is always possible to calculate the length of one side of a right triangle if the other sides are known. Some other noteworthy rules concerning triangles follow.

- The sum of the three angles in a triangle must be 180 degrees.
- If all three sides of a triangle are equal, then all three angles are the same, 60 degrees. This is called an *isosceles* triangle.
- If the two legs of a right triangle are the same length, then the two angles opposite each leg are equal and 45 degrees.

The circle is another shape that occurs in construction; for example, in a brick patio. Before we review the formula for the area of a circle it will be helpful to define the various parts of the circle, and some constant relationships.

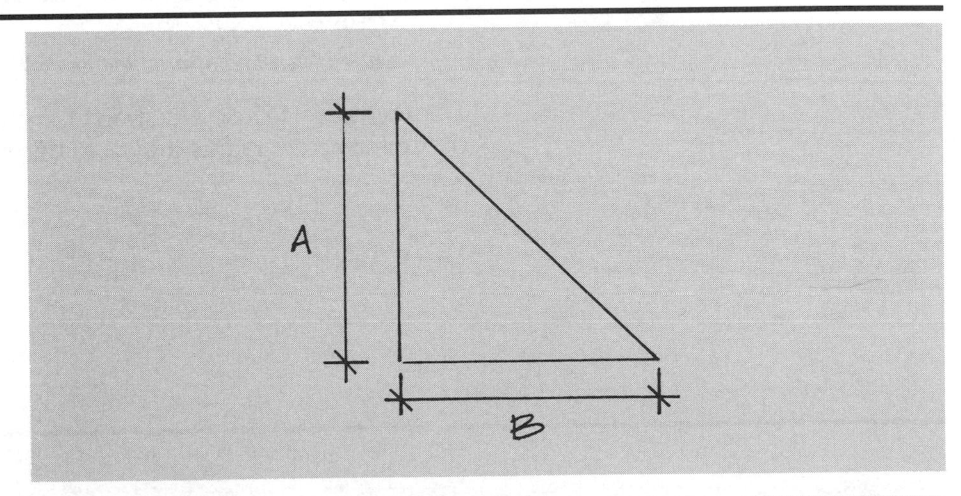

Figure 2.2 Area of a Triangle

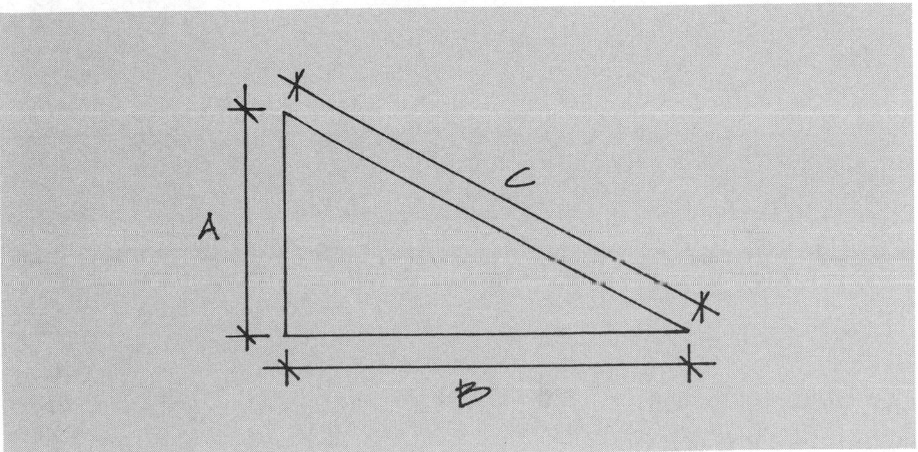

Figure 2.3 Right Triangle

The *circumference* is the perimeter of the circle.

The *diameter* is defined as a line drawn through the center of the circle, beginning and ending on the circumference. Any number of diameters drawn on a circle should render two equal halves of that circle. All diameters of the same circle are equal.

A *radius* of a circle is a line from the center point within a circle to a point on the circumference. All radii of the same circle are equal in length. The radius, by definition, is one-half of the diameter. Knowing this relationship, we can establish the following formulas:

$D = 2 \times R$, or $R = 1/2 \times D$
where D = diameter and R = radius.

Figure 2.4 illustrates the various parts of a circle.

The circumference of a circle has a constant relationship to the diameter of the same circle. That constant is the number 3.1416, and has been given the name of "pi" and the corresponding symbol, π.

$C = pi \times D$, or $C = 2 \times pi \times R$
where C = circumference, pi = 3.1416, D = diameter, and R = radius.

A *chord* is a straight line connecting two points on the circumference, without passing through the center of the circle.

An *arc* is any portion of the circumference of the circle. A circle has 360 degrees. Therefore, if the radius and the interior angle between the radii are known, the length of the arc can be calculated using the formula:

$N/360 \times 2 \times pi \times R$
where N is the central angle, pi = 3.1416, and R = radius.

The *tangent* of a circle is a straight line touching only one point on the circumference. A radius drawn to this point is at 90 degrees to the tangent.

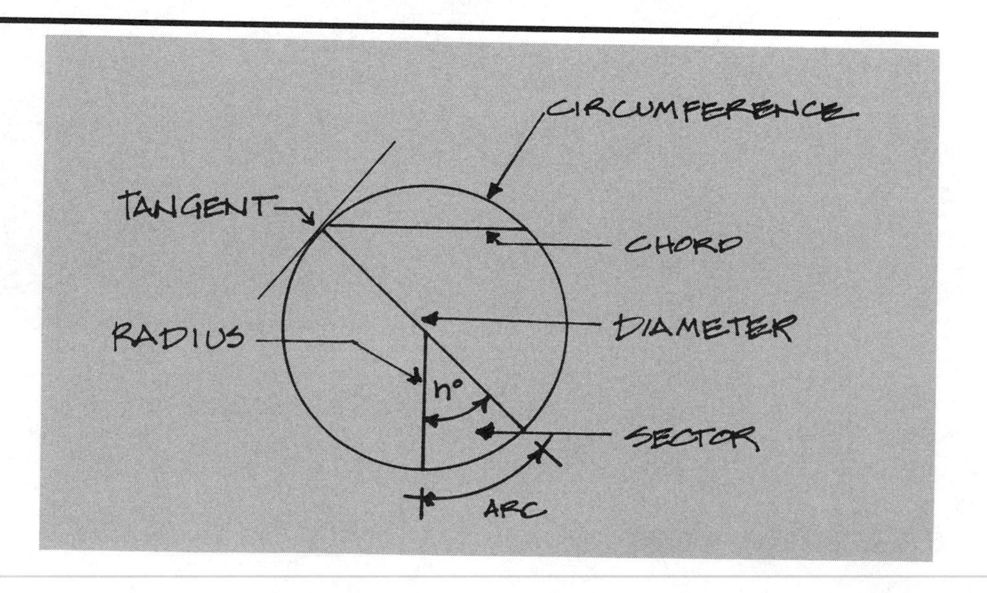

Figure 2.4 Parts of a Circle

By using the mathematical relationships expressed above, it is possible to calculate the area of a circle if certain information about the circle is known.

The area of a circle is the circumference multiplied by 1/2 the radius, or pi times the radius squared. As a formula:

$$A = C \times R/2, \text{ or } A = pi \times R^2$$

It is also possible to calculate the area of a portion of a circle. If one cuts a pie-shaped piece out of a circle—in other words, two radii with an angle in between—given the radius and the angle between it, we can calculate that area, which is called a *sector*. As we discovered in the formula for the arc, the length of the arc is a fraction of the total circumference. A similar deduction can be used to devise a formula for the area of the sector:

$$A_s = \frac{n}{360} \times pi \times R^2$$

where A_s *= area of a sector,* n *= angle in degrees between the radii,* $pi = 3.1416$*, and* R *= radius.*

Contractors are often required to calculate the areas of more complex polygons such as hexagons, octagons, trapezoids, and other irregular shapes. Rather than confuse the calculations with complicated formulas, it is best to divide the area into more common shapes and either add or subtract the pieces to arrive at the total. For most construction applications, the accurate approximation of the area of an irregular shape is sufficient.

Volume and Cubic Measure

In contrast to area, which has only two dimensions, volume has a third dimension, *depth*. The depth of a shape can also be called its *thickness* or *height*. Once a shape takes on this third dimension, it is no longer planar, but becomes a solid. The term *cubic* refers to the volume of a solid, whereas *square* accounts only for its area. The standard units of cubic measure are *cubic inches, cubic feet, and cubic yards.*

If we were to visualize the shape in Figure 2.5 as a surface area with a height, it would be called a *prism*. Examples of prisms in construction are pilecaps, footings, or precast concrete tanks. If the dimensions of length, width, and height were the same, it would be a cube.

If we take the formula for the area of that planar surface and add the new dimension, we arrive at the formula for the volume of a prism.

$$V = A \times h$$

where V *= volume of the prism,* A *= area of the base, and* h *= height.*

This formula applies only to shapes whose ends and opposite sides are parallel.

To further expand this formula:

$$V = l \times w \times h$$

where, again, l *= length,* w *= width, and* h *= height.*

The contractor will encounter an endless variety of shapes that are one form or another of the prism. The triangular prism is a shape that has a triangular surface area and a height. The rule of base area multiplied by height still applies. The volume of a triangular prism is as follows:

$$V = 1/2 \times l \times w \times h$$

A less common, more sophisticated shape is the *cylinder*. The formula for the volume of a cylinder is essential in calculating the amount of concrete to fill a sonotube, or a round column form. The volume of a cylinder is the area of its circular base multiplied by its height.

$V = pi \times R^2 \times h$

where V = volume of a cylinder, pi = 3.1416, R = radius, and h = height.

Figure 2.6 illustrates a typical cylinder and its parts.

Summary

These are by no means all the shapes that may be encountered in a quantity takeoff, but they are the most common ones. Figure 2.7 lists additional formulas for less common shapes. Also included is a table of conversions for linear, square, and cubic measure.

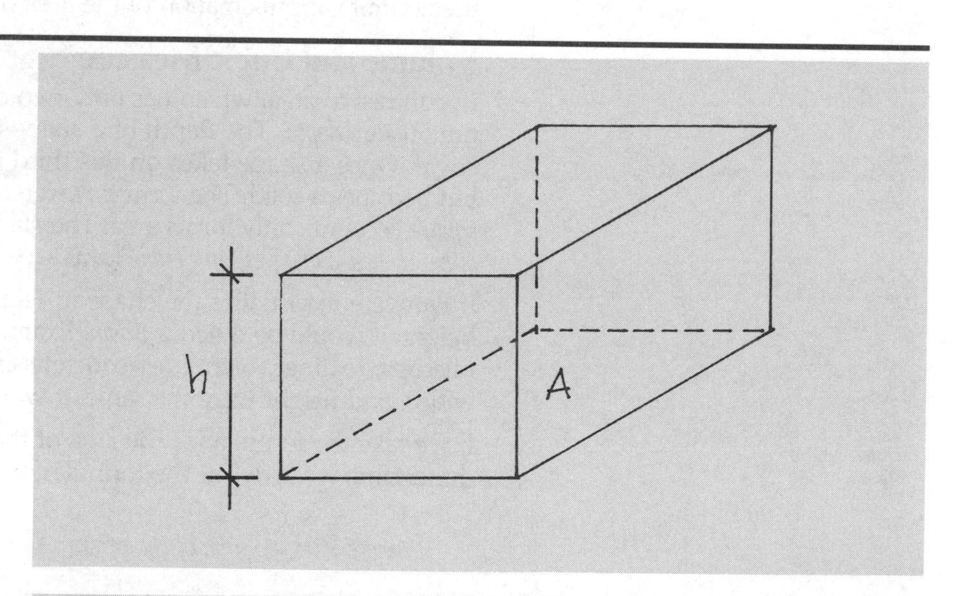

Figure 2.5 Prism

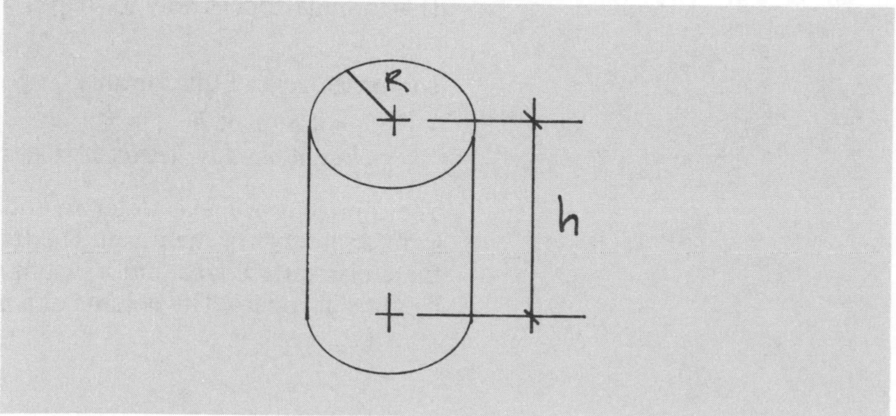

Figure 2.6 Cylinder

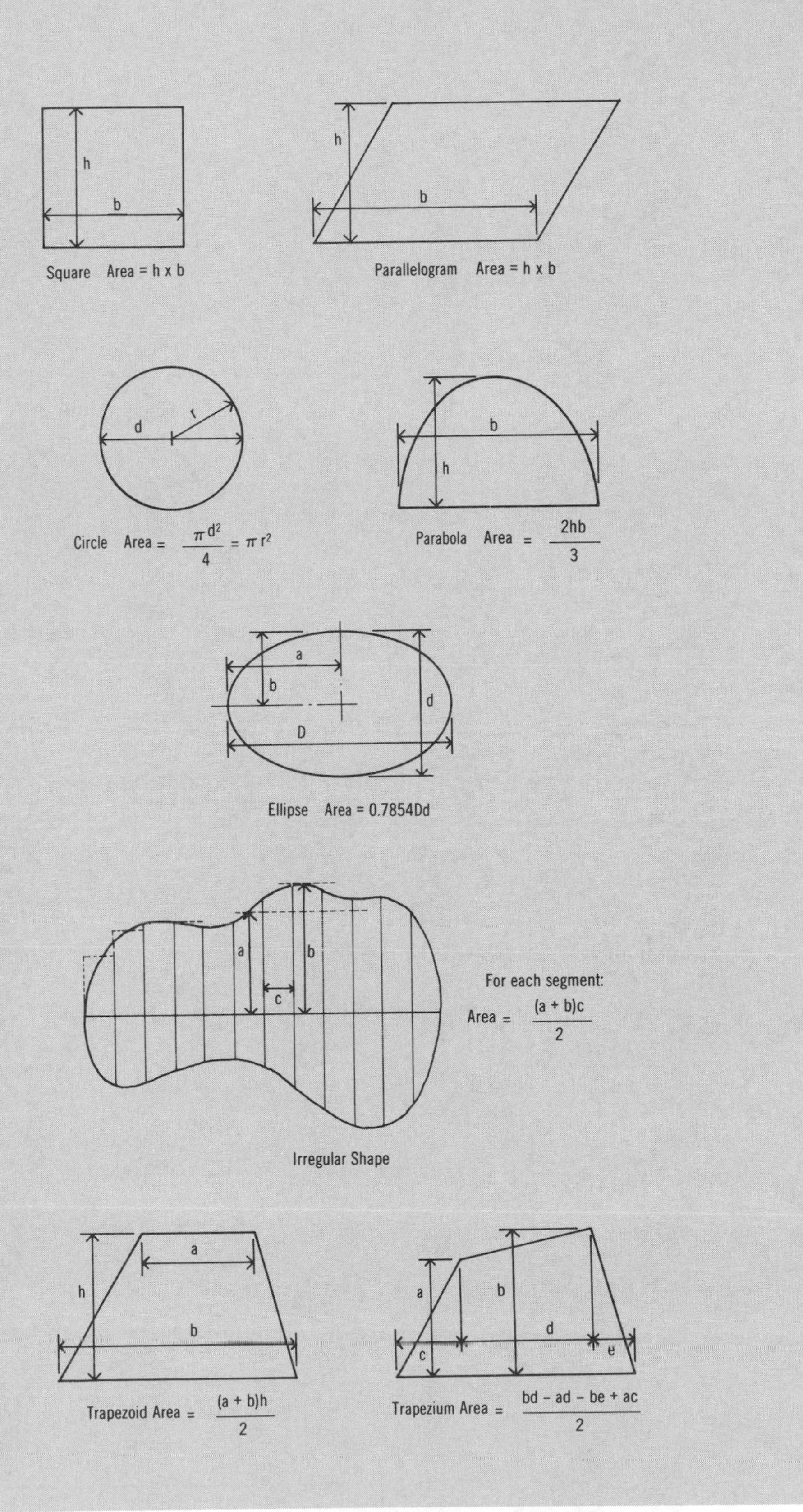

Figure 2.7 Formulas for Less Common Shapes and Standard Weights and Measures

Volume Calculations

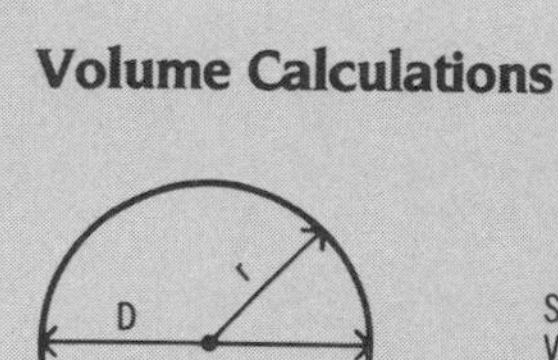

Sphere
Volume = $0.5236\ D^3$
Surface Area = $4\pi r^2$ ($\pi = 3.1416$)

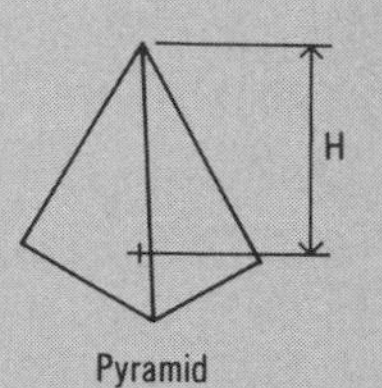

Pyramid

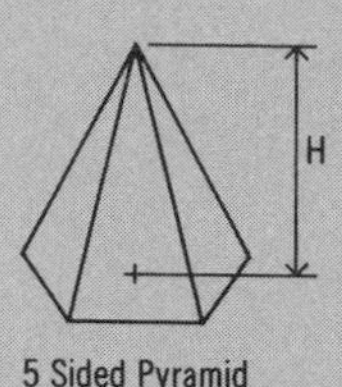

5 Sided Pyramid

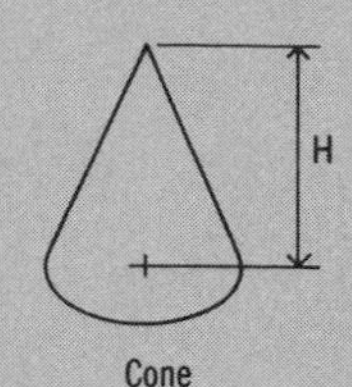

Cone

Volume = Area of Base x 1/3 Height (H)

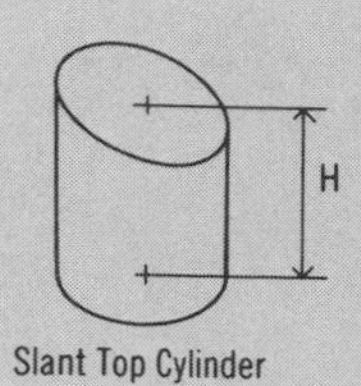

Slant Top Cylinder

Oblique Cylinder

Volume = Area of the Base x Height (H)

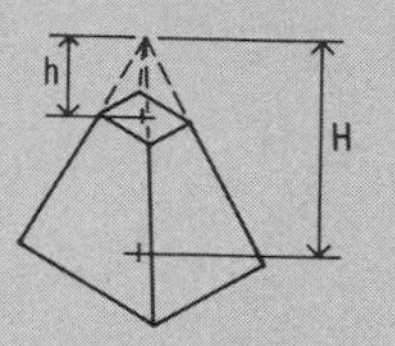

Truncated Pyramid

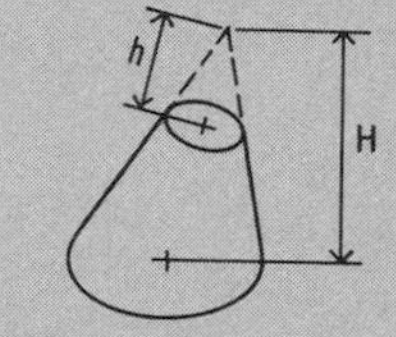

Truncated Oblique Cone

h = Height of Cut-Off
H = Height of Whole

Volume = Volume of the Whole Solid Less Volume of the Portion Cut Off

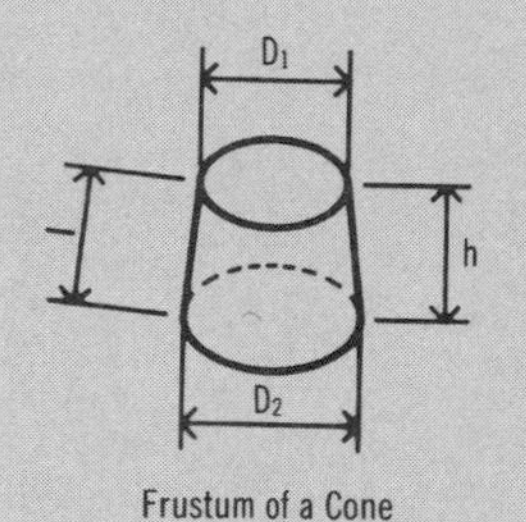

Frustum of a Cone

Volume =

$$\frac{D_1 + D_2 + (\text{Area of Top} + \text{Area of Base}) \times h \times 0.7854}{3}$$

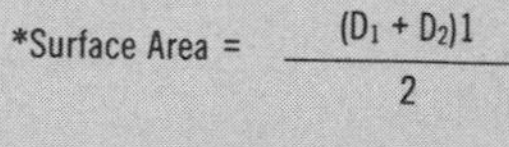

*Surface Area = $\dfrac{(D_1 + D_2)1}{2}$

*Excludes Top & Base Areas

Volume and Capacity

Units and Equivalents	
1 cu. ft. of water at 39.1° F	= 62.425 lbs.
1 United States gallon	= 231 cu. in.
1 imperial gallon	= 277.274 cu. in.
1 cubic foot of water	= 1728 cu. in.
	= 7.480519 U.S. gallons
	= 6.232103 imperial gallons
1 cubic yard	= 27 cu. ft. = 46.656 cu. in.
1 quart	= 2 pints
1 gallon	= 4 quarts
1 U.S. gallon	= 231 cu. in.
	= 0.133681 cu. ft.
	= 0.83311 imperial gallons
	= 8.345 lbs.
1 barrel	= 31.5 gallons = 4.21 cu. ft.
1 U.S. bushel	= 1.2445 cu. ft.
1 fluid ounce	= 1.8047 cu. in.
1 acre-foot	= 43,560 cu. ft.
	= 1,613.3 cu. yds.
1 acre-inch	= 3,630 cu. ft.
1 million U.S. gallons	= 133,681 cu. ft.
	= 3.0689 acre-ft.
1 ft. depth on 1 sq. mi.	= 27,878,400 cu. ft.
	= 640 acre-ft.
1 cord	= 128 cu. ft.

Standard Weights and Measures

Linear Measure	
1000 mils =	1 inch
12 inches =	1 foot
3 feet =	1 yard
2 yards =	1 fathom 6 feet
5-1/2 yards =	1 rod 16-1/2 feet
40 rods =	1 furlong 660 feet
8 furlongs =	1 mile 5280 feet
1.15156 miles =	1 nautical mile, or knot 6080.26 feet
3 nautical miles =	1 league 18,240.78 feet

Square Measure	
144 square inches =	1 square foot
9 square feet =	1 square yard
30-1/4 square yds. =	1 square rod 272-1/4 square feet
160 square rods =	1 acre 43,560 square feet
640 acres =	1 square mile 27,878,400 square feet

A circular mil is the area of a circle 1 mil, or 0.001 inch in diameter.

1 square inch =	1,273,239 circular mils
A circular inch is the area of a circle 1 inch in diameter =	0.7854 square inches.
1 square inch =	1.2732 circular inches

Dry Measure	
2 pints =	1 quart
8 quarts =	1 peck
4 pecks =	1 bushel 2150.42 cubic in. 1.2445 cubic feet

Weight—Avoirdupois or Commercial	
437.5 grains =	1 ounce
16 ounces =	1 pound
112 pounds =	1 hundredweight
20 hundredweight =	1 gross, or long ton 2240 pounds
2000 pounds =	1 net, or short ton
2204.6 pounds =	1 metric ton
1 lb. of water (39.1°F) =	27.681217 cu. in.
=	0.016019 cu. ft.
=	0.119832 U.S. gallon
=	0.453617 liter

Figure 2.7 Formulas for Less Common Shapes and Standard Weights and Measures (continued)

Chapter Three

THE SPECIFICATIONS

Chapter Three

THE SPECIFICATIONS

Technical information pertaining to the quality of materials and workmanship is not always incorporated on the drawings themselves. In most cases, working drawings are issued with a set of specifications. Even the simplest projects have specifications, whether incorporated on the drawings or issued as a separate document, to guide the contractor and subcontractors.

The person preparing the specifications, which are more commonly referred to as the *specs*, makes every effort to cover all of the items or segments of work shown on the working drawings. The specifications serve as a guideline for bidding and performing the work. In the event of a discrepancy between the plans and the specifications, it is generally accepted that the specifications take precedence over the drawings.

The CSI MasterFormat

The most widely accepted system for arranging construction specifications is the CSI MasterFormat, developed by the Construction Specifications Institute. As this is the system of coding and arrangement most commonly used in daily practice, our discussion will center around MasterFormat. In addition to organizing specifications, the MasterFormat system is used for classifying data and filing manufacturers' literature for products and services.

The CSI format categorizes the information presented in specifications into four major categories:

- Bidding Requirements
- Contract Forms
- General Conditions
- Specifications (Technical)

The technical specifications are further divided into 16 divisions. Each of these divisions is a grouping of similar work numerically organized into "subdivisions," or units of work. Figure 3.1 shows the project information, categorized by division, and its relationship to the overall project documents.

These 16 divisions were determined based on relationships established in the actual construction process, and roughly follow the natural order of the construction of a building. The divisions are used in Chapters 4–18 of this book, which cover the various phases of construction.

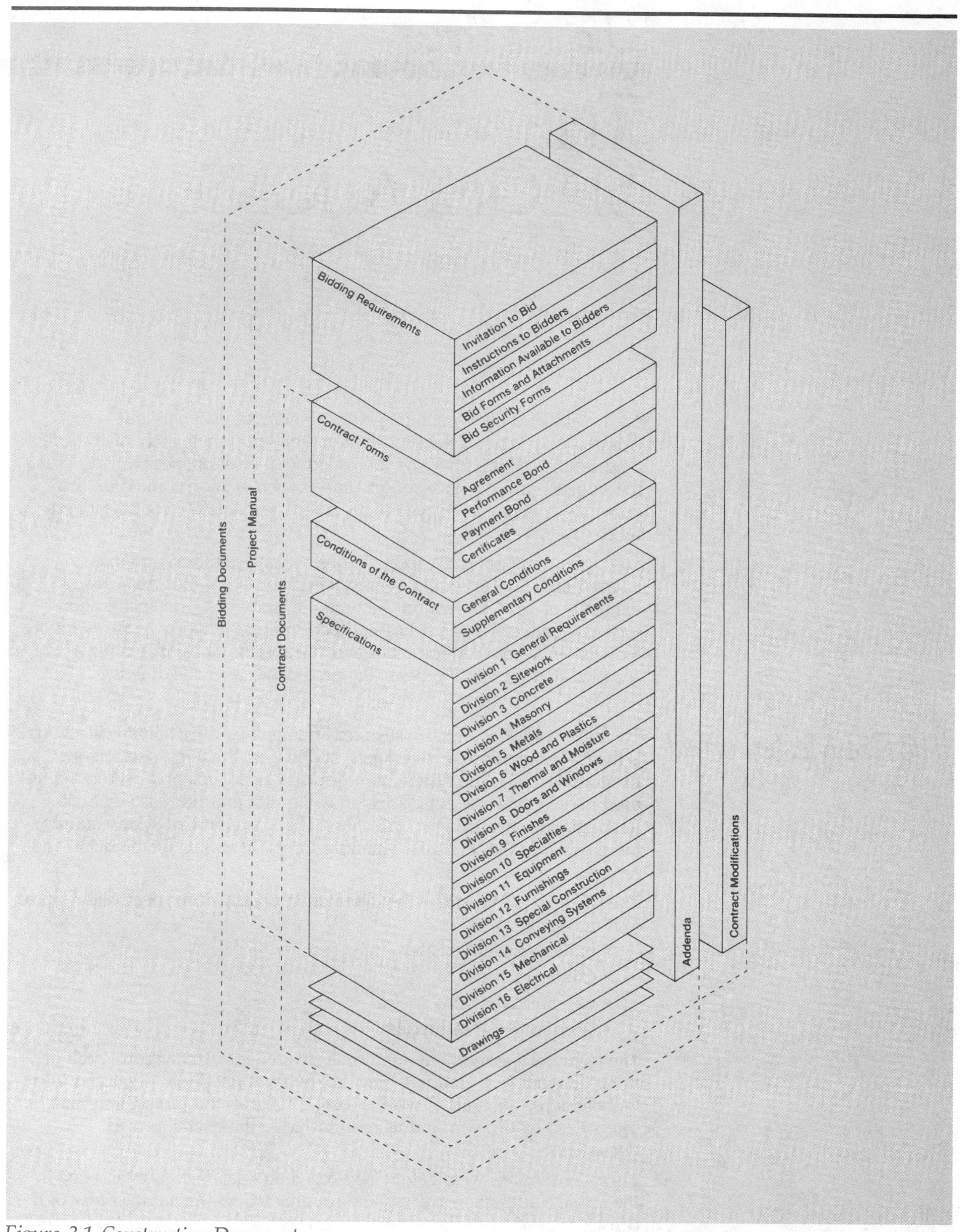

Figure 3.1 Construction Documents

Reprinted from the Construction Specifications Institute (CSI) ***Manual of Practice,*** *(FF Module, 1992 edition) with permission from CSI.*

The 16 specification divisions and a general summary of their contents are as follows:

Division 1–General Requirements includes a brief summary of the work, as well as the definitions and standards for the project, the project coordination, project meetings, schedules, reports, testing, samples, submittals, shop drawings, project close-out, cleanup, quality control, and temporary facilities. In addition, it addresses pricing concerns such as unit prices, alternates, and allowances.

Division 2–Site Work includes clearing of the site, earthwork, site drainage, utilities, roads, walks, pavings, curbing, general site improvements, subsurface investigations, landscaping, and heavy site work such as shoring, pile driving, and caissons.

Division 3–Concrete includes formwork, reinforcing, precast and cast-in-place concrete, and cementitious decks.

Division 4–Masonry includes brick, block, stone, mortar, anchors and reinforcement, and masonry restoration and cleaning.

Division 5–Metals includes structural steel, joists, metal decking, light-gauge framing, and ornamental and miscellaneous metals.

Division 6–Wood and Plastics includes rough and finish carpentry, millwork, casework, and plastic laminate work.

Division 7–Thermal and Moisture Protection includes waterproofing, dampproofing, insulation, roofing, siding, caulking, and sealants.

Division 8–Doors and Windows includes metal and wood doors and frames, windows, glass and glazing, and finish hardware.

Division 9–Finishes includes gypsum wallboard systems, board and plaster systems, painting and wallcoverings, flooring, acoustical ceiling systems, and ceramic and quarry tile.

Division 10–Specialties includes demountable partitions, toilet partitions, toilet accessories, fire extinguishers, postal specialties, flagpoles, lockers, signage, and retractable partitions.

Division 11–Equipment includes specialized equipment for banks, gymnasiums, schools, churches, laboratories, residences, prisons, libraries, and hospitals.

Division 12–Furnishings includes cabinetry, rugs, tables, seating, artwork, and window treatments.

Division 13–Special Construction includes greenhouses, clean rooms, swimming pools, integrated ceilings, incinerators, and sound and vibration controls.

Division 14–Conveying Systems includes elevators, escalators, lifts, dumbwaiters, cranes, and hoists.

Division 15–Mechanical Systems includes fire protection systems, plumbing, heating, air-conditioning, ventilating, gas piping, special piping, refrigeration, and controls.

Division 16–Electrical Systems includes electrical service and distribution, wiring devices, fixtures, communications, and power.

Each of the 16 CSI divisions is numerically organized with a five-digit code. The first two numbers of the code identify the division. The next three digits further define the type of work within the division. Additional numbers not already specified by CSI may be added for your own use. For example, 000 can be used for Alternates.

The contractor should be familiar with major division codes by number. Figure 3.2 is a detailed listing of the 16 MasterFormat divisions and subdivisions, reprinted with permission of the Construction Specifications Institute.

Analysis of the Specifications

Bidding Requirements

The bidding requirements begin with a solicitation for bids or proposals. This solicitation can be in the form of an *Invitation for Bid, Request for Proposals*, or, in the case of public work, an *Advertisement for Bid*. All are similar in the information conveyed to the bidder.

The Request for Proposals, or RFP, invites prospective qualified general contractors and subcontractors to submit proposals for a particular project. It clearly defines the date, time, and location for bids to be submitted. It identifies the authority responsible for award and whether the bids will be publicly or privately opened. In the case of taxpayer-funded projects, the bids are opened publicly and made available for the inspection of the general public.

The RFP contains any required prequalification or eligibility criteria to eliminate bidders that could later be considered unacceptable. In the case of private bidding, the Invitation to Bid may be all that is required. In some states, publicly-bid projects require formal qualification forms and a summary of the contractor's performance record. If a pre-bid conference or site inspection is scheduled, the date, time, and location are also stated. In addition, the RFP defines the various forms and amount of bid security that will be accepted. Some publicly-funded projects require that certain subcontractors submit their proposals separately prior to the general contractor's bid date. This practice is called *filed sub-bidding*. The required filed sub-bidders are listed according to their MasterFormat codes and respective trades. The RFP states the date, time, and location for the submission of filed sub-bids.

The date, time, and location for procuring a set of Contract Documents is stated, as is the cost to the bidder. Other pertinent information such as the time frame for award or rejection, special wage rates, tax exempt status, or legal rights of the awarding authority is also provided. In some cases the RFP contains sufficient information for the contractor to decide whether the project is worth bidding.

Contract Forms

The next major section of the specifications deals with contract forms. These forms include those needed for both bidding and award. The bidding portion of the contract includes those documents used in the submission of the general contractor's and subcontractors' bids. All bid forms are similar in terms of the information they contain. The purpose of using a pre-printed form is to keep comparisons between bidders uniform. If each bidder submitted his or her bid on a different type of form, it would be difficult to analyze the bid for conformity and completeness. As the bid form is a legal document, it must be filled out in accordance with the instructions or run the risk of being qualified as nonresponsive.

BIDDING REQUIREMENTS, CONTRACT FORMS, AND CONDITIONS OF THE CONTRACT

00010 PRE-BID INFORMATION
00100 INSTRUCTIONS TO BIDDERS
00200 INFORMATION AVAILABLE TO BIDDERS
00300 BID FORMS
00400 SUPPLEMENTS TO BID FORMS
00500 AGREEMENT FORMS
00600 BONDS AND CERTIFICATES
00700 GENERAL CONDITIONS
00800 SUPPLEMENTARY CONDITIONS
00900 ADDENDA

Note: The items listed above are not specification sections and are referred to as "Documents" rather than "Sections" in the Master List of Section Titles, Numbers, and Broadscope Section Explanations.

SPECIFICATIONS

DIVISION 1 -- GENERAL REQUIREMENTS

01010 SUMMARY OF WORK
01020 ALLOWANCES
01025 MEASUREMENT AND PAYMENT
01030 ALTERNATES/ALTERNATIVES
01035 MODIFICATION PROCEDURES
01040 COORDINATION
01050 FIELD ENGINEERING
01060 REGULATORY REQUIREMENTS
01070 IDENTIFICATION SYSTEMS
01090 REFERENCES
01100 SPECIAL PROJECT PROCEDURES
01200 PROJECT MEETINGS
01300 SUBMITTALS
01400 QUALITY CONTROL
01500 CONSTRUCTION FACILITIES AND TEMPORARY CONTROLS
01600 MATERIAL AND EQUIPMENT
01650 FACILITY STARTUP/COMMISSIONING
01700 CONTRACT CLOSEOUT
01800 MAINTENANCE

DIVISION 2 -- SITEWORK

02010 SUBSURFACE INVESTIGATION
02050 DEMOLITION
02100 SITE PREPARATION
02140 DEWATERING
02150 SHORING AND UNDERPINNING
02160 EXCAVATION SUPPORT SYSTEMS
02170 COFFERDAMS
02200 EARTHWORK
02300 TUNNELING
02350 PILES AND CAISSONS
02450 RAILROAD WORK
02480 MARINE WORK
02500 PAVING AND SURFACING
02600 UTILITY PIPING MATERIALS
02660 WATER DISTRIBUTION
02680 FUEL AND STEAM DISTRIBUTION
02700 SEWERAGE AND DRAINAGE
02760 RESTORATION OF UNDERGROUND PIPE
02770 PONDS AND RESERVOIRS
02780 POWER AND COMMUNICATIONS
02800 SITE IMPROVEMENTS
02900 LANDSCAPING

DIVISION 3 -- CONCRETE

03100 CONCRETE FORMWORK
03200 CONCRETE REINFORCEMENT
03250 CONCRETE ACCESSORIES
03300 CAST-IN-PLACE CONCRETE
03370 CONCRETE CURING
03400 PRECAST CONCRETE
03500 CEMENTITIOUS DECKS AND TOPPINGS
03600 GROUT
03700 CONCRETE RESTORATION AND CLEANING
03800 MASS CONCRETE

DIVISION 4 -- MASONRY

04100 MORTAR AND MASONRY GROUT
04150 MASONRY ACCESSORIES
04200 UNIT MASONRY
04400 STONE
04500 MASONRY RESTORATION AND CLEANING
04550 REFRACTORIES
04600 CORROSION RESISTANT MASONRY
04700 SIMULATED MASONRY

DIVISION 5 -- METALS

05010 METAL MATERIALS
05030 METAL COATINGS
05050 METAL FASTENING
05100 STRUCTURAL METAL FRAMING
05200 METAL JOISTS
05300 METAL DECKING
05400 COLD FORMED METAL FRAMING
05500 METAL FABRICATIONS
05580 SHEET METAL FABRICATIONS
05700 ORNAMENTAL METAL
05800 EXPANSION CONTROL
05900 HYDRAULIC STRUCTURES

DIVISION 6 -- WOOD AND PLASTICS

06050 FASTENERS AND ADHESIVES
06100 ROUGH CARPENTRY
06130 HEAVY TIMBER CONSTRUCTION
06150 WOOD AND METAL SYSTEMS
06170 PREFABRICATED STRUCTURAL WOOD
06200 FINISH CARPENTRY
06300 WOOD TREATMENT
06400 ARCHITECTURAL WOODWORK
06500 STRUCTURAL PLASTICS
06600 PLASTIC FABRICATIONS
06650 SOLID POLYMER FABRICATIONS

DIVISION 7 -- THERMAL AND MOISTURE PROTECTION

07100 WATERPROOFING
07150 DAMPPROOFING
07180 WATER REPELLENTS
07190 VAPOR RETARDERS
07195 AIR BARRIERS
07200 INSULATION
07240 EXTERIOR INSULATION AND FINISH SYSTEMS
07250 FIREPROOFING
07270 FIRESTOPPING
07300 SHINGLES AND ROOFING TILES
07400 MANUFACTURED ROOFING AND SIDING
07480 EXTERIOR WALL ASSEMBLIES
07500 MEMBRANE ROOFING
07570 TRAFFIC COATINGS
07600 FLASHING AND SHEET METAL
07700 ROOF SPECIALTIES AND ACCESSORIES
07800 SKYLIGHTS
07900 JOINT SEALERS

Figure 3.2

*Reprinted from the Construction Specifications Institute (CSI) and Construction Specifications Canada (CSC), **MasterFormat,** (1988 edition), with permission from CSI.*

DIVISION 8 – DOORS AND WINDOWS

08100 METAL DOORS AND FRAMES
08200 WOOD AND PLASTIC DOORS
08250 DOOR OPENING ASSEMBLIES
08300 SPECIAL DOORS
08400 ENTRANCES AND STOREFRONTS
08500 METAL WINDOWS
08600 WOOD AND PLASTIC WINDOWS
08650 SPECIAL WINDOWS
08700 HARDWARE
08800 GLAZING
08900 GLAZED CURTAIN WALLS

DIVISION 9 – FINISHES

09100 METAL SUPPORT SYSTEMS
09200 LATH AND PLASTER
09250 GYPSUM BOARD
09300 TILE
09400 TERRAZZO
09450 STONE FACING
09500 ACOUSTICAL TREATMENT
09540 SPECIAL WALL SURFACES
09545 SPECIAL CEILING SURFACES
09550 WOOD FLOORING
09600 STONE FLOORING
09630 UNIT MASONRY FLOORING
09650 RESILIENT FLOORING
09680 CARPET
09700 SPECIAL FLOORING
09780 FLOOR TREATMENT
09800 SPECIAL COATINGS
09900 PAINTING
09950 WALL COVERINGS

DIVISION 10 – SPECIALTIES

10100 VISUAL DISPLAY BOARDS
10150 COMPARTMENTS AND CUBICLES
10200 LOUVERS AND VENTS
10240 GRILLES AND SCREENS
10250 SERVICE WALL SYSTEMS
10260 WALL AND CORNER GUARDS
10270 ACCESS FLOORING
10290 PEST CONTROL
10300 FIREPLACES AND STOVES
10340 MANUFACTURED EXTERIOR SPECIALTIES
10350 FLAGPOLES
10400 IDENTIFYING DEVICES
10450 PEDESTRIAN CONTROL DEVICES
10500 LOCKERS
10520 FIRE PROTECTION SPECIALTIES
10530 PROTECTIVE COVERS
10550 POSTAL SPECIALTIES
10600 PARTITIONS
10650 OPERABLE PARTITIONS
10670 STORAGE SHELVING
10700 EXTERIOR PROTECTION DEVICES FOR OPENINGS
10750 TELEPHONE SPECIALTIES
10800 TOILET AND BATH ACCESSORIES
10880 SCALES
10900 WARDROBE AND CLOSET SPECIALTIES

DIVISION 11 – EQUIPMENT

11010 MAINTENANCE EQUIPMENT
11020 SECURITY AND VAULT EQUIPMENT
11030 TELLER AND SERVICE EQUIPMENT
11040 ECCLESIASTICAL EQUIPMENT
11050 LIBRARY EQUIPMENT
11060 THEATER AND STAGE EQUIPMENT
11070 INSTRUMENTAL EQUIPMENT
11080 REGISTRATION EQUIPMENT
11090 CHECKROOM EQUIPMENT
11100 MERCANTILE EQUIPMENT
11110 COMMERCIAL LAUNDRY AND DRY CLEANING EQUIPMENT
11120 VENDING EQUIPMENT
11130 AUDIO-VISUAL EQUIPMENT
11140 VEHICLE SERVICE EQUIPMENT
11150 PARKING CONTROL EQUIPMENT
11160 LOADING DOCK EQUIPMENT
11170 SOLID WASTE HANDLING EQUIPMENT
11190 DETENTION EQUIPMENT
11200 WATER SUPPLY AND TREATMENT EQUIPMENT
11280 HYDRAULIC GATES AND VALVES
11300 FLUID WASTE TREATMENT AND DISPOSAL EQUIPMENT
11400 FOOD SERVICE EQUIPMENT
11450 RESIDENTIAL EQUIPMENT
11460 UNIT KITCHENS
11470 DARKROOM EQUIPMENT
11480 ATHLETIC, RECREATIONAL, AND THERAPEUTIC EQUIPMENT
11500 INDUSTRIAL AND PROCESS EQUIPMENT
11600 LABORATORY EQUIPMENT
11650 PLANETARIUM EQUIPMENT
11660 OBSERVATORY EQUIPMENT
11680 OFFICE EQUIPMENT
11700 MEDICAL EQUIPMENT
11780 MORTUARY EQUIPMENT
11850 NAVIGATION EQUIPMENT
11870 AGRICULTURAL EQUIPMENT

DIVISION 12 – FURNISHINGS

12050 FABRICS
12100 ARTWORK
12300 MANUFACTURED CASEWORK
12500 WINDOW TREATMENT
12600 FURNITURE AND ACCESSORIES
12670 RUGS AND MATS
12700 MULTIPLE SEATING
12800 INTERIOR PLANTS AND PLANTERS

Figure 3.2 (continued)

DIVISION 13 -- SPECIAL CONSTRUCTION

13010 AIR SUPPORTED STRUCTURES
13020 INTEGRATED ASSEMBLIES
13030 SPECIAL PURPOSE ROOMS
13080 SOUND, VIBRATION, AND SEISMIC CONTROL
13090 RADIATION PROTECTION
13100 NUCLEAR REACTORS
13120 PRE-ENGINEERED STRUCTURES
13150 AQUATIC FACILITIES
13175 ICE RINKS
13180 SITE CONSTRUCTED INCINERATORS
13185 KENNELS AND ANIMAL SHELTERS
13200 LIQUID AND GAS STORAGE TANKS
13220 FILTER UNDERDRAINS AND MEDIA
13230 DIGESTER COVERS AND APPURTENANCES
13240 OXYGENATION SYSTEMS
13260 SLUDGE CONDITIONING SYSTEMS
13300 UTILITY CONTROL SYSTEMS
13400 INDUSTRIAL AND PROCESS CONTROL SYSTEMS
13500 RECORDING INSTRUMENTATION
13550 TRANSPORTATION CONTROL INSTRUMENTATION
13600 SOLAR ENERGY SYSTEMS
13700 WIND ENERGY SYSTEMS
13750 COGENERATION SYSTEMS
13800 BUILDING AUTOMATION SYSTEMS
13900 FIRE SUPPRESSION AND SUPERVISORY SYSTEMS
13950 SPECIAL SECURITY CONSTRUCTION

DIVISION 14 -- CONVEYING SYSTEMS

14100 DUMBWAITERS
14200 ELEVATORS
14300 ESCALATORS AND MOVING WALKS
14400 LIFTS
14500 MATERIAL HANDLING SYSTEMS
14600 HOISTS AND CRANES
14700 TURNTABLES
14800 SCAFFOLDING
14900 TRANSPORTATION SYSTEMS

DIVISION 15 --MECHANICAL

15050 BASIC MECHANICAL MATERIALS AND METHODS
15250 MECHANICAL INSULATION
15300 FIRE PROTECTION
15400 PLUMBING
15500 HEATING, VENTILATING, AND AIR CONDITIONING
15550 HEAT GENERATION
15650 REFRIGERATION
15750 HEAT TRANSFER
15850 AIR HANDLING
15880 AIR DISTRIBUTION
15950 CONTROLS
15990 TESTING, ADJUSTING, AND BALANCING

DIVISION 16 -- ELECTRICAL

16050 BASIC ELECTRICAL MATERIALS AND METHODS
16200 POWER GENERATION - BUILT-UP SYSTEMS
16300 MEDIUM VOLTAGE DISTRIBUTION
16400 SERVICE AND DISTRIBUTION
16500 LIGHTING
16600 SPECIAL SYSTEMS
16700 COMMUNICATIONS
16850 ELECTRIC RESISTANCE HEATING
16900 CONTROLS
16950 TESTING

Figure 3.2 (continued)

Reprinted from the Construction Specifications Institute (CSI) and Construction Specifications Canada (CSC), ***MasterFormat,*** *(1988 edition), with permission from CSI.*

Nonresponsive is the term applied to a bidder who has incorrectly filled out or inadvertently left out information on the bid form, thereby rendering that bidder ineligible for award.

Public and private bid forms share some common information. The bid form must address the specific project, and the bid must be in accordance with the Contract Documents. It must also acknowledge any changes, known as Addenda, to the original Contract Documents. Any alternates or unit prices must be listed clearly. Naturally, the base bid price must be listed. The last part of the form provides space for the bidder's company information and the signature of the authorized representative of the bidder's company. Other information specific to the project or to the format of the awarding authority bid form may be required, such as the names of filed sub-bidders, a breakdown of costs by division, or special authorization forms.

Additional forms are included for review by the prospective bidders, such as the *Bid Bond, Performance Bond,* and *Labor and Material Payment Bond.*

The Owner and Contractor Agreement is also provided so that the prospective bidder can see the contract that will be executed when the project is awarded. Contractors should review the proposed contract immediately, for conformity to company policy. The American Institute of Architects (AIA) has published a series of contract documents that are frequently used by owners. Some contracts written by architects and/or owners tend to favor the owner or architect to the point that the risks involved greatly outweigh the chance for success or profit, and bidding would be a waste of time and resources. Bidders should be aware of this and make every effort to understand the contract and any related information. In fact, bidders should either accept the contract language or decline to bid the project. Changing unfavorable contract language after the contract has been awarded (or the bidder selected) is unlikely.

General Conditions

The Contract, in the case of the AIA agreements, refers to the *General Conditions of the Contract for Construction.* The General Conditions are a set of rules and regulations on which the contract is based. Private companies and agencies of the U.S. Government have drafted versions of their own General Conditions. All are similar in content and address the 14 basic articles found in AIA A201. Those articles are listed below in order of appearance in the AIA A201:

1. General Provisions
2. Owner
3. Contractor
4. Administration of the Contract
5. Subcontractors
6. Construction by Owner/Separate Contractors
7. Changes in the Work
8. Time
9. Payments and Completion
10. Protection of Persons and Property
11. Insurance and Bonds
12. Uncovering and Correction of Work
13. Miscellaneous Provisions
14. Termination or Suspension of the Contract

It is imperative that contractors become familiar with the AIA A201, as it is regularly used. Bidders who encounter projects with unique versions of the General Conditions may want to consult an attorney to assess the risk of exposure and to determine whether bidding the project is in the best interest of the contractor.

Addressing the needs of individual projects is done in the *Supplementary General Conditions*. These amendments or supplements to the General Conditions can have an enormous impact on the original articles. The contractor should carefully review these changes, item by item.

The final General Conditions section is comprised of contract conditions. These include wage rate schedules, minority business participation, EEOC requirements, and Affirmative Action requirements, which the contractor should review with great care. All of the above-mentioned contract conditions have a direct effect on the bid price. If they are not complied with, they may have an adverse effect on the project outcome.

Technical Specifications

The last of the four categories is the *Technical Specifications*, which deal with the actual products and execution of the individual trades or segments of work. The Technical Sections are divided into three parts: Part 1 is *General*, Part 2 is *Products*, and Part 3 is *Execution.*

Part 1, the General section of the specifications, refers to the scope or limits of the work in that particular trade section. It makes the correlation between the technical specs and the General Conditions and Supplementary General Conditions of the Contract. This is where the administrative portion of the work for that trade would be found. For example, the submittals or shop drawings for that category of work are described in detail in the General section. It also describes the related work of other trades by specification section. It may note specific items to be furnished by others and installed in this particular section. The General section lists any special requirements for that work.

In Part 2, products may be spelled out by name and model number, or characteristics of the product may be described. Some of the products that are not specified by name can be identified generically by reference to a particular ASTM testing number or a Federal Specification number. The contractor should use great caution when pricing materials by a product's conformance with an ASTM number, as there could be several different grades of one product with sharply different prices. This is also the area in which products "or an approved substitute" would be specified. The contractor should remember that the designer decides whether a product is actually equal, and that the bidder should be able to prove equality with cold, hard facts and evidence. Failure to do so could make the bidder responsible for providing the specified item, even if the bid was based on a lower-priced item.

Part 3, Execution, deals exclusively with the method, techniques, and quality of the workmanship. This section makes clear the allowable tolerances of the workmanship. Tolerances refer to plumb, straight, level, or true. This section should also describe any preparation to the existing surfaces in order to accommodate the new work, as well as a particular technique or method for executing the work. The contractor should take off the work in accordance with this method, as other methods may render the work unacceptable.

Any discrepancies between the contract drawings and the specifications should be addressed to the designer or owner immediately. When the drawings and specifications differ as to the quality of a particular product, the standard rule of thumb is that the product of higher quality will be expected.

One final note concerning the specifications: Each page of a particular section should be sequentially numbered, including the MasterFormat code number. This helps the contractor to know that all the pages are intact and the complete information for that section is available.

Addenda

Any changes to the contract documents during the bidding period (the time period beginning the date the drawings are issued and ending on bid day) in the form of modification, clarifications, or revisions for any reason, are called *Addenda*. These addenda can be issued only by the designer. Addenda must be in writing and will automatically become part of the Contract Documents. There are special places on bid forms for the acknowledgment of Addenda. Failure to acknowledge the Addenda could, again, render the bid nonresponsive.

It is the responsibility of the contractor to evaluate how each addendum will affect the bid price. As Addenda can affect the bids of all parties involved, it is the responsibility of the general contract bidder to make the subcontractors and materials suppliers aware of any Addenda so they can adjust their bids accordingly.

Alternates

On some projects, the owner may want to see how a change in materials, method of construction, or addition or subtraction of work will affect the price. Such requests are presented in the form of additions or deletions to the base price. Typically the Alternate is listed at the end of the specification section that is affected by it, and also in Division 1, *Alternates*. The Alternate must include the increase or decrease in cost for all work, including all taxes, labor burden, overhead, and profit. The contractor must further calculate the total consequence of the Alternate. For example:

Alternate 1

Delete door, frame, and hardware for Door #3 in its entirety.

DELETE $_______

The contractor must not only delete the cost of the door, frame, and hardware, but must now include the additional wall materials, wood, gypsum board, paint, etc., to fill in the area Door #3 originally occupied in the base bid. The actual price of the Alternate is the difference between the two. In some cases, the addition or deletion of large scopes of work by Alternate can have a tremendous effect on the project's duration, thereby increasing or decreasing overhead and other time-sensitive costs of the project. Projects with limited budgets often include a series of Alternates as a way of choosing how to spend the money.

Allowances

Occasionally, as the Contract Documents are ready to be issued, certain items have yet to be finalized and are not ready for inclusion. Rather than leaving the item out altogether, the designer includes a cash allowance. The allowance is a fixed lump sum such as "$10,000 for the purchase and delivery of sod and plantings." The allowance can also be in the form of a unit price, such as "an allowance of $450 per M for brick, including delivery to the job site." Typically, it is clearly stated just what the allowance is for: materials only, materials and labor, or the entire item of work. If there is any doubt, it is up to the contractor to request clarification.

At the completion of the project, the actual cost is computed for items included as allowances, savings are returned to the owner, and overages are added to the contract price.

Unit Prices

In the course of design for some projects, architects or engineers are sometimes unable to pinpoint an exact quantity of a certain material. The best example of this is excavation of rock or unsuitable fill materials. The architect or engineer is aware of what needs to be done and the techniques or quality required, but is unable to determine the exact amount of rock or unsuitable materials to be removed. In an effort to at least predict the cost of this work, unit prices are requested and submitted as part of the bid form or proposal. Unit prices are included on the bid for each item, by a unit of measure. For example,

Excavation of unsuitable materials $ _______ per CY

The contractor should try to make an educated calculation as to the approximate quantity. The reason for this is that unit prices tend to decrease as the quantity increases. This is not always true, but is a point to consider. The unit price should always include markups for taxes, insurance, overhead, and profit.

Summary

While the project plans are a source of quantitative information, the specifications provide the qualitative requirements. Combined, these documents provide the contractor with the needed resources to proceed to the next step—an accurate quantity takeoff.

Chapter Four

GENERAL REQUIREMENTS

(Division 1)

Chapter Four

GENERAL REQUIREMENTS (Division 1)

The requirements set forth in the specifications in Division 1 are appropriately named the *General Requirements*, and are primarily concerned with how the project will be administered. The General Requirements as typically outlined in Division 1 and the General Conditions as discussed in Chapter 3 are different, but interrelated. The General Conditions define the contractual relationships that will bind the parties involved. The General Requirements specify the exact requirements for the particular project in such a way that the contractor can assign a monetary value to each.

While pricing of the items in Division 1 is often a matter of experience and judgment and should therefore be handled by a seasoned estimator, it is important to be aware of the kinds of items that will need to be addressed in the takeoff for Division 1. Summarizing these items is frequently done on a Project Overhead Summary sheet (see Figure 4.1 for an example). The contractor should determine which items on this sheet are pertinent to a particular project while reviewing the specifications. A simple "X" next to the item will be a reminder that this item is needed as part of the estimate.

Many of the items that comprise the General Requirements, such as scaffolding and the hookup and dismantling of temporary power, are not production-related and sometimes are mistaken as having no real cost. A review of the plans will help to identify and account for these items in the quantity takeoff. Any experienced contractor can testify to the fact that there is a definite cost to the administration of a project. Most of this cost is time-sensitive, and therefore increases proportionately with the duration of the project.

To quantify General Requirements items, it is essential that the contractor define the anticipated duration of each activity to determine the overall project duration.

The Project Schedule

Drafting a schedule is a necessity, whether for a simple residential addition or a complex multi-story tower. Often several revisions are required as the contractor becomes more familiar with the project. Drafting and revising a schedule are the only means of determining both how long the project will last and what the time-sensitive General Requirements expenses will be.

Means Forms

PROJECT OVERHEAD SUMMARY

PROJECT

LOCATION

ARCHITECT

QUANTITIES BY:

PRICES BY:

EXTENSIONS BY:

SHEET NO.

ESTIMATE NO.

DATE

CHECKED BY:

DESCRIPTION	QUANTITY	UNIT	MATERIAL/EQUIPMENT		LABOR		TOTAL COST	
			UNIT	TOTAL	UNIT	TOTAL	UNIT	TOTAL
Job Organization: Superintendent								
Project Manager								
Timekeeper & Material Clerk								
Clerical								
Safety, Watchman & First Aid								
Travel Expense: Superintendent								
Project Manager								
Engineering: Layout								
Inspection/Quantities								
Drawings								
CPM Schedule								
Testing: Soil								
Materials								
Structural								
Equipment: Cranes								
Concrete Pump, Conveyor, Etc.								
Elevators, Hoists								
Freight & Hauling								
Loading, Unloading, Erecting, Etc.								
Maintenance								
Pumping								
Scaffolding								
Small Power Equipment/Tools								
Field Offices: Job Office								
Architect/Owner's Office								
Temporary Telephones								
Utilities								
Temporary Toilets								
Storage Areas & Sheds								
Temporary Utilities: Heat								
Light & Power								
Water								
PAGE TOTALS								

Page 1 of 2

Figure 4.1 Project Overhead Summary Sheet

Means Forms

DESCRIPTION	QUANTITY	UNIT	MATERIAL/EQUIPMENT		LABOR		TOTAL COST	
			UNIT	TOTAL	UNIT	TOTAL	UNIT	TOTAL
Totals Brought Forward								
Winter Protection: Temp. Heat/Protection								
Snow Plowing								
Thawing Materials								
Temporary Roads								
Signs & Barricades: Site Sign								
Temporary Fences								
Temporary Stairs, Ladders & Floors								
Photographs								
Clean Up								
Dumpster								
Final Clean Up								
Punch List								
Permits: Building								
Misc.								
Insurance: Builders Risk								
Owner's Protective Liability								
Umbrella								
Unemployment Ins. & Social Security								
Taxes								
City Sales Tax								
State Sales Tax								
Bonds								
Performance								
Material & Equipment								
Main Office Expense								
Special Items								
TOTALS:								

Page 2 of 2

Figure 4.1 Project Overhead Summary Sheet (continued)

There are many different methods for scheduling, from the simple bar chart to the complex Critical Path Method. The bar or GANTT chart is the simplest, but not the most efficient. Nevertheless, it is the choice of most contractors in the residential/light commercial market. Bar charts are easy to understand for those with and without construction experience.

The typical arrangement of a bar chart construction schedule roughly follows the order of the MasterFormat, with some modification for the way the work will actually progress. Each activity is shown as a separate bar, which seems to suggest that each is independent and not related to the others. In reality, delays or changes to one activity have the potential to disrupt the entire project. The bar chart does not illustrate the interdependence of activities.

The basic principle of the bar chart, like any scheduling method, is that the work is sequentially connected; in other words, it follows a specific order of events. An example of the sequencing of the events is that the concrete foundation work must be completed before the first-floor deck is framed, and the roof must be sheathed before the roofing system is installed. It would make construction a lot simpler if the activities actually progressed this way, one starting after the other is finished. Unfortunately, even a simple residence would take a great deal of time to complete using this approach. That is why certain activities can and should occur simultaneously without interference, thereby shortening the duration of the overall project. When preparing a preliminary schedule for bidding, the contractor tries to schedule as many concurrent activities as possible without becoming counterproductive.

In the bar chart in Figure 4.2, each activity or item of work is shown as a solid line with a starting and completion date. The overall schedule is from March 26 through May 30.

The most important information is the duration of the project: approximately 10 weeks. Other information, such as when certain activities will occur, are important for calculating weather-related expenses or projecting weather-related delays. For example, if this project were in a northern climate where precipitation and cold temperatures are a factor, it would make sense to leave some margin of flexibility in the start and completion dates of the masonry work to allow for weather conditions. Another option and expense, in this case, would be "closing the project in" and providing heat to do the masonry work.

Bar charts tend to show the work on a general basis. The main advantage is that they illustrate the work as it occurs in relation to calendar dates. The contractor must develop at least a rough schedule for the purpose of evaluating time-sensitive project and office overhead.

The Superintendent and Other Field Personnel

The contractor must predict the cost for the salary of a superintendent or foreman to manage the project. Some projects require that the same superintendent be present from commencement to completion, to provide a sense of continuity. In that case there is little creativity involved in determining the cost. In other situations, a superintendent specializing in site work may supervise until that scope of work is complete; then a working foreman (with a lower salary) may handle the balance of the project.

Additional field or office personnel—such as security, secretarial clerks, assistant superintendents, engineers, and timekeepers—may be indicated

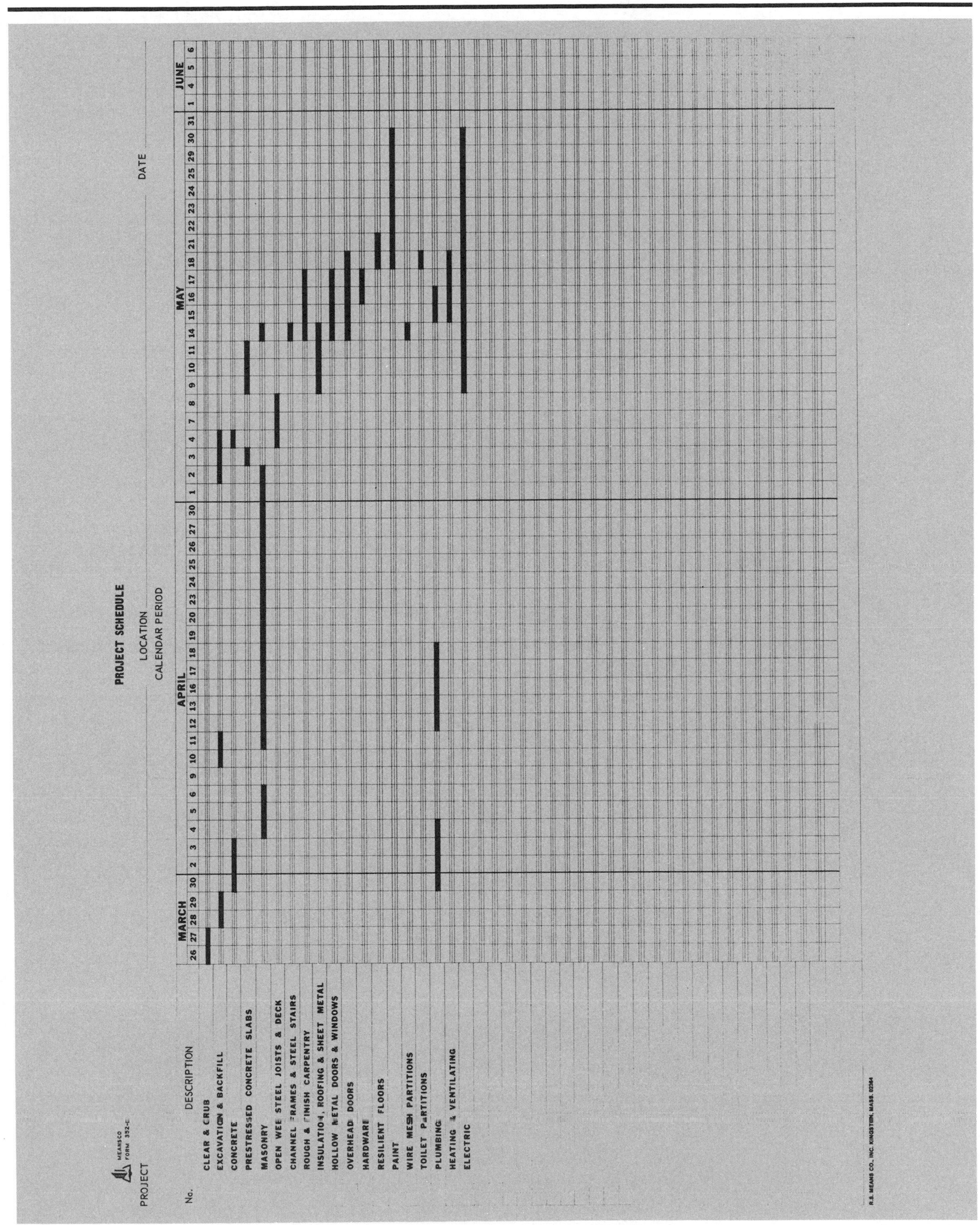

Figure 4.2 Bar Chart

in the specifications. Expenses such as travel, lodging, and certain benefits may be part of the cost of such personnel.

Note: The engineering staff on a project perform a specific function, such as establishing line and grade. Some projects require a registered land surveyor to do the initial site layout. These two costs for engineering services should not be confused.

Temporary Facilities

Temporary facilities include costs such as those associated with a field office, particularly the expenses incurred in the set-up, removal, and rental of a trailer. Also in this category are monthly or weekly expenses for equipment such as telephone, FAX, copy machines, and blueprinting machines. The contractor must also calculate the cost of such services as temporary light and power, heat, and water services for both the field office and the project itself. This involves determining how long the field office will be needed, and at what point the permanent utilities will be available to the building. Other items classified as temporary facilities are portable toilets; temporary barricades; noise, dust, rodent, and pest control; staging and platforms; cranes and hoists; storage trailers; and a project sign. The cost of most of these items is related to the amount of time needed and can be predicted by reviewing the schedule.

Distribution and Handling

The cost of handling and distribution can be included as part of the unit price for a particular item, or listed separately as a total price for handling and distribution. For example, door hardware in a school or office building would typically be installed by carpenters, but the distribution of the correct hardware set for each door would be done by someone else.

The contractor must evaluate the cost of handling and distribution based on the quantity and complexity of items.

Testing and Inspections

Many specifications require the testing and inspection of completed work. Testing often takes place in the field as the work is in progress. Soil and concrete testing are examples of such testing services. The cost of these services is commonly assumed by the contractor. The contractor should secure a price, either per day or per unit of inspection, from an independent agency versed in these services. The cost for the services can be calculated based on the frequency of inspection as indicated in the specs or as stipulated by code requirements.

The contractor might also include in the takeoff any delays or down time resulting from such inspections.

Submittals and Samples

Specifications often require that literature and samples be submitted to the designer for approval. This may seem redundant, since the submittals are materials specified by the designer, but it is simply one item in a series of checks and balances that ensure the project is built as designed. The contractor must take into account the cost of purchasing required samples. Because most contractual agreements make the contractor responsible for the correctness and compliance of the submittals, the main cost is for the contractor's personnel who will review and organize submittals.

Some projects may require a "mock-up"—a model of a system or assembly that shows construction details, strength, and appearance. An example is a mock-up panel of masonry brick, block, or stone. The mock-up is created so that the designer and owners can see the item prior to construction. The cost of the mock-up involves materials, labor, and equipment.

Other submittals include construction schedules, shop drawings, and "as-built" or record drawings.

Permits and Fees

The category of permits and fees deals almost exclusively with the costs imposed by the local municipality or state agency. Included are such items as building and occupancy permits; assessments for the connection of sewer, water, or storm drainage lines to the municipal services; street opening fees; municipal inspection services; and plumbing and electrical permits. Each locality has different rules for calculating the cost of the permits and fees. It is the contractor's responsibility to contact local building officials to determine the costs of such fees for bidding purposes.

Other permit and fee charges could be levied by the local utility company for connections of temporary services.

Cutting and Patching

On many projects the general contractor is responsible for cutting and patching holes or openings for the installation of work by other trades. The contractor must base the cost of such work on the quantity or occurrence of the opening or patch, and either assign it as part of a subcontract or carry the cost as a project overhead item.

Project Meetings

Project meetings typically include the cost of time expended by personnel attending and administering periodic job-site meetings. The general contractor normally attends all project meetings for the duration of the project. Subcontractors, on the other hand, should take into account the cost for only the time period they are present on the project.

Cleanup and Debris Removal

There are two basic classifications of cleanup—*general*, which is ongoing for the duration of the project, and *final*, which entails the final cleaning of the premises prior to occupancy by the owners. The contractor must calculate the man-hours and resources needed to perform the general cleanup over the length of the project, including such costs as dumpster rental and removal, and storage of the dumpster in the street or on public property (a common occurrence in cities). It is important to note the associated costs of constructing and dismantling such items as ramps and chutes.

Final cleaning tends to be more specialized, requiring buffing and vacuuming equipment not usually considered part of the general contractor's typical work, and may be performed by a company that normally provides these services.

Project Closeout

Project closeout is bringing the project to final completion. It includes the submission of record drawings (sometimes called "as-built" drawings), warranties, maintenance and operational instructions and literature; the removal of contractor equipment; and the punch list. All of these items represent a definite cost to the contractor and should be carried in the bid. These items tend to be manpower-intensive and should not be overlooked. The punch list is a list of items that remain to be repaired, replaced or completed, as identified by the owner and owner's representatives in a "final" walk-through of the job site with the contractor. Even the smallest of projects requires provisions for the correction of punch list work. There are no formulas for projecting the cost of punch list items, but the company's prior record for performance of these tasks should give the contractor a good idea of what to expect.

Winter Protection

Many projects occur during the winter months in cold climates. The contractor must determine by the schedule when temperature-sensitive work will occur, and provide adequate closure and heat in order to perform the work in accordance with the specifications and building codes. The rental of heaters and the provision of fuels, tarps, and moisture barriers should be considered, in addition to erection and dismantling of any enclosure.

Summary

While General Requirements items are not directly referred to on the plans, they are an inherent part of the construction process. A thorough review of the plans is required in order to determine the quantities associated with General Requirements items, both construction aids such as scaffolding and equipment rental, and administrative overhead.

The takeoff sheets in Figure 4.3 show the quantities for the work of Division 1, General Requirements, as they relate to the sample project. Additional items may be required for specific conditions that relate to the individual site or to the contractor performing the work. The following checklist may be helpful for making decisions regarding other items that may not be included on the takeoff sheet.

Division 1 Checklist

The following checklist can be used to ensure all items for this division have been accounted for.

Project Overhead Costs

Personnel

- ☐ Superintendent
- ☐ Field engineer
- ☐ Foreman
- ☐ Clerk or field secretary
- ☐ Security personnel

Temporary Facilities

- ☐ Field office rental
- ☐ Equipment trailer rental
- ☐ Set-up and removal of trailers
- ☐ Temporary power and light
- ☐ Temporary heat or cooling
- ☐ Temporary water
- ☐ Toilets
- ☐ Telephone
- ☐ Fencing
- ☐ Barricades
- ☐ Project sign
- ☐ Access roads

- ☐ Staging and platforms
- ☐ Cranes and hoists
- ☐ Noise control
- ☐ Dust control
- ☐ Pest and rodent control

Handling and Distribution

- ☐ Materials handling
- ☐ On-site storage
- ☐ Distribution and transportation

Quality Controls

- ☐ Soils testing
- ☐ Concrete testing
- ☐ Structural testing

Submittals

- ☐ Shop drawings
- ☐ Submittals and samples
- ☐ Construction schedules
- ☐ Photographs

Cleanup/Debris Removal

- ☐ General cleanup
- ☐ Dumpster fees
- ☐ Landfill permit or fees
- ☐ Final cleaning

Project Meetings

Project Closeout

- ☐ Operations and maintenance manuals
- ☐ Operations and maintenance instruction
- ☐ Record drawings
- ☐ Punch list

Bonds

- ☐ Bid
- ☐ Performance and payment
- ☐ Roof or maintenance

Insurance

- ☐ General liability
- ☐ Workers' compensation
- ☐ Vehicular
- ☐ Fire
- ☐ Property damage
- ☐ Unemployment and Social Security

Taxes and Benefits

- ☐ Sales tax
- ☐ Union benefits

Permits and Fees
- ☐ Building and occupancy
- ☐ Plumbing and gas
- ☐ Electrical
- ☐ Sewer and water
- ☐ Licenses
- ☐ Betterment fees

Cutting and Patching

Engineering
- ☐ Layout with own forces
- ☐ Registered land surveying

Winter Protection
- ☐ Heat
- ☐ Enclosure
- ☐ Snow plowing
- ☐ Thawing materials

Equipment
- ☐ Small tools and equipment
- ☐ Ladders
- ☐ Pumping
- ☐ Equipment maintenance
- ☐ Fuel and oil

Main Office Overhead
- ☐ Personnel
- ☐ Rent
- ☐ Telephone and utilities
- ☐ Postage and supplies
- ☐ Legal and accounting
- ☐ Vehicular
- ☐ Miscellaneous expenses

Means Forms

QUANTITY SHEET

PROJECT: SAMPLE PROJECT

SHEET NO. 1-1/1

ESTIMATE NO. 1

LOCATION:

ARCHITECT: HOME PLANNERS

DATE:

TAKE OFF BY: WJD

EXTENSIONS BY: WJD

CHECKED BY: KF

DESCRIPTION	NO.	DIMENSIONS				UNIT		UNIT		UNIT		UNIT
DIVISION 1: GENERAL REQUIRMENTS												
BUILDING PERMIT	1	LS			1	LS					1	LS
FIELD LAYOUT FOR FOUNDATION LOCATION	1	LS			1	LS					1	LS
TEMPORARY ELECTRIC SERVICE	1	EA			1	EA					1	EA
GENERAL CLEANUP DURING CONSTRUCTION	100	HR			100	HR					100	HR
30 CY DUMPSTER FOR DEBRIS	2	EA			2	EA					2	EA

Figure 4.3

Chapter Five

SITE WORK

(Division 2)

Chapter Five

SITE WORK
(Division 2)

Site work takeoff can often be the most difficult part of an estimate. One reason is the number of unpredictable factors. Even with subsurface investigation and a careful site inspection, it is not always possible to know what lies beneath the surface. The site contractor must carefully study all available information. A site inspection should be conducted only after the drawings and specifications have been examined.

Most site work is shown on the civil drawings. The contractor will use the site plan or site grading plan to determine the cut and fill quantities to bring the site to the new grade. Utilities drawings will be used to take off trench excavation and backfill for the various utilities that will service the building, as well as the piping and related work such as manholes. Drainage drawings will provide the contractor with the information needed to take off the drainage work and the limits of the excavation and backfill. Site plans also provide the contractor with information such as the removal or relocation of existing site features such as trees or benches, and general site improvements such as paving and curbs.

Architectural drawings must be reviewed for details that would be shown in section, such as foundation drains. Structural drawings, specifically the foundation plans and details, give the site contractor the limits of the excavation for the foundation. They also show interior details such as footings, or depressed areas in the slab that will require excavation, backfill, and compaction.

Mechanical and electrical drawings must be reviewed for details that show any trenching for piping or conduits that are in the basement or under the slab-on-grade. This work is often overlooked because it is not shown in the civil drawings.

Finally, the specifications must be reviewed for the type and quality of the product and installation. The specifications may offer additional information, such as Subsurface Investigation Reports.

Interpreting Subsurface Investigations

Architects often retain the services of geotechnical engineers to investigate the existing conditions below the surface of the soil prior to the design of many foundations on commercial structures. The purpose of the investigation is to define the conditions of the soil that will support the structure, and to provide bidders with an idea of the conditions that will be encountered during excavation.

Several methods can be used to sample the soil beneath the surface. The simplest is the *test pit,* which allows a visual inspection of the soils. Information such as soil contents, stratification, water table height, and cohesiveness can be obtained through visual inspection. Unfortunately, the test pit is restricted by the reach of the excavating equipment, and most structures require analysis of soils at far greater depths. A fairly common method for reaching greater depths is *test boring*. This method provides an actual sample of the materials as they occur in the soil, including the location of the water table. The soil samples are analyzed, and the results are interpreted by the geotechnical engineer and distributed in the form of a report. The report typically provides a plan that locates each boring with respect to the proposed structure. The actual report contained in the specs is in the form of a simple chart providing the physical description of the soil sample as it occurs vertically, and the corresponding depth. The location of water, if any was encountered, should be clearly noted, as should any major obstructions.

The contractor should make careful note of the disclaimer that accompanies each report. It explicitly states that the information contained in the boring reports is for the convenience of the contractor and, further, that the geotechnical firm assumes no responsibility for their representation of the soil conditions of the site as a whole. Figure 5.1 is an example of a boring report.

Clearing and Grubbing

Clearing of the site refers to the removal of brush, trees, topsoil, and other structures on the surface. *Grubbing* refers to the removal of tree stumps. Most of this work is done with power equipment, with some hand work. Occasionally, clearing and grubbing require the use of specialized equipment. The contractor should inspect the site to identify work that can be done by equipment versus work that must be done by hand. The depth of the topsoil can be determined from the test boring report or a visual inspection. Topsoil is typically stockpiled on the site for reuse in planting and lawn areas.

Units for Takeoff

Stripping of topsoil is calculated and listed in the takeoff as an area multiplied by the depth and expressed as cubic yards. When calculating the area of topsoil to be removed, it should be enlarged to accommodate any clearance needed. Grubbing and removal of stumps are taken off by the piece. Takeoff quantities should include the count of stumps to be removed, as well as trucking and fees for disposal. These should be listed by the truckload in the takeoff.

Demolition

Demolition, by definition, is the dismantling and removal of unwanted existing work. Demolition can include a wide variety of tasks and is usually labor- and equipment-intensive. Again, the contractor must decide what demolition work can be accomplished by equipment and what requires handwork. The scope of the demolition work should include the transporting and loading of dumpsters or trucks for the removal of debris.

Units for Takeoff

Frequently, the units of demolition are difficult to label. Demolition tasks can be quantified by the linear foot, square foot, cubic foot, cubic yard, or

DEPTH (FT)	CASING (BLOWS/FT)	SAMPLE No.	SAMPLE PEN/REC.	SAMPLE DEPTH (ft.)	SAMPLE BLOWS/6"		SAMPLE DESCRIPTION Classification	REMARKS	STRATUM DESCRIPTION
0									TOPSOIL
		J-1	24/18	0-2	3	4	Loose, fine SAND, trace inorganic, silt, dark brown, wet, poorly graded (SP)		
					4	5			.5
2									
		J-2	24/8	2-4	5	7	Medium dense, fine SAND, trace silt, dark brown, wet, poorly graded (SP)		
					9	7			
4									
		J-3	24/13	4-6	5	4	Loose, fine SAND, trace silt, dark brown, wet, poorly graded (SP)		
					3	3			FINE
6									SAND
		J-4	24/10	6-8	4	4	Loose, fine SAND, trace silt, trace coarse gravel, dark brown, wet, poorly graded (SP)		
					3	4			
8									
		J-5	24/16	8-10	4	4	Loose, fine SAND, trace silt, brown, dry, poorly graded (SP)		
					4	7			
13									13'
		J-6	24/18	13-15	7	3	Medium dense, fine to coarse SAND, some fine to coarse gravel, brown, dry, well graded (SW)		
					15	23			

GRANULAR SOILS BLOWS/FT	DENSITY	COHESIVE SOILS BLOWS/FT.	DENSITY
		2	V. SOFT
0-4	V. LOOSE	2-4	SOFT
4-10	LOOSE	4-8	M. STIFF
10-30	M. DENSE	8-15	STIFF
30-50	DENSE	15-30	V. STIFF
>50	V. DENSE	>30	HARD

REMARKS:

Figure 5.1 Boring Report

each. There are some tasks or segments of work that contain more than one operation, but are better left as a whole for the purposes of estimating. This work is sometimes classified as a *lump sum* (LS). An example might be removing a metal window and frame from an existing concrete block wall, cutting the jambs to the floor, and removing the debris. It might be wise to consider a lump sum because of the variety of tasks to be performed.

Other Takeoff Considerations

Sometimes demolition work requires the erection of barriers, ramps, chutes, or some protection for surrounding work. The contractor must include this as well. In addition to protection, the cost of disposal by means of a dumpster or trucking to a landfill must be included. The contractor has the option of including these costs as part of the demolition work or as part of Division 1 (see Chapter 4). In either case the takeoff for disposal should be a separate item so that the contractor can analyze various options.

Excavation

Simply put, excavation is the digging of a hole in the earth. In construction, excavation is performed for the purpose of erecting a building or other site improvement. Excavation, and earthwork in general, is a volume calculation.

Units for Takeoff

Regardless of the method used, excavation is taken off as a calculation of the length, width, and depth of the area to be excavated in CF and converted to CY, where 1 CY equals 27 CF (see "Units of Measure" in Chapter 2). The contractor should allow for additional digging to allow room to work within the excavated area.

Other Takeoff Considerations

It is the nature of soil to expand once it has been excavated. This expansion refers to the increased volume the fill will occupy once it is excavated. This is called the *swell* of the material. Swell is expressed as a percentage over and above the original volume. The inverse is true when the volume of a material is compacted. It tends to shrink or lose volume. This is called *compaction*. The actual percentage of swell or compaction will depend on the conditions of the material being excavated. It is safe to say that the more finely graded the materials, the less swell and compaction can be expected. Figure 5.2 lists soil characteristics and factors for calculating the volume of soils in various conditions.

There are several different types of excavation. For estimating purposes, we can group them into two important classifications: bulk excavation by cross-section, and trench or general excavation.

Bulk Excavation by Cross-Section

Moving large masses of earth to establish new grades for parking lots, roads, or buildings is referred to as *bulk excavation*. The contractor must examine the site closely and determine the amount of earth that will have to be handled in order to transform the existing grades to the proposed grades.

One of the more frequently used methods is called the *cross-section method*. It involves the tabulation of cuts and fills for small increments of the total parcel. *Cutting* refers to the removal of earth in order to achieve the desired grade. *Filling* is the addition of earth to raise the existing conditions to meet the desired grade (see Figure 5.3). The cross-section method divides the area in question into a series of smaller, equal areas. This is done with a grid drawn on the grading plan. It is helpful if both the existing and the new proposed grades are shown in the form of contours. Contours are lines that indicate the same horizontal elevation. Existing contours are shown on the grading or site plan as dashed lines, with the new or proposed grades shown as solid lines. Figure 5.4 illustrates the cross-section grid in place over the existing grading plan.

Weights and Characteristics of Materials

Approximate Material Characteristics*				
Material	Loose (Lbs./C.Y.)	Bank (Lbs./C.Y.)	Swell (%)	Load Factor
Clay, dry	2,100	2,650	26	0.79
Clay, wet	2,700	3,575	32	0.76
Clay and gravel, dry	2,400	2,800	17	0.85
Clay and gravel, wet	2,600	3,100	17	0.85
Earth, dry	2,215	2,850	29	0.78
Earth, moist	2,410	3,080	28	0.78
Earth, wet	2,750	3,380	23	0.81
Gravel, dry	2,780	3,140	13	0.88
Gravel, wet	3,090	3,620	17	0.85
Sand, dry	2,600	2,920	12	0.89
Sand, wet	3,100	3,520	13	0.88
Sand and gravel, dry	2,900	3,250	12	0.89
Sand and gravel, wet	3,400	3,750	10	0.91

*Exact values will vary with grain size, moisture content, compaction, etc. Test to determine exact values for specific soils.

Typical Soil Volume Conversion Factors				
Soil Type	Initial Soil Condition	Converted to: Bank	Loose	Compacted
Clay	Bank	1.00	1.27	0.90
	Loose	0.79	1.00	0.71
	Compacted	1.11	1.41	1.00
Common earth	Bank	1.00	1.25	0.90
	Loose	0.80	1.00	0.72
	Compacted	1.11	1.39	1.00
Rock (blasted)	Bank	1.00	1.50	1.30
	Loose	0.67	1.00	0.87
	Compacted	0.77	1.15	1.00
Sand	Bank	1.00	1.12	0.95
	Loose	0.89	1.00	0.85
	Compacted	1.05	1.18	1.00

$$\text{Swell (\%)} = \left(\frac{\text{Wt./bank C.Y.}}{\text{Wt./loose C.Y.}} - 1 \right) \times 100$$

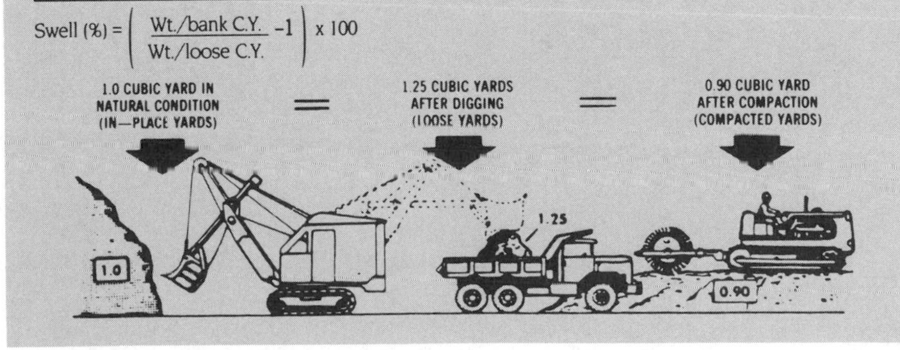

Figure 5.2 Soil Characteristics

The size of the grid can be whatever is convenient. The important thing is that the size be uniform throughout the plan and that the grids be square for ease of calculation. Also, the grid spacing really depends on the area to be cross-sectioned. If the lot slopes very gradually, the spacing can be spread out, but if the lot has dramatic changes in elevation (a concentration of contours), closer spacing (smaller grids) may prove to be more accurate in the takeoff. Again, whatever the grid size, it should remain constant throughout the takeoff.

Each of the smaller areas is numbered. The horizontal lines are numbered, and the vertical lines numbered and lettered so that each intersection can be referenced. The contractor must now calculate the elevation of the intersection of each of these lines by interpolating between the contours. *Interpolation* is an approximation of the elevation of a particular point between two known points. The existing grade is placed in the upper right-hand quadrant, the proposed grade in the upper left-hand quadrant. The difference between the two is noted as either a cut or fill, in the lower right or left quadrant, respectively. Each grid area is examined to see if it changes from fill to cut or vice versa. These must be calculated separately. For grids that are all cut or all fill, the contractor calculates the average cut or fill for the grid, multiplies it by the area of the grid, and divides by 27 to convert CF to CY. The total of each grid is kept in separate columns on a cut/fill sheet as seen in Figure 5.5.

For grids that change from cut to fill, the contractor must determine what portion of the area is cut and what portion is fill. The same principles are applied to arrive at the cut and fill portion for each. Once all the grids have been calculated, the total cut and fill for each column is calculated. From this information the contractor can determine quantities of earth to be exported or imported to the site, along with the total CY of material to be handled.

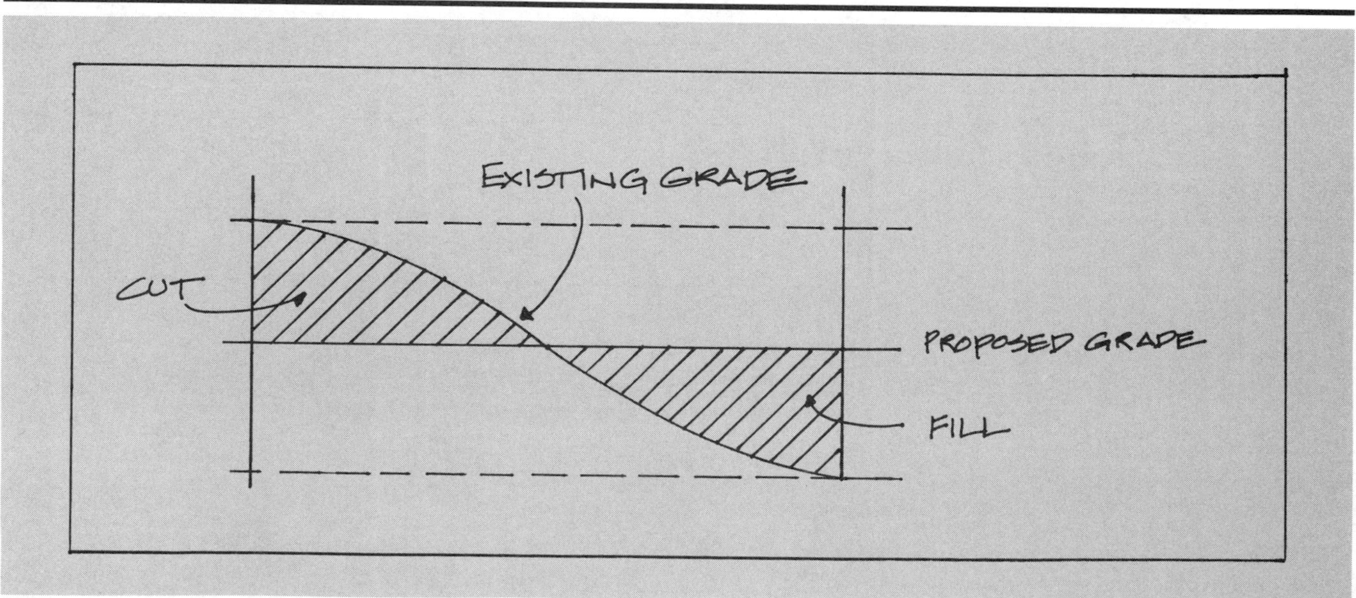

Figure 5.3 Filling to Meet Grade

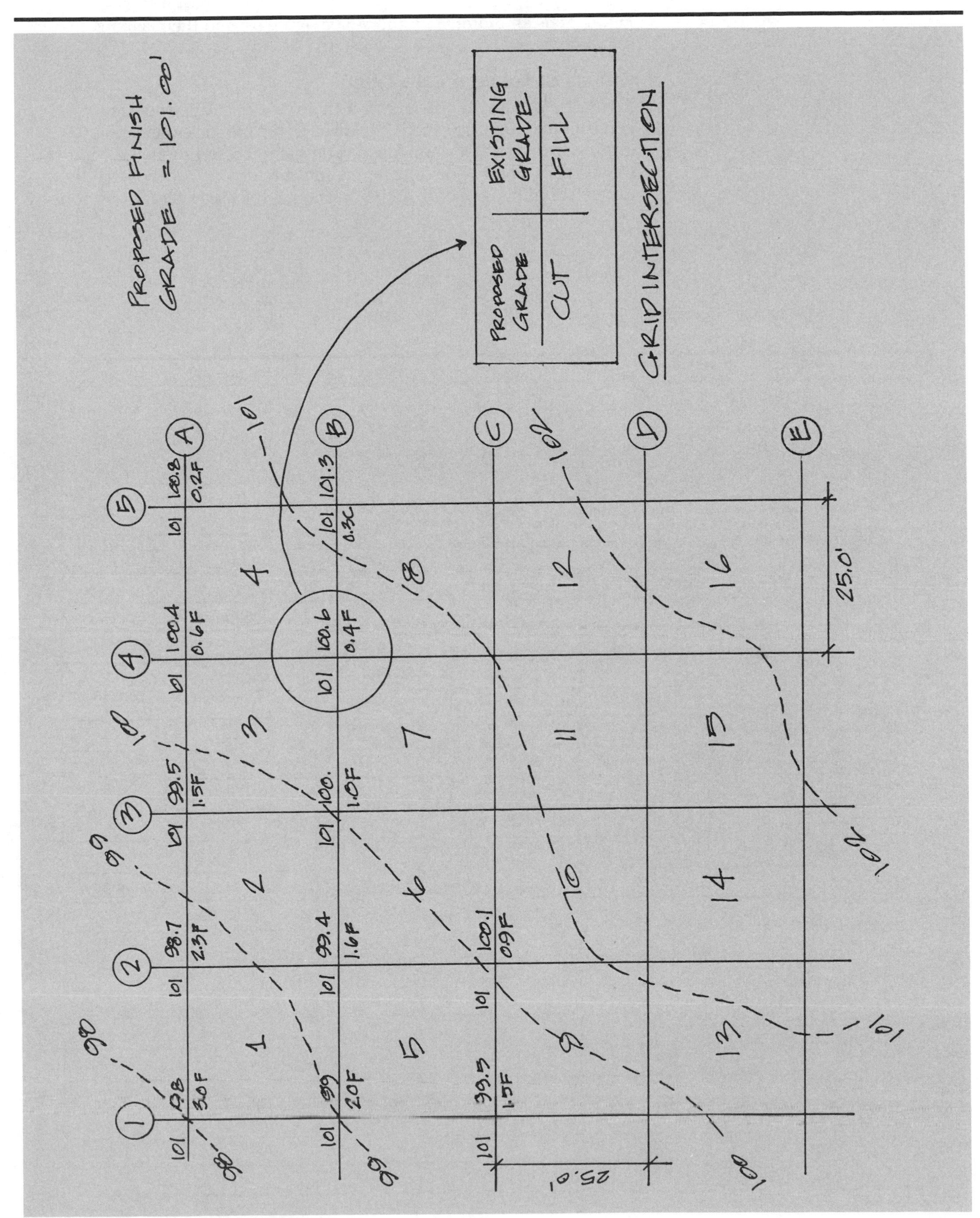

Figure 5.4 Cross-Section Grid Layout

Figure 5.6 is an example of grid #5 of Figure 5.4 and the resulting calculations.

Trench or General Excavation

A large part of earthwork takeoff includes the calculations of trench or general excavation, classified as excavation for specific engineering improvements such as foundations and site utilities or drainage. Most of this work is done with digging equipment such as backhoes and excavators, although some of the dressing of the excavated area is done by

CUT & FILL ESTIMATE SHEET

PROJECT ____________

DATE ____________
BY ____________
CHK. ____________

GRID	CUT			FILL		
	AREA(SF)	AVE. CUT(FT)	VOLUME(CY)	AREA(SF)	AVE FILL(FT)	VOLUME(CY)
1	625			625	2.23	51.62
2	625			625	1.6	37.04
3	625			625	.88	20.37
4	625			625		
5	625			625		
6						
7						
8						
9						
10						
11						
12						
13						
14						
15						
16						

Figure 5.5 Cut/Fill Sheet

hand. The contractor should keep the quantities for machine and hand excavation separate, as there is a considerable difference in pricing.

When determining the limits of the excavation, the contractor must take into account the actual size of the basement foundation, then add a sufficient buffer to provide for the access of workers and materials to do the work. In addition to the buffer for workers, the contractor must include overdigging to stabilize the slope. Overdigging allows the soils to stabilize naturally, to prevent earth from sliding into the excavated area. Different soil compositions tend to stabilize at different angles, as measured from the horizontal plane at the bottom of the excavation. This is called the *angle*

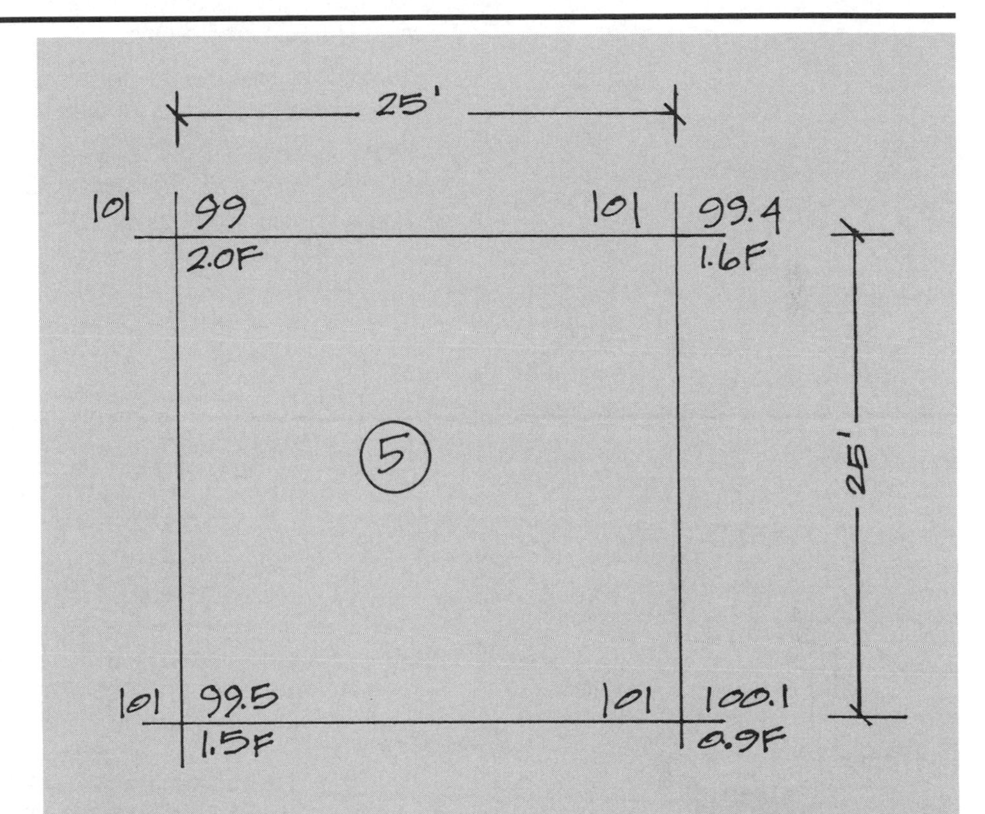

In analyzing Grid #5 from the Cross Section Grid in Figure 5.4, it is determined that the grid is a "fill" grid, because all four intersections show that fill is required to reach the Proposed Grade. The total amount of fill required is the average depth of fill multiplied by the area to be filled. In this case the area to be filled is the area of Grid #5—25′ x 25′ or 625 square feet.

To find the average fill required; Add the fill required for each intersection and divide by 4 to find the average.

$$\frac{2.0 + 1.6 + 0.9 + 1.5}{4} = 1.5 \text{ ft. (average depth of fill)}$$

1.5 ft. x 625 SF = 937.5 CF divided by 27 CF per CY = 34.7 CYs

Figure 5.6 Calculation for Grid #5

of repose. The more cohesive the soil, the steeper the angle of repose. The less cohesive the soil, the shallower the angle of repose. The term *cohesive* refers to soils with a large clay content. Noncohesive soils are represented by sand or gravel materials. An extreme example of noncohesive soils is finely-graded beach sand. Regardless of the care taken in digging, it always seems to cave right back in. At the other end of the scale, many of us have seen excavated trenches that stabilize vertically at the limits of the bucket on the excavator. This is a cohesive soil. Figure 5.7 and the accompanying table illustrate the angle of repose for various soils.

Once the angle of repose and the required size of the excavation have been determined, it is fairly easy to calculate the cross-sectional area of the trench and, subsequently, its volume. Figure 5.8 shows the cross-sectional area of a sample trench. Its corresponding calculations follow.

> *Calculate the volume of earth to be excavated for a trench 4'-0" deep × 100'-0" long. The bottom of the trench is 4'-0" wide including access. The angle of repose is 45 degrees.*
>
> *An angle of repose of 45 degrees is a ratio of 1:1; that is, for every foot of depth, the excavation is 1'-0" wide as measured from the vertical plane.*
>
> *Therefore, the area of A1 = 4' × 4' = 8 SF. Since A3 is symmetrical with A1, then A3 = 8 SF. The area of A2 = 4' × 4' = 16 SF.*
>
> *Adding all three areas, A1 + A2 + A3 = 32 SF.*

GRADE

ANGLE OF REPOSE

SOIL TYPE	ANGLE OF REPOSE (DEGREES)	
	DRY	DAMP
SAND	25-45	30-50
GRAVEL	25-40	20-30
CLAY	45-90*	30-45

* IT SHOULD BE NOTED THAT SOME CLAYS OR COHESIVE SOIL WILL MAINTAIN A VERTICAL ANGLE OF REPOSE WHEN EXCAVATED.

Figure 5.7 Angle of Repose

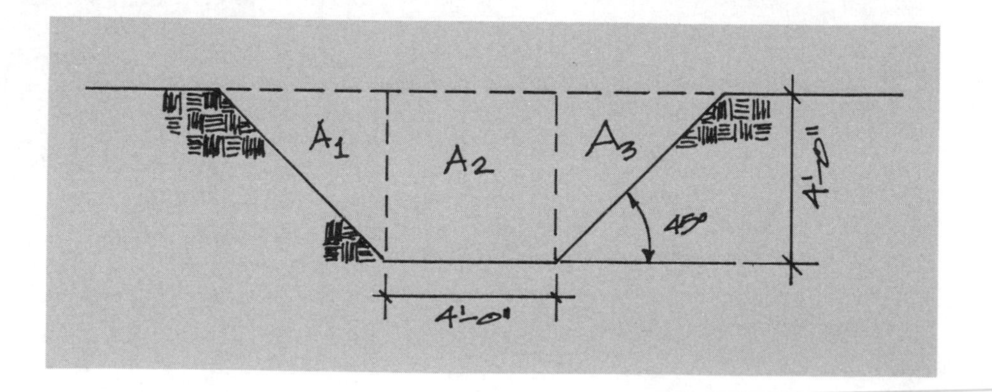

Figure 5.8 Sample Trench

If we use the formula for volume from Chapter 3,

$$V = A \times l$$

$$V = 32\ SF \times 100.0' = 3200\ CF$$

$$\text{Then convert to cubic yards: } \frac{3200\ CF}{27\ CF/CY} = 118.52\ CY$$

Shoring

Shoring may be required either because of safety considerations or because there is insufficient room to stabilize the slope naturally. Shoring can be in the form of wood, steel, or concrete sheet piling. Other similar artificial means, such as trench boxes, are available. The contractor must calculate the area to be retained or held back. The area to be shored will require overdigging to provide sufficient access to install the shoring and bracing.

Units for Takeoff

The unit of takeoff for shoring and bracing is the SF. Units are calculated by multiplying the perimeter of the area to be shored by the height of the shoring from the bottom of the trench to the top of the excavation. The vertical dimension should include additional length to embed the shoring below the bottom of the trench level. The takeoff should include the materials as well as the labor and equipment to install and remove the shoring. The takeoff for shoring should be separate from the work it is meant to protect.

Hauling Earth On/Off Site

During the excavation process, materials frequently need to be moved greater distances than economically feasible with the excavating equipment. Moving the material to an alternate location on or off site is accomplished by the use of trucks.

Units for Takeoff

The quantity of earth to be transported by trucks is calculated by the CY. The contractor must calculate the amount of material to be excavated and moved. Factors for the increased volume resulting from swell must be added in the form of additional truckloads. Estimates of the distance to be traveled and the approximate time to load and dump must be calculated to determine the cost per yard. Additional costs for dumping of excess material and handling at the dump site may be warranted. Alternate pricing can be by the truckload, based on the quantity the type of truck will carry.

Backfill and Compaction

The contractor must be familiar with the requirements for backfilling and compaction as mandated in the specifications. Special attention should be paid to allowable thickness of the layers of backfill, the materials required, and the density of the materials after compaction. In the absence of written specifications governing the placement and compaction of backfill, the contractor can refer to sections on the architectural drawings that show the thickness of the compacted materials under the slab. Trench excavation for utilities or drainage, as well as general excavation for foundations, require backfill and compaction in some capacity. Figure 5.9 illustrates a section of the foundation that requires backfill and compaction under the slab on grade.

Units for Takeoff

The unit of takeoff for backfill and compaction is the CY, although some contractors prefer to separate the backfill from the compaction because backfill is placed by machine and the compaction is done by manpower and a plate compactor or roller. Backfill should still be quantified by CY, and compaction by SF or SY.

Other Takeoff Considerations

It may be necessary to add the cost of delays for testing of the compacted soils between layers.

Grading

Grading can be accomplished by equipment, by hand, or a combination of the two. Quantities for each should be kept separate. Movement of large quantities of earth should not be labeled as grading, but as excavation. Grading should be restricted to the dressing of previously filled or cut areas.

The contractor can determine the area to be graded by the site improvements shown on the site drawings as well as architectural drawings as preparation for slabs-on-grade. Areas paved with bituminous concrete, brick walkways, and concrete walks or pads require a graded sub-base as preparation for the work.

Units for Takeoff

Hand grading is taken off by the SF or SY. Machine grading is done by bulldozer or similar equipment and supplemented by manpower for hand

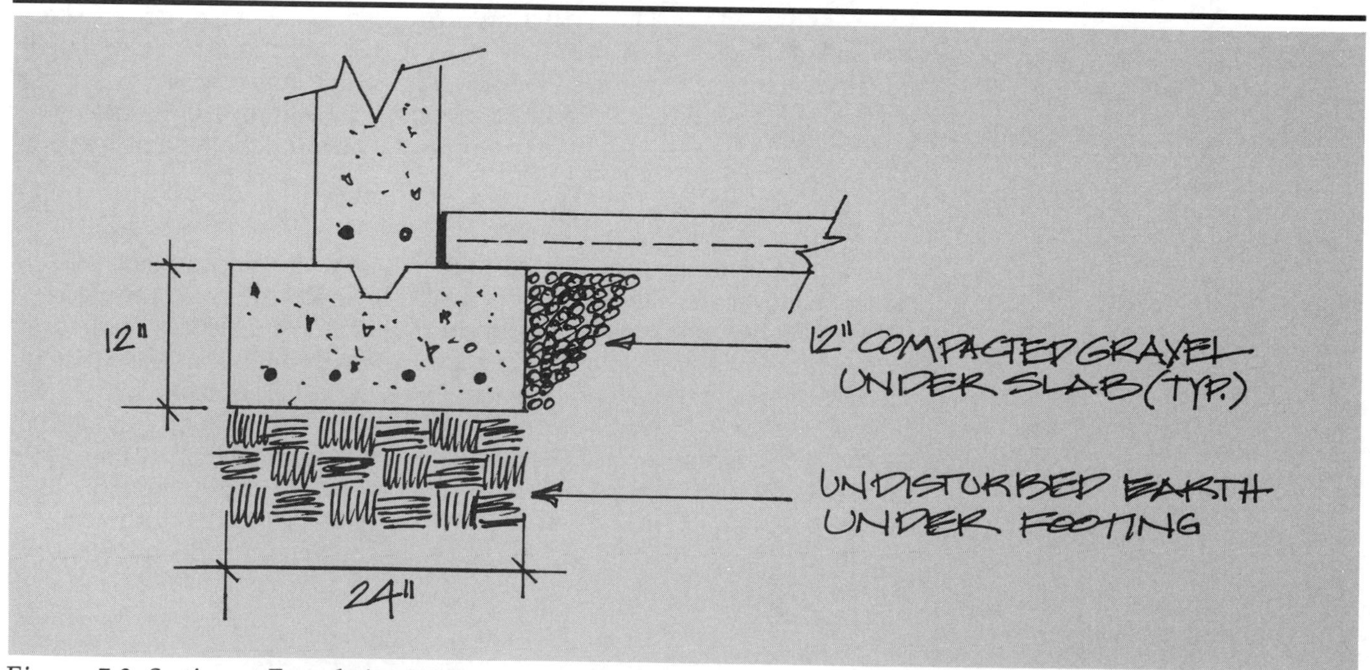

Figure 5.9 Section at Foundation Wall

grading in areas close to structures. Typically, machine grading is taken off by the SF and converted to a larger unit, such as SY or acres.

Other Takeoff Considerations

The cost of grading varies in accordance with the degree of accuracy required. The contractor should use caution in selecting the appropriate equipment for the task, considering maneuverability and proximity of buildings or site improvements.

The contractor may want to include costs for additional equipment to "spot" fill in low areas, or removal of excess materials in "high" areas.

Utilities and Drainage

Drainage and utilities require the installation of piping and related fittings to carry services to and from the building and site. The contractor should refer to the site plans to determine the limits of the drainage and utility work. Plans such as the site drainage plans and the utilities plan will show the location of the various types of piping and any changes in direction or elevation. Sections of the piping may be shown on site detail plans. The contractor should refer to the specifications for particular information concerning the type of materials used and the method of installation. This information is important for the takeoff and pricing of the work. Additional items such as catch basins and manholes are also shown on the site plans, with elevations for inverts and rims. Sections or details for these items are referred to in order to clarify the construction and materials to be included in the takeoff. Figure 5.10 illustrates a partial site plan showing the sewer line and the corresponding details necessary for the takeoff.

Units for Takeoff

All pipe is taken off by the linear foot. Depending on the material and size of the pipe, costs for cutting to specific lengths in the field must be included. The cost for cutting a thin-wall PVC pipe with a handsaw is negligible. The cost for the cutting of larger-diameter pipe that requires special cutting tools, however, is not negligible. The quantity of cuts for such pipe should be taken off by each piece (EA). Quantities for different types and sizes of pipe should be kept separate.

The connection of pipe lengths should also be kept separate, especially if it involves labor-intensive tasks such as welding, fusion, or installing mechanical-type fittings. Typical bell-and-spigot or glued PVC couplings are not classified as a separate labor task. Special fittings such as valves, bends, tees, or wyes should be taken off separately and listed according to type and size, with the unit of EA. Any other tasks for the connection of various pipe, such as the mortaring of joints in reinforced concrete pipe, should be noted separately, taking into account the pipe's diameter and accessibility.

Many types of piping can be installed by hand, as in the case of smaller-diameter PVC pipe. Larger-diameter pipe, such as ductile iron, corrugated drain pipe, or reinforced concrete pipe, requires the use of equipment for lifting and placing. This should be included as part of the unit cost for installation.

Certain types of piping work require the use of precast concrete structures such as manholes, handholes, catch basins, and tanks. Each of these structures should be taken off individually, separated by type, size, capacity, or use. Similar items whose sizes are constant should be

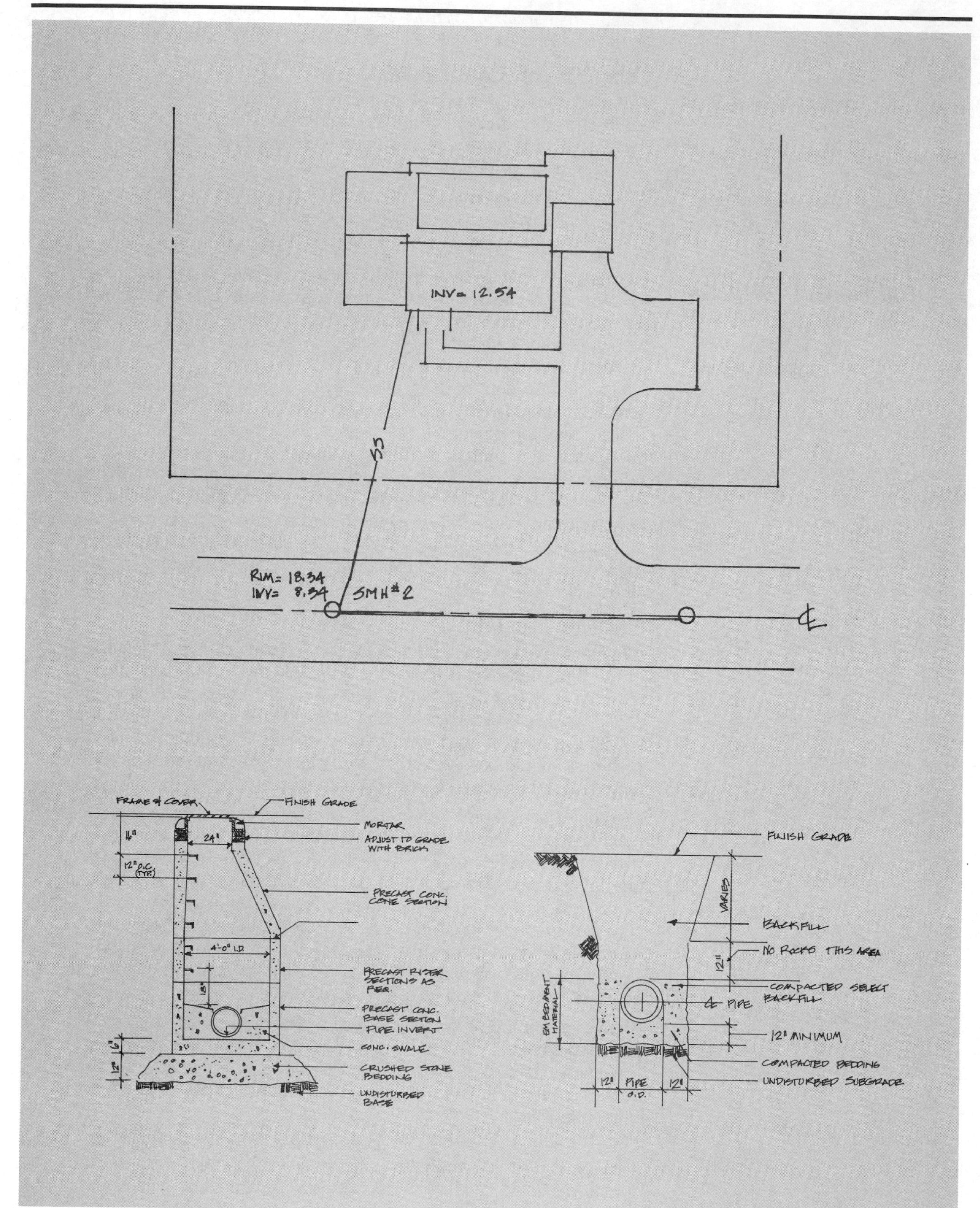

Figure 5.10 Partial Site Plan

quantified as EA (e.g., 1,000 gal. septic tanks, 4′ catch basins). Items that vary in size, such as the vertical heights of manholes, must be separated and quantified by EA or VLF (vertical linear foot). Additional pieces to complete the structure, such as manhole frames and covers, catch basin frames and grates, metal steps, and special swales or inverts should be quantified separately.

The contractor should secure quotes for the testing of installed piping services, as often this is specified to be done by an independent agency. The cost of testing should be carried as an independent contractor price. Typically, the cost of testing the pipe should be included in Division 2 because it is considered as part of the actual installation work.

Related work, such as concrete thrust blocks for the water service, should be calculated on an individual basis, including the CY of material needed and any specified or required formwork. Special work for the support of existing pipe should be taken off and priced separately, and may be quantified as a LS (lump sum).

Other Takeoff Considerations

Takeoff for excavation and backfill for site drainage and utilities should be kept separate from the actual piping and appurtenances. The contractor should use caution when quantity surveying piping on plans of large engineering scale (1″ = 50′ or greater), as the chance for error increases dramatically. Piping is sold in specific lengths; the takeoff should be rounded to the nearest full length required. Special bedding materials such as sand and stone may also be required as part of the installation. The quantities of bedding materials should be kept separate, and should be included as part of the backfill work.

In most jurisdictions, the pipe within the confines of the foundation itself, such as underground waste lines for sanitary sewer services or conduit piping for underground electrical work, is the responsibility of the respective trade and not the work of the site contractor. Similarly, the installation of piping for underground gas services or site lighting may be the responsibility of the gas utility company or licensed electrician. These practices may not always be apparent on the drawings, but should be researched by the contractor.

Underpinning

When the excavation of new work comes too close to, or goes below, the foundation of an existing structure, it must be underpinned. While underpinning is not always required by the specifications, it is a matter of safety and responsibility. Most contracts hold the contractor liable for damage to adjacent structures. Underpinning requires the existing foundation support to be brought to the depth of the proposed structure. This is often an expensive task and may require the services of a specialized contractor. It can be quantified by a variety of units, and should be selected according to the application.

Paving and Curbing

Paving is the surfacing of a subbase, typically compacted gravel or stone with a course of materials such as bituminous concrete, brick, or concrete to provide a wearing surface for vehicular or pedestrian traffic. The most common examples are parking areas, walks, and patios. The perimeters of parking areas and walkways are often surrounded by curbing. Curbing is available in a variety of materials and methods of installation. Some of the more common types are bituminous concrete, cast-in-place concrete, and precast concrete.

The contractor should refer to the site plans to determine the areas and types of paving and curbing. In addition to plan view drawings, sections and details showing sizes, thicknesses, and composition of the items should be included. Reference to the specs to determine the quality and method of installation is also required.

Paving and Surfacing

Units for Takeoff

The rough grading or preparation of the subbase is categorized under the grading scope of work. Fine grading and rolling of the subbase just prior to installation is typically done by the paving contractor. Takeoff units for this task are by the SF and can be extended to SY for ease of pricing. Typically the work is performed by equipment, with labor allocated for minor hand work.

Additional work for tying in new surfaces to existing surfaces may be required and should be listed separately. It can be taken off by the LF of abutting surfaces.

Bituminous concrete or asphaltic paving is usually applied in two courses. The first course is called the binder or base course. The top course is called the wearing course. Each course should be taken off and listed separately. Trenches that require temporary patching should be taken off by the SY and listed separately. Paving contractors often have minimum charges for such things as temporary patches.

The placing of concrete for walkways or aprons involves the calculation of the concrete needed. The contractor should include a small percentage for waste resulting from spillage while handling. The quality and strength of the concrete, including any additives or hot water, should be noted as it will affect the unit cost. Concrete should be quantified by the CY. Placement and finishing of the concrete surface should be taken off by the SF.

Special curing compounds or protection from rapid hydration should be listed separately and taken off by the SF. In the case of liquid compounds, the contractor can convert the square footage into coverage of gal./SF and to the manufacturer's units, such as 5-gallon containers. Polyethylene barriers to contain moisture can be calculated similarly by the SF and converted to the convenient size roll. Additional labor may be required to manually assure proper hydration. (Refer to "Concrete Curing" in Chapter 6.)

Most concrete placed in walkways requires some type of formwork at the perimeter of the pour, usually called *edge forms*. Edge forms are listed by the LF, and pricing should include the cost for stripping the forms. Materials for edge forms often have multiple uses, so the materials cost may be divided by the number of expected uses. (Refer to "Formwork" in Chapter 6.) Reinforcing for concrete walkways in the form of welded-wire fabric is calculated by the SF, with allowance for overlap and waste resulting from the size of the walkway. Reinforcing in the form of steel rebar will be discussed at length in Chapter 6.

Other Takeoff Considerations

Takeoff should be an area calculation in SF, extended to SY for pricing. Large areas such as parking lots or roadways should be kept separate from walks. The cost of application varies with the size of the paved area and the equipment used.

Curbing

Units for Takeoff

Curbing is available in many types and compositions; each type is taken off and listed separately. Curbing manufactured off site, such as granite or precast concrete, often requires the mortaring of joints between each section. This should be included as part of the installation cost. Radii for precast concrete or granite curb are listed separately and grouped according to radius. These are more expensive than straight pieces. The cost of materials should include any delivery charges. Because of its weight, manufactured curbing requires the use of equipment for setting and poured concrete to hold the curb in place. These should be included in the cost. Ready-mix concrete for holding the precast curb in place should be taken off by the CY.

Cast-in-place concrete curb or extruded curb and gutter will require the calculation of the concrete needed in CY. The contractor may need to include waiting time for ready-mix concrete trucks, as curb tends to pour more slowly than flatwork. Cast-in-place curb requires the use of forms. The unit cost should include forming, placing, stripping, and rubbing of the exposed surfaces. Cast-in-place curb is taken off by the LF.

The cost for bituminous concrete curb extruded by machine is based on production quantity and type of material used and is taken off by the LF.

Other Takeoff Considerations

The cost for extruded curb and gutter is based on the production (quantity) rate, setup and moves, small quantities, and the accessibility of the work, all of which affect unit costs.

Pumping and Dewatering

High levels of groundwater during excavation may require the use of pumping or dewatering equipment. Simple dewatering may require the use of a localized pump, in which case the cost would include fuel, pump rental, and the labor to operate the pump. This could be calculated based on the time required to dewater the excavation, and is quantified as an allowance or lump sum. More elaborate means of dewatering, such as a well point system, may be required. Factors that can affect the cost of dewatering are location, season, and the amount of precipitation.

Units for Takeoff

Simple dewatering is quantified by the amount of time the pumping operation will take. It can be listed in the takeoff by the day, week, or month.

Other Takeoff Considerations

Because of its specialized nature, the dewatering of large areas requires the services of a contractor specializing in this type of work. The contractor would be advised to secure the price of an independent contractor for this work.

Landscaping

Landscaping includes a variety of components to improve the appearance of the site. Projects of a more sophisticated nature often include a landscaping plan as part of the site plans. The landscaping drawing is in plan view, and often includes sections or details for plant material installation, decorative stone walls, or general improvements of an ornamental nature. As part of the landscaping drawings, a planting schedule lists the plantings by species, size, and variety. (Refer to the "Landscaping Plan" section in Chapter 1.)

Landscaping work requires the spreading, raking, and fine grading of topsoil as a base for sod or grass seed. The installation of plantings, seed or sod, and bark mulch are part of the landscaping scope of work.

Units for Takeoff

Plantings should be taken off by counting the plantings shown and listing them according to species, size, and variety of the planting. The contractor can check the quantities with the planting schedule as a means of verification.

Sod and seeded areas are taken off by the SF. The contractor should convert the SF area of seed to the rate required in the specifications. Rate refers to pounds per 100 SF.

Spreading, fine grading, and raking of topsoil is taken off by the SF. The quantity of topsoil needed is calculated by multiplying the area by the required depth and extending the amount to the CY.

Grading and distribution can be accomplished with small equipment, such as a bobcat or backhoe. Areas that are inaccessible by machine are done by hand and should be listed separately. Grading, raking, and rolling of topsoil is quantified by the SF. Additional topsoil may be needed for raised areas in planting beds and is also calculated by the CY.

Other landscaping tasks may involve the placing of stone, or timbers, as a focal point. Such tasks should be figured separately. Units will vary in accordance with the task.

Additional costs for landscaping work include fertilizers converted from lbs./SF to lbs., and wood stakes and wire for staking trees, taken off by the piece (EA).

The spreading of bark mulch, wood chips, or similar materials is taken off by the SF or SY. The materials are calculated in CY, based on the depth of the material.

Other Takeoff Considerations

Cost of the plantings and sod is affected by availability, season, species, and size.

Allowances should be made for compaction resulting from rolling of the topsoil.

When calculating the cost of labor for planting trees and shrubs, the contractor may want to include equipment for placing items too big to be handled by hand.

Frequently the specifications call for a maintenance period that includes watering of the materials until growth has been established. The cost for such work is contingent upon the frequency and the season. For example, watering and mowing are required more often during peak growing seasons.

Site Improvements

Site improvements include a wide range of miscellaneous items and tasks. These can be shown on the site plan, or on a separate plan called the site improvement plan. The units used are as follows.

Fencing, wood or chain link	LF
Site irrigation	per Head or SF covered
Playground equipment	EA (piece)
Trash receptacles	EA
Fountains	EA
Planters and furniture	EA (piece)

Other Takeoff Considerations

Some site improvements may require assembly or shipping costs. This should be noted in the takeoff for accurate pricing.

Rock

Special types of excavating require the removal of ledge or large formations of rock. This can be accomplished by various methods. Two common methods are drilling and blasting, both of which require the use of specialized equipment and techniques.

Units for Takeoff

Drilling or blasting of rock is taken off by its in-ground quantity expressed in CY, and should be listed separately from common methods of excavation once the rock has been broken up. Rock can exceed its in-ground volume by as much as 60%.

Other Takeoff Considerations

Rock excavation is affected by a variety of factors. The contractor should quantify rock excavation after careful consideration of the following factors.

- Classification of rock—hard, medium, or soft
- Proximity of adjacent structures (blasting)
- Depth of drilling
- Quantity of rock to be drilled/blasted
- Type of explosive required
- Special permits, insurance, or safety requirements

Minor rock drilling may be done by jackhammer with carbide bits.

Because of its unique nature, the price of drilling and blasting should be secured from a contractor specializing in this work.

Pile Foundations

The type of pile foundations varies extensively with location, type of pile (wood or steel), depth driven, quantity of piles, and type of soil. Costs for layout, moving and transporting the rig, cutting off piles, and test piles must be included.

Units for Takeoff

Piles are taken off by the VLF of each pile, multiplied by the quantity of piles.

Other Takeoff Considerations

Pile driving is a specialized form of work requiring heavy equipment and considerable experience. The contractor should secure an independent contractor for the work whenever possible.

Miscellaneous

Miscellaneous items that should be accounted for in the site work takeoff and estimate are:

- Special fees or permits
- Charges for transporting equipment
- Temporary protection for open excavations
- Engineering (layout and grades)
- Police details for street work (traffic control)
- Storage or trailer charges for large items
- Sweeping between paving courses

Summary

The takeoff sheets in Figure 5.11 are the quantities for the work of Division 2–Site Work for the sample project. It should be noted that the takeoff proceeds (generally) in the order that the work would occur during the construction process. This helps eliminate errors or omissions that may be caused by not following the natural order of the activities.

Division 2 Checklist

The following list can be used to ensure that all items in the takeoff for this division have been accounted for.

Demolition

- ☐ Above-ground structures
- ☐ Below-ground structures
- ☐ Removal

Clear and Grub

- ☐ Clearing brush and trees
- ☐ Chipping
- ☐ Stockpiling topsoil
- ☐ Removing and disposing of stumps

Excavation

- ☐ Bulk excavation
- ☐ Trench excavation for foundations
- ☐ Trench excavation for utilities and drainage
- ☐ Miscellaneous hand excavation

Shoring and Bracing

Hauling Materials On/Off Site

Grading

- ☐ Site grading by machine
- ☐ Hand grading trenches
- ☐ Hand grading at subbases for walks/slabs

Backfill and Compaction

- ☐ General backfill
- ☐ Trench backfill
- ☐ Special backfill or bedding
- ☐ Machine and hand equipment compaction
- ☐ Swell and shrinkage
- ☐ Excess and borrow

Utilities and Drainage

- ☐ Gas distribution
- ☐ Water distribution
- ☐ Sewer or septic systems
- ☐ Electric and communications
- ☐ Drainage
- ☐ Manholes, catch basins, precast tanks

Underpinning

Pumping and Dewatering

Paving and Curbing

- ☐ Bituminous concrete paving
- ☐ Concrete paving
- ☐ Curbing
- ☐ Rolling and fine grading
- ☐ Special paving surfaces

Landscaping

- ☐ Plantings
- ☐ Lawns
- ☐ Loam
- ☐ Maintenance
- ☐ Warranty
- ☐ Miscellaneous

Rock

Pile Foundations

Site Improvements

- ☐ Fencing and gates
- ☐ Retaining walls
- ☐ Miscellaneous

Miscellaneous

- ☐ Permits and fees
- ☐ Temporary protection and barricades
- ☐ Water taps
- ☐ Traffic control
- ☐ Transportation and hauling
- ☐ Storage
- ☐ Layout and engineering

Means Forms

QUANTITY SHEET

PROJECT: SAMPLE PROJECT
SHEET NO.: 2-1/3
ESTIMATE NO.: 1
LOCATION:
ARCHITECT: HOME PLANNERS
DATE:
TAKE OFF BY: WJD
EXTENSIONS BY: WJD
CHECKED BY: KF

DESCRIPTION	NO.	DIMENSIONS				UNIT		UNIT		UNIT		UNIT
DIVISION 2: SITEWORK												
CLEAR & GRUB SITE APPROX 25'-0" AROUND PERIMETER OF HOUSE (HOUSE 35'x60')	1	85	110		9350	SF					9350	SF
STRIP & STOCK PILE TOPSOIL (6") @ CLEARED AREA	1	85	110	.5	4675	CF					174	CY
EXCAVATION FOR MAIN HOUSE FOUNDATION 3'-0" OVER DIG 1:1 SLOPE RATIO @ 8'-0" BELOW GRADE MAIN EXCAV.	1	38	36	8	10944	CF						
SLOPE	76	8	8	.5	2432	CF						
	72	8	8	.5	2304	CF						
					15680	CF					581	CY
HAND GRADE AND COMPACT @ MAIN EXCAVATION	1	38	36		1368	SF					1368	SF
4" GRAVEL @ INTERIOR OF FOOTING	1	31.33	29.33	.33	304	CF					12	CY
ADDITIONAL MATERIAL FOR 15% COMPACTION	1	304	.15		46	CF					2	CY
HAND GRADE & MACHINE COMPACT FILL @ FOOTING INTERIOR	1	31.33	29.33		919	SF					919	SF

Figure 5.11

Means Forms

QUANTITY SHEET

SHEET NO. 2-2/3

PROJECT SAMPLE PROJECT

ESTIMATE NO. 1

LOCATION

ARCHITECT HOME PLANNERS

DATE

TAKE OFF BY WJD

EXTENSIONS BY: WJD

CHECKED BY: KF

DESCRIPTION	NO.	DIMENSIONS			UNIT	UNIT	UNIT	UNIT
BACK FILL MAIN HOUSE FOUNDATION (LESS FOUND. VOLUME)	1	581 CY			581 CY			
	1	30	32	8	(285 CY)			
					296 CY			296 CY
ADD FOR COMPACTION 15% ADDITIONAL FILL	1	296	.15		45 CY			45 CY
TRUCK EXCESS FILL OFF SITE	1	240 CY			240 CY			240 CY
MACHINE COMPACTOR PERIMETER	1	341 CY			341 CY			341 CY
EXCAVATE CRAWL SPACE GARAGE FOUNDATION & COVERED PORCH FOUND. TO -4'-0" W/ 2'-0" OVERDIG & NO SLOPE	2	16	5	4	640 CF			
	1	17.33	5	4	347 CF			
	1	49.67	4.67	4	928 CF			
	1	5	5	4	100 CF			
	1	7	11.33	4	318 CF			
	1	23.67	4.67	4	443 CF			
					2776 CF			103 CY
HAND GRADE & COMPACT TRENCH BOTTOM FOR CRAWL, GARAGE & COVERED PORCH	2	16	5		160 SF			
	1	17.33	5		87 SF			
	1	49.67	4.67		232 SF			
	1	5	5		25 SF			
	1	7	11.33		80 SF			
	1	23.67	4.67		111 SF			
					695 SF			695 SF

Figure 5.11 (continued)

Means Forms

QUANTITY SHEET

SHEET NO. 2-3/3

PROJECT: SAMPLE PROJECT

ESTIMATE NO. 1

LOCATION:

ARCHITECT: HOME PLANNERS

DATE:

TAKE OFF BY: WJD

EXTENSIONS BY: WJD

CHECKED BY: KF

DESCRIPTION	NO.	DIMENSIONS				UNIT		UNIT		UNIT		UNIT
BACKFILL @ FOUNDATION												
DEDUCT FOR CMU												
& CONCRETE	1	103 CY			103	CY						
	(1	19 CY)			(19	CY)						
					84	CY					84	CY
ADDITIONAL MATERIAL												
15% FOR COMPACTION	1	84	15		13	CY					13	CY
REMOVE 12" MATERIAL												
@ CRAWL SPACE TO												
MAINTAIN 2'-0" MIN												
BELOW JOISTS	1	13	16	1	208	CF					8	CY
ROUGH GRADE @												
PERIMETER OF HOUSE	1	85	110		9350	SF						
DEDUCT FOR HOUSE	(1	27	14)		(378	SF)						
	(1	30	32)		(960	SF)						
	(1	7.33	5)		(37	SF)						
	(1	14	21)		(294	SF)						
					7681	SF					7681	SF
SPREAD TOPSOIL & RAKE												
FINE GRADE @ SAME												
AREA TO 4" DEPTH	1	7681	.33		2534	CF					94	CY
ADD 25% FOR COMPACTION												
OF TOPSOIL	1	94	.25		24	CY					24	CY
TRUCK ADDITIONAL												
MATERIAL OFFSITE	1	56			56	CY					56	CY

Figure 5.11 (continued)

Chapter Six

CONCRETE

(Division 3)

Chapter Six

CONCRETE (Division 3)

Concrete as defined by *Means Illustrated Construction Dictionary* is "a composite material consisting of sand, coarse aggregate, cement and water. When mixed and allowed to harden, it forms a stone-like material." Concrete, because of its high compressive strength, durability, ability to withstand the weather, and relative ease in handling and shaping, is a versatile material for widespread use in the construction industry.

When used in conjunction with steel reinforcing, concrete takes on the ability to withstand elongation, called *tensile strength*, as well as its own compressive strength. This added characteristic makes reinforced concrete an ideal product for use as a structural component.

Because of its versatility, concrete can be used in many parts of a project. To prepare an accurate takeoff, the contractor must review all drawings in the set for concrete work. Division 3, Concrete, is found in the structural drawings, specifically the foundation plan and sections showing details of footings, walls, piers, and slabs.

Other drawings, such as mechanical/electrical, should also be studied for items like boxouts for piping and conduit or equipment. Architectural drawings are reviewed for indications of surface finish treatments.

The specifications and drawings should be reviewed to determine the strength of the mix or any particular additive required. This is necessary to properly separate the different mixes for accurate pricing.

Refer to Chapter 1 for symbols and abbreviations that relate to concrete work.

Takeoff of concrete work can be organized into the categories discussed in the following sections. This chapter provides basic information about the properties and typical units of measure used to take off concrete.

Concrete Materials

While concrete has many applications in construction, its main function is as a structural component. Concrete has an extremely high *compressive strength*; that is, the ability to resist a crushing force, usually imposed by the weight of the structure it supports.

With water as a main ingredient, mixed concrete is greatly affected by temperature and weather extremes. When placing concrete in cold weather, the poured concrete must be protected against freezing. Concrete that freezes prior to curing will never achieve its full strength. Similarly, pouring concrete on hot, dry days may evaporate essential water used for the curing process before it has a chance to set. Both conditions should be avoided. Evaluating the requirements for placing concrete is covered later in this chapter.

Units for Takeoff

Concrete is taken off by the cubic foot and converted to cubic yards, the accepted units for both estimating and purchasing concrete. One CY of concrete contains 27 CF (see Figure 6.1).

The quantities of different types of concrete should be listed separately in the takeoff because a variety of factors will affect the price per CY.

Other Takeoff Considerations

The following factors affect the price of the concrete work, and should therefore be considered in the takeoff:

- The strength of the mix, specified as the concrete's *compressive strength per square inch* after curing (e.g., 3000 psi, 3500 psi, 4000 psi).
- The use of additives that accelerate the time for reaching full strength. For example, High Early Strength Portland cement will achieve the same strength in 72 hours that other types normally achieve in 7 days.
- The size and type of the coarse aggregate used in the batching; for example, gravel, peastone, stone, and/or local aggregates such as slag.
- The percentage of air incorporated into the mix, in the form of tiny air bubbles. This is known as *air entrainment*. Air entrainment

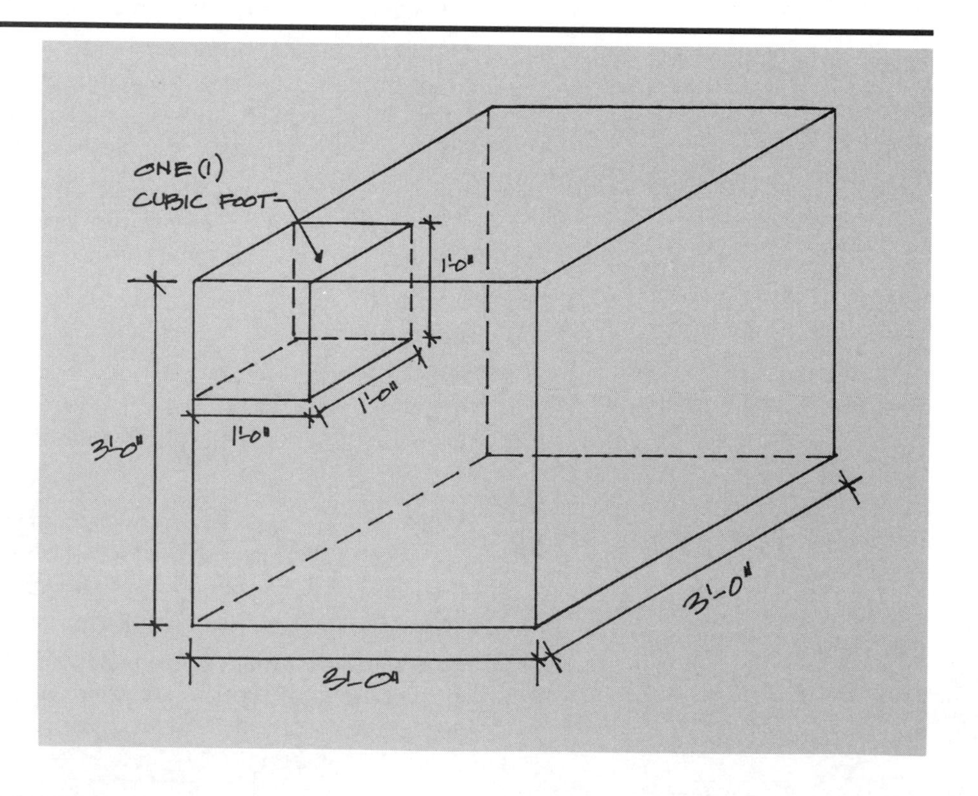

Figure 6.1 One Cubic Yard

increases the concrete's workability and resistance to weathering and salts. The standard air entrainment is between 3% and 5% and is accomplished by the use of an additive included in the mix.

- Chemicals added to the mix, such as calcium chloride, which acts as a drying agent and decreases setting time for pouring in cold weather. Other chemicals are available that increase the curing time.
- The use of lightweight aggregates to reduce weight per CY. Materials such as perlite and vermiculite are the most common and add considerable cost per CY. They are used to reduce the weight of the concrete, thereby reducing the overall structural load.

Other factors affecting the price, such as a requirement for hot water (common practice in cold climates during the winter months), may be related to localized batching practices and should be reviewed on an individual basis.

As a result of significantly better quality control, the use of ready-mixed concrete has all but eliminated the job-site batching of concrete in any large quantity. *Ready-mixed concrete* refers to concrete that is batched at an off-site location, then transported to the site in mixers.

When calculating the unit price of ready-mixed concrete, the contractor should consider any small quantities, typically under 5 CY, commonly referred to as "short loads." Most ready-mixed concrete companies have a 3–5 CY minimum charge for short loads unless prior arrangements have been made.

Difficulty in pouring concrete may also affect the unit price in the form of a charge for "waiting time" accrued by a mixer that is delayed in unloading the concrete. The contractor should list separately the quantities of concrete for short loads and pours that take extra time.

A waste factor of approximately 3% can be applied for an average concrete pour. Excessive handling or transporting of the concrete after it has left the mixer may require a 5% waste factor.

Formwork

Because of its fluid-like consistency during placement, all cast-in-place concrete must be contained in some type of formwork. Formwork varies in size and composition, but is most often constructed from a wood facing applied over a steel or wood frame. Simpler forms, such as those used in forming a footing, may be no more than a plank anchored by stakes and straps.

The quantities of all formwork should include the erection and bracing of forms until the concrete has hardened, and stripping and cleaning of forms. The contractor may separate the quantities of each of these tasks for pricing or price the work as a single process.

The takeoff of formwork is based on the actual area of the form that comes in contact with the concrete. It is listed as *square foot of contact area (SFCA).* Other methods include the takeoff and pricing of formwork for walls by the linear foot, where contact area is less critical than the size of the form panels used.

Formwork is typically listed in separate categories based on each application. Some of the more common types of formwork and their respective takeoff units are described in the following paragraphs.

Footing Forms

There are two main types of footings: *continuous strip footings*, on which walls will be erected, and *isolated spread footings*, which are used for supporting interior columns.

Strip Footings

To better distribute the load imposed by the structure, strip footings are wider than the walls they support. Strip footings follow the shape and perimeter of the wall. They are typically formed on both sides and braced on the top with temporary wood braces at 2′ to 3′ intervals. Bottoms of the footing are braced with a perforated metal strap at approximately 2′ to 4′ intervals. The forms used for footings are rough planking, similar to staging planks approximately 2″ x 12″, in varying lengths. Common sizes for footings are 20″ to 36″ in width by 12″ to 18″ in depth. The materials, with the exception of the perforated straps, are reuseable (see Figure 6.2).

Units for Takeoff

Linear feet are used for taking off continuous strip footings because most strip footings remain constant in height. Because of the conditions of the soil after excavation, changes in elevation at the bottom of the footings may

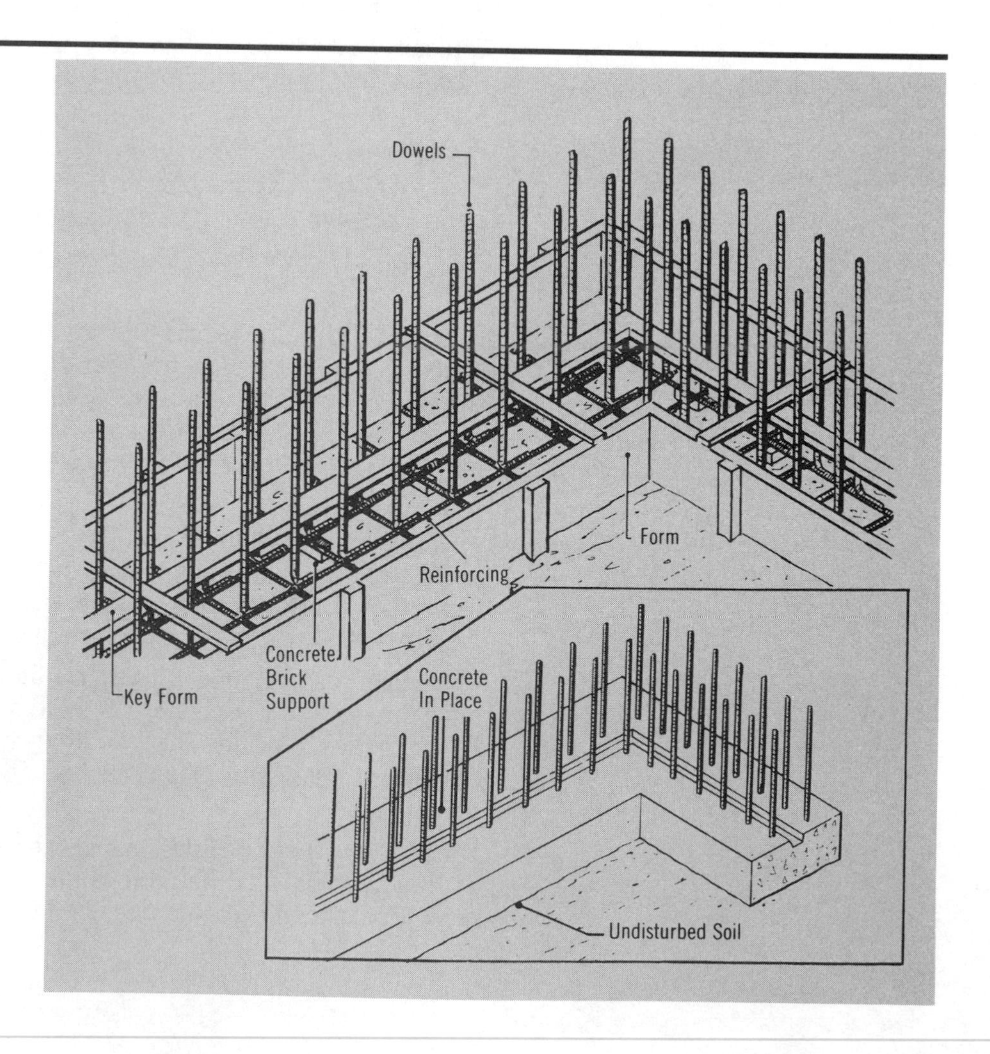

Figure 6.2 Strip Footings

be required to ensure that the footing will rest on suitable soil. The contractor should note any changes in elevation that may require stepping the footings. Productivity is reduced for stepped footings, so they should be listed separately. Takeoff is still per LF, but the height of the steps should be noted. Figure 6.3 shows an example of a stepped footing.

To reduce the lateral movement of the wall to be placed on the continuous strip footing, a small trough, called a *keyway,* is formed by using a tapered 2" x 4" embedded in the top surface of the wet concrete in the footing. Takeoff units for the keyway are listed by the LF. Figure 6.4 illustrates the keyway form and the resulting trough after the forms have been stripped.

Spread Footings

Spread footings are isolated masses of concrete, often square or rectangular in shape, with thicknesses varying from 12" to 24". Their main purpose is to support point loads from the columns that rest on them. Their actual size

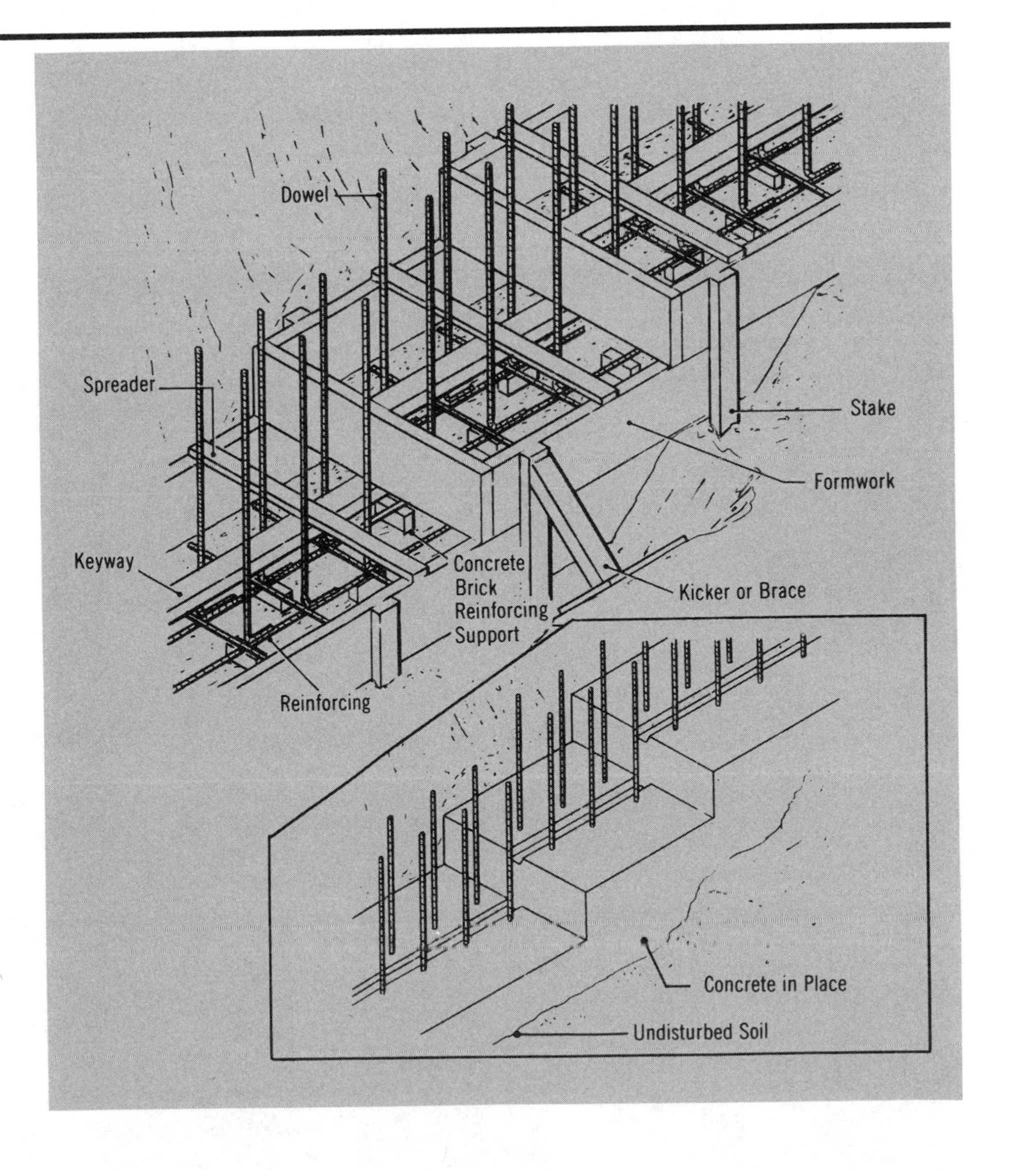

Figure 6.3 Stepped Footings

is contingent upon the load carried and the soil's bearing capacity. A typical spread footing is shown in Figure 6.5.

The form material may be planks or panels, depending on the thickness of the footing. Footing forms are erected and braced in a manner similar to that used for strip footings.

Units for Takeoff

Spread footings are typically taken off by the piece, and labeled EA (each). Quantities of same-size footings should be grouped together. Foundations with various-sized footings are typically listed in a footing schedule shown on the structural drawings.

In addition, the spread footing may require the use of a template for embedded anchor bolts to support the column. This can be listed separately from the formwork, as it requires the layout of the template to be exact and often requires the services of a site engineer. See the "Embedded Items" section near the end of this chapter.

Walls and Piers

Concrete foundation walls are cast-in-place. Their function is to support the structure above. They are supported by footings and are mostly below grade. In a structure with a basement, foundation walls act to retain, or hold back, the outside soil.

A *pier* is a short column of plain or reinforced concrete used to support a concentrated load. Piers are used as components in foundation walls or as isolated, separate members.

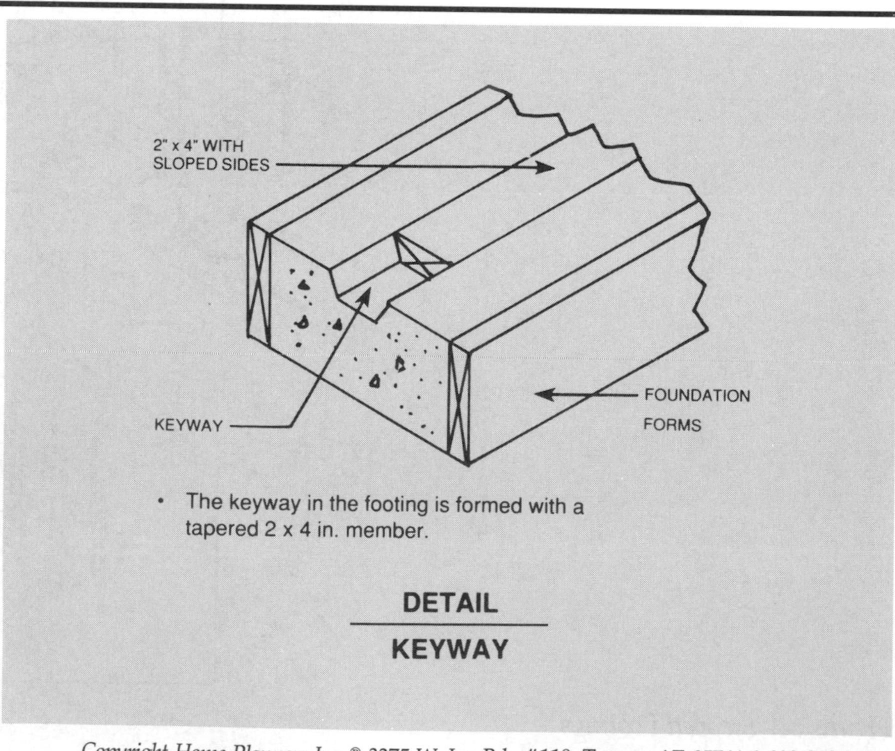

Figure 6.4

Formwork for foundation walls is constructed of smooth wood sheathing applied to a 2″ x 4″ wood or steel frame. Formwork is built in modular sizes starting at approximately 8″ in width and increasing to 16″ widths in 2″ increments. Larger panels for longer straight runs are in 24″ and 48″ widths. Standard panel heights are 48″, 72″, and 96″.

Foundation walls are formed by erecting and fastening modular panels side by side on top of the strip footing. Foundation wall formwork requires the "doubling up" of panels to create a narrow box to hold the concrete until it has hardened. The panels are held apart at a predetermined space using metal ties. This predetermined space is ultimately the thickness of the wall. Figure 6.6 illustrates formwork for a typical wall.

Ties are usually spaced at 24″ on center both horizontally and vertically, though they may require closer spacing for greater loads imposed by the wet cast-in-place concrete. In addition, the panels are braced on the exterior (against the hydrostatic pressure caused by the wet concrete) by a series of horizontal wood or metal braces known as *walers*.

Units for Takeoff

The takeoff units of foundation formwork are either square foot of contact area (see definition of SFCA earlier in this chapter) or linear foot of formed wall. Linear feet are used most often by residential contractors. Because the price of forming a wall is affected more by the size (height) of the panel than the amount of concrete placed in it, the contractor should separately list formwork of different sizes (heights); specifically 4′, 6′, and 8′ heights. Walls to be formed in excess of 8′ in height should be listed separately in vertical increments of 2′, because of the premium cost for such formwork. Odd-shaped or round forms should also be listed separately.

In addition, the contractor must make provisions for coating the forms with a release agent to break the bond between the wood and the concrete

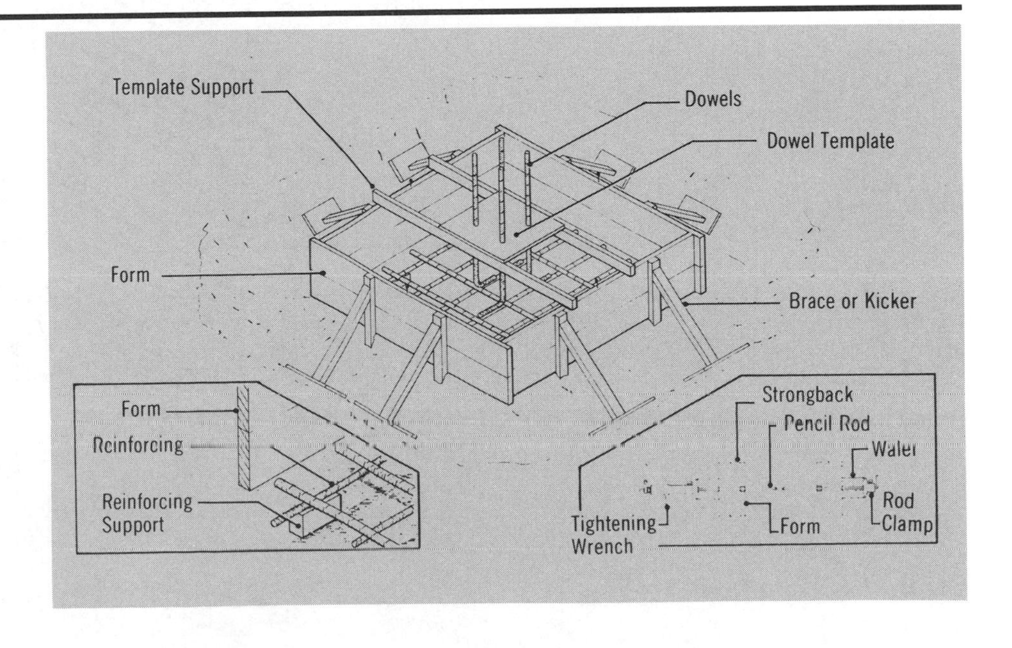

Figure 6.5 Spread Footings

during the curing process. Releasing agents are liquid and are converted from SF of area to be covered to gallons per SF, and finally, to the gallons needed, including 10 – 15% waste.

Takeoff units for wall ties are by the piece (EA), converted to the common method of purchase, per hundred count.

Formwork for Grade Beams and Elevated Slabs

Grade beams are horizontal beams supported at the ends, as opposed to foundation walls that are supported by footings on the ground. The structure's load is carried along the grade beam and transmitted through the end supports (piers) to the soil below.

Units for Takeoff

Grade beams are taken off by the square foot of contact area (SFCA). Grade beams differ from wall formwork in that they sometimes require the forming of the bottom of the grade beam as well as the sides (see Figure 6.7).

If custom-made or one-time-use forms are required for certain applications, the contractor should list them separately.

Elevated cast-in-place slabs are often integrated with concrete beams, similar to grade beams. Again, the unit of takeoff is SFCA. Formed horizontal areas should be listed separately because they require considerably more bracing to support the weight of the concrete they contain. Figure 6.8 is an illustration of an elevated cast-in-place slab.

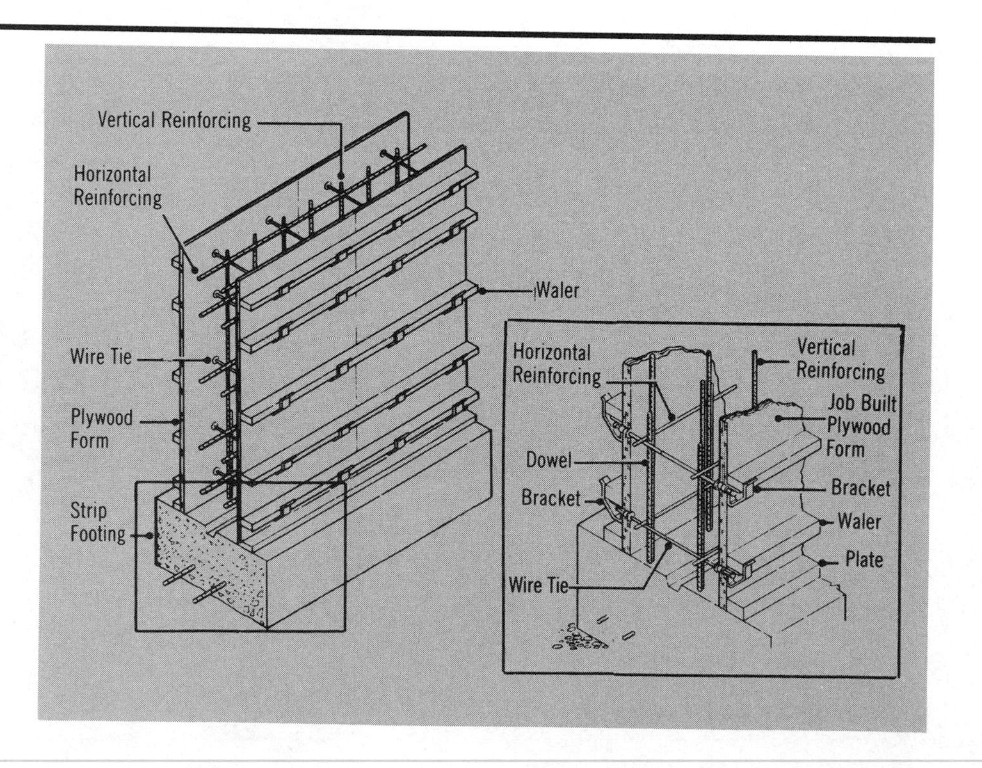

Figure 6.6 Foundation Wall

Another type of elevated slab involves placement of concrete on corrugated metal decking supported by bar joists. In this case, the metal decking work should not be included as part of the concrete formwork, and will be discussed in Chapter 8.

Edge Forms

The simplest type of form, called the edge form, is most commonly used to contain shallow pours of concrete for slab-on-grade, walks, or pads. Edge form materials are typically rough-grade lumber in the dimension required by the depth of the pour, such as 2″ x 4″, 2″ x 6″, 1″ x 4″. The actual edge form is held in place by wood or metal stakes, driven into the ground at spacing as needed to support the work and prevent bowing.

Units for Takeoff

Edge forms are taken off by the LF. Quantities of straight edge forms should be listed separately from curved edge forms. The contractor should also note any vertical surfaces that may be used as edge forms. For example, using a foundation wall as an edge when pouring a basement slab may eliminate most of the edge form required.

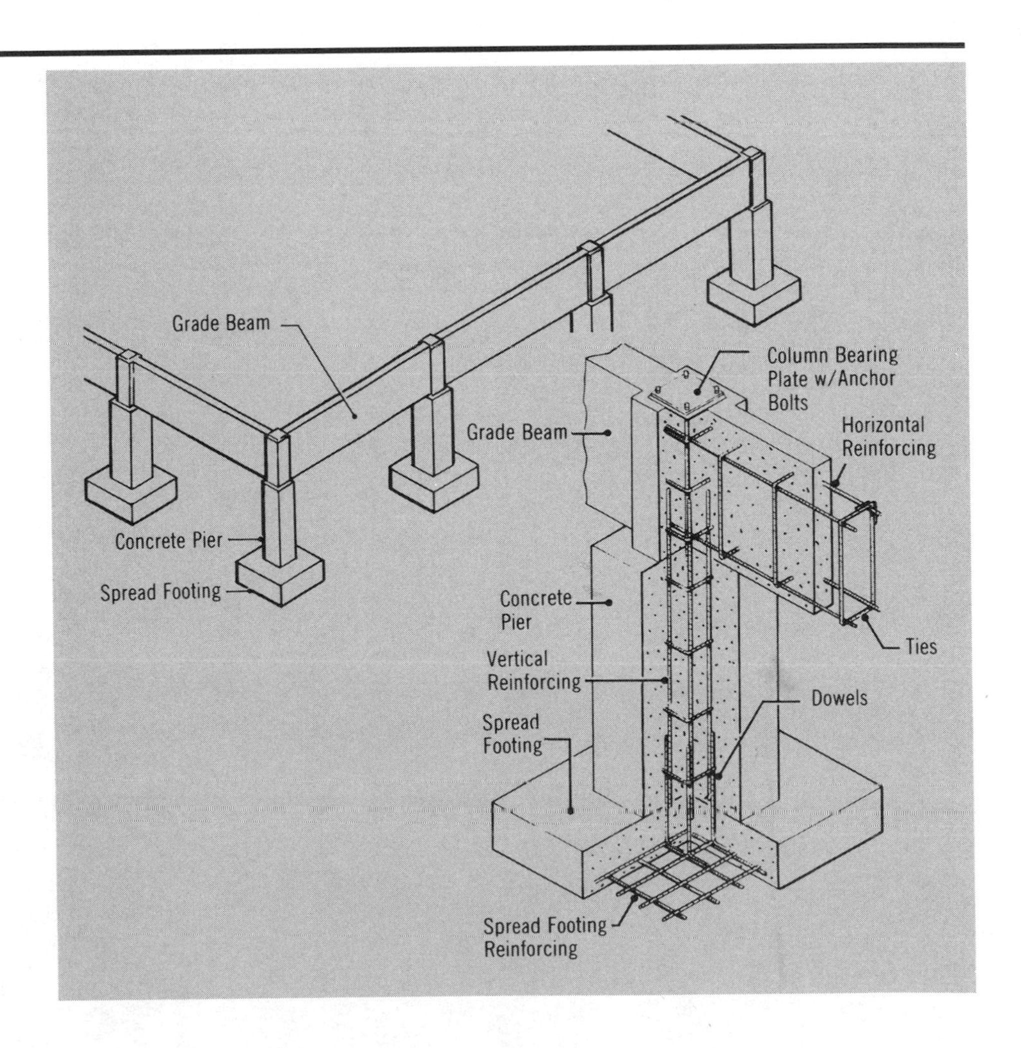

Figure 6.7 Grade Beams

Expansion and Control Joints

Expansion Joints

Concrete, like other construction materials, expands and contracts with temperature changes. To allow for safe expansion and contraction without defects to the work, certain precautions must be taken during construction. For example, a premolded joint filler (which compresses as the concrete expands) allows room for expansion. These asphalt-impregnated fibrous boards come in a variety of widths, the most common of which are 4″ and 6″. Thicknesses are 3/8″, 1/2″, and 3/4″.

Expansion joints typically occur at the perimeter of a concrete slab where it terminates at a masonry or concrete wall. Figure 6.9 shows the use of a

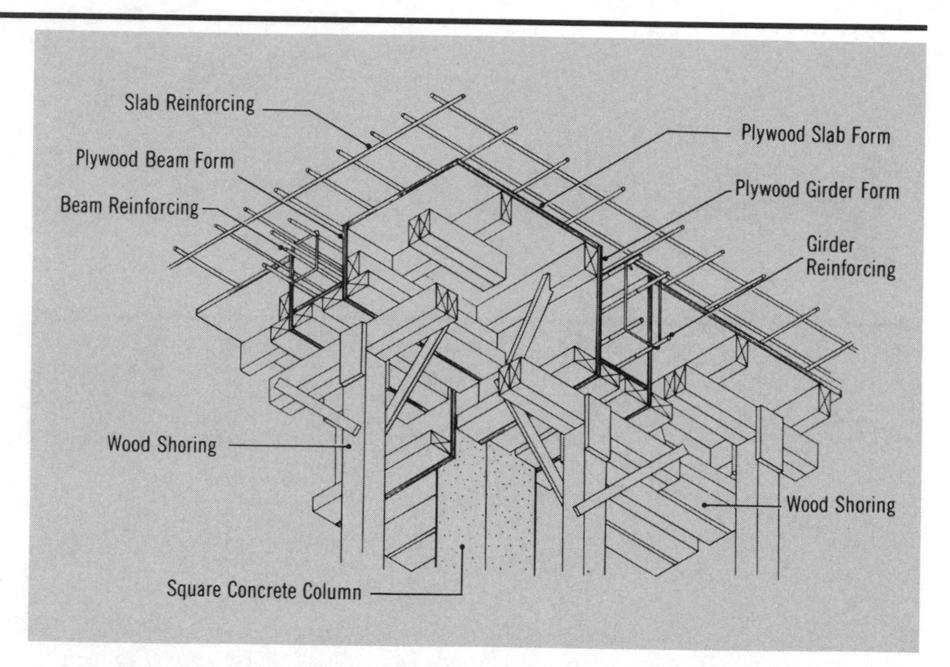

Figure 6.8 Elevated Slab

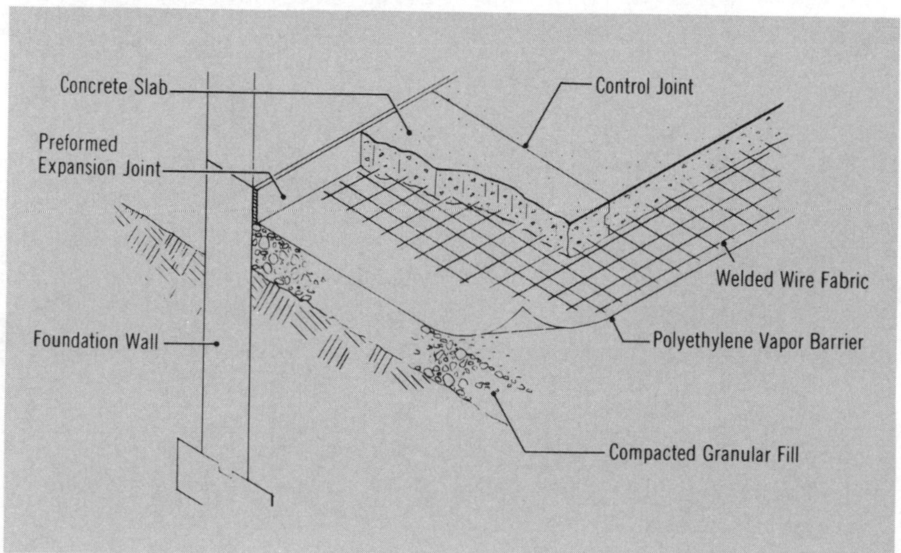

Figure 6.9 Use of Premolded Joint Filler

premolded joint filler at the perimeter of a slab that meets a foundation wall.

Units for Takeoff

Expansion joints are taken off by the LF. The contractor should list each quantity separately by size and thickness.

Control Joints

A control joint is a formed, sawed, or tooled groove in a concrete surface. Its purpose is to create a weakened plane to regulate the location of cracking that results from the dimensional changes in large volumes of poured concrete.

Units for Takeoff

Each method of providing control joints involves taking off quantities by the LF. It is not uncommon to have a combination of all three methods on the same project. The first, *tooled joints*, is accomplished by the use of a hand-held tool, and is typically used for walkways. The grooves are cut perpendicular to the length of the walk at intervals of approximately 5′ during the finishing process. The second method, saw cutting, is done after the surface is hardened, but before final strength of the concrete has been achieved. Saw cutting is done with a diamond blade set to a specific depth, commonly 1/4″ to 1/2″. Saw-cut control joints occur at column lines in large slabs. Finally, *formed control joints* are created by edge-forming areas to be poured at different times. (The slab is edge-formed in a checkerboard fashion and alternate squares are poured.) Formed joints are typically created at the intersection of a column's base and the surrounding slab. Figure 6.10 shows a control joint at the base of a column.

Concrete Reinforcement

Concrete reinforcement is the placing of steel bars or wire within the formwork prior to the placing of concrete. The concrete and the steel reinforcing are designed to act as a single unit, providing both compressive

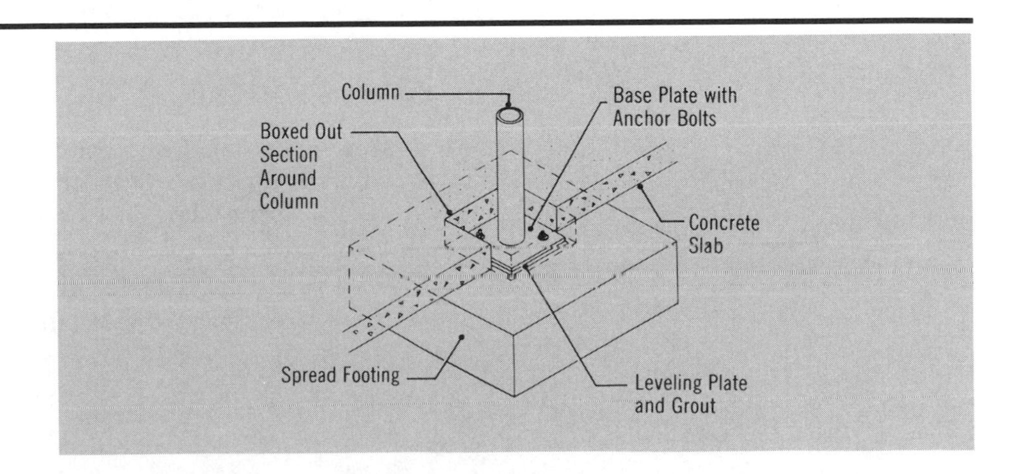

Figure 6.10 Control Joint at Base of Column

and tensile strength in resisting the forces caused by the weight and mass of the structure. There are two basic types of reinforcing: *welded wire fabric* and *steel reinforcing bars.*

Welded Wire Fabric (WWF)

Welded wire fabric, also called *welded wire mesh,* is a series of longitudinal and transverse wires of various gauges arranged at right angles to each other and welded at all points of intersection. Welded wire fabric is used in concrete slabs both to provide reinforcement against contraction and to reduce cracking.

Units for Takeoff

WWF, taken off by the square foot, is manufactured in sheets and rolls, and is sold based on the price per SF. Rolls are 250′ x 5′ wide, and flat sheets are 5′ x 10′. Figure 6.11 is a table of specifications for common styles of WWF and an illustration of its parts.

The contractor should list separately the different sizes of WWF required for a job and, in the absence of a specified overlap, should add 10% for overlap and waste.

Reinforcing Bars

Steel reinforcing bars, commonly refered to as *rebar,* are deformed or knurled round bars of high-grade steel. Rebar is available in stock lengths of 20′, or can be cut, formed, or bent into any required shape. Bars are designated by a number that refers to the nominal diameter of the bar in eighths of an inch. Standard bar designation numbers are 3, 4, 5, 6, 7, 8, 9, 10, 11, 14, and 18. Therefore, the diameter of a #3 bar is 3/8″. Figure 6.12 is a table of standard rebar weights and measures.

Units for Takeoff

Rebar placement in walls or footings is horizontal and/or vertical. Horizontal bars are taken off by total length, multiplied by the number of bars shown. Vertical bars are taken off by dividing the total length of wall or footing by the spacing, and multiplying by their height or length.

In the case of rebar in slabs, the length and width of the slab are divided by the longitudinal and transverse spacing respectively, to arrive at a quantity for each. The quantity is multiplied by the length of each, and a total linear footage is tallied. For example:

> *Calculate the LF of #4 bar in a 20′ × 25′ slab where the bars are spaced at 12″ O.C. EW (each way).*
>
> *Longitudinal (length): 25′/12″ O.C. = 25 pcs × 20′ = 500 LF*
>
> *Transverse (width): 20′/12″ O.C. = 20 pcs × 25′ = 500 LF*
>
> *Total of #4 bar = 1,000 LF*

Rebar is taken off by the LF and converted to pounds (lbs.) or, in the case of large quantities, tons (tns). One ton is equal to 2,000 pounds. Figure 6.13 lists symbols and abbreviations used on reinforcing steel drawings.

The contractor should list separately each size (by bar designation) and shape in order to calculate weights for correct pricing. Using the table in Figure 6.12, we can calculate the weight of a specific quantity of rebar:

> *Calculate the weight of 2,956 LF of #5 rebar.*
>
> *2,956 LF × 1.043 lbs./LF = 3,083.11 lbs. / 2,000 lbs./tn = 1.54 tns*

When the length of continuous reinforcing exceeds the length of stock bars, overlap becomes necessary to maintain the structural integrity of the member. Most specifications indicate the amount of overlap as a multiple

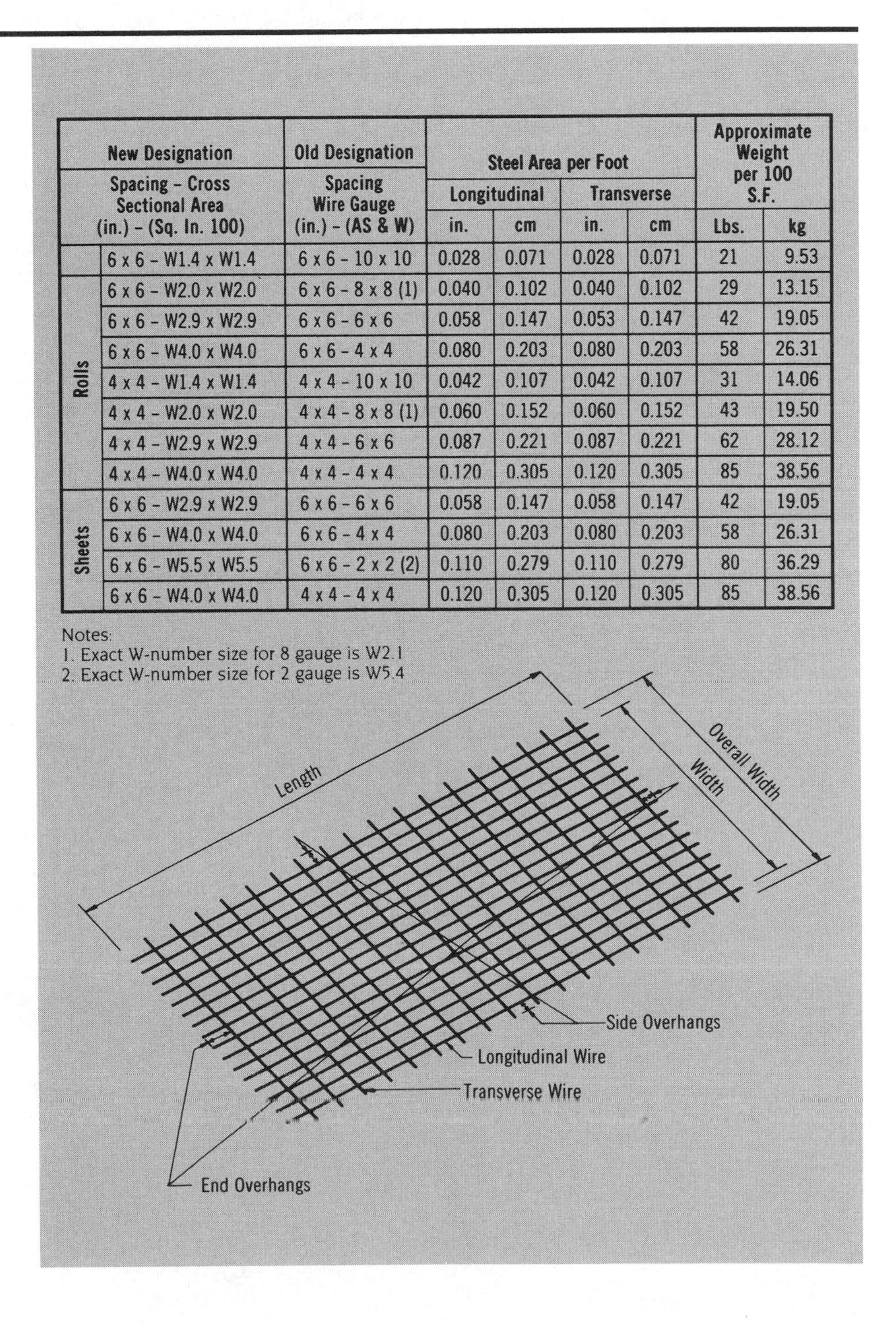

	New Designation	Old Designation	Steel Area per Foot				Approximate Weight per 100 S.F.	
	Spacing – Cross Sectional Area (in.) – (Sq. In. 100)	Spacing Wire Gauge (in.) – (AS & W)	Longitudinal		Transverse			
			in.	cm	in.	cm	Lbs.	kg
	6 x 6 – W1.4 x W1.4	6 x 6 – 10 x 10	0.028	0.071	0.028	0.071	21	9.53
Rolls	6 x 6 – W2.0 x W2.0	6 x 6 – 8 x 8 (1)	0.040	0.102	0.040	0.102	29	13.15
	6 x 6 – W2.9 x W2.9	6 x 6 – 6 x 6	0.058	0.147	0.053	0.147	42	19.05
	6 x 6 – W4.0 x W4.0	6 x 6 – 4 x 4	0.080	0.203	0.080	0.203	58	26.31
	4 x 4 – W1.4 x W1.4	4 x 4 – 10 x 10	0.042	0.107	0.042	0.107	31	14.06
	4 x 4 – W2.0 x W2.0	4 x 4 – 8 x 8 (1)	0.060	0.152	0.060	0.152	43	19.50
	4 x 4 – W2.9 x W2.9	4 x 4 – 6 x 6	0.087	0.221	0.087	0.221	62	28.12
	4 x 4 – W4.0 x W4.0	4 x 4 – 4 x 4	0.120	0.305	0.120	0.305	85	38.56
Sheets	6 x 6 – W2.9 x W2.9	6 x 6 – 6 x 6	0.058	0.147	0.058	0.147	42	19.05
	6 x 6 – W4.0 x W4.0	6 x 6 – 4 x 4	0.080	0.203	0.080	0.203	58	26.31
	6 x 6 – W5.5 x W5.5	6 x 6 – 2 x 2 (2)	0.110	0.279	0.110	0.279	80	36.29
	6 x 6 – W4.0 x W4.0	4 x 4 – 4 x 4	0.120	0.305	0.120	0.305	85	38.56

Notes:
1. Exact W-number size for 8 gauge is W2.1
2. Exact W-number size for 2 gauge is W5.4

Figure 6.11 Welded Wire Fabric

of the bar diameter. For example, "#4 bars will be overlapped by 20d," where 20d refers to 20 times the bar diameter. In the case of #4 bars:

$20d = 20 \times 4/8'' = 10''$

Other Takeoff Considerations

The contractor must provide the additional LF to satisfy the overlap requirement. In the absence of a specified overlap, a 10% factor for overlap and waste is acceptable. Preparation and review of shop drawings, and storage, handling, and protection of rebar once it is delivered to the project, should also be planned and accounted for.

Note: Because steel is an international product, projects that are taxpayer-funded may be bound by the *Buy American Act.* This means that the steel must be of domestic origin. The contractor should account for the rebar accordingly.

Cast-In-Place Concrete

The term *cast-in-place concrete* refers to concrete that is poured into forms at its final location, as opposed to concrete that is formed and poured off site and transported. This section addresses the various methods used to place and finish concrete—the labor and equipment portion only. Concrete itself and the curing of concrete have been addressed separately.

Units for Takeoff

In brief, the placement of wet or unhardened concrete is taken off by the CY when concrete is placed in erected formwork (as opposed to slab work). The unit CY is used because concrete is ordered, sold, delivered, and handled by the CY. The only exception to this rule is the pouring of concrete flatwork, or slab work, which also includes leveling and smoothing the surface of fresh concrete, commonly referred to as *finishing.*

Reinforcing Steel Weights and Measures

		U.S. Customary Units				SI Units		
		Nominal Dimensions*				Nominal Dimensions*		
Bar Designation No.**	Nominal Weight, Lb./Ft.	Diameter in.	Cross Sectional Area, in.²	Perimeter in.	Nominal Weight kg/m	Diameter, mm	Cross Sectional Area, cm²	Perimeter mm
3	0.376	0.375	0.11	1.178	0.560	9.52	0.71	29.9
4	0.668	0.500	0.20	1.571	0.994	12.70	1.29	39.9
5	1.043	0.625	0.31	1.963	1.552	15.88	2.00	49.9
6	1.502	0.750	0.44	2.356	2.235	19.05	2.84	59.8
7	2.044	0.875	0.60	2.749	3.042	22.22	3.87	69.8
8	2.670	1.000	0.79	3.142	3.973	25.40	5.10	79.8
9	3.400	1.128	1.00	3.544	5.059	28.65	6.45	90.0
10	4.303	1.270	1.27	3.990	6.403	32.26	8.19	101.4
11	5.313	1.410	1.56	4.430	7.906	35.81	10.06	112.5
14	7.65	1.693	2.25	5.32	11.384	43.00	14.52	135.1
18	13.60	2.257	4.00	7.09	20.238	57.33	25.81	180.1

*The nominal dimensions of a deformed bar are equivalent to those of a plain round bar having the same weight per foot as the deformed bar.

**Bar numbers are based on the number of eighths of an inch included in the nominal diameter of the bars.

Figure 6.12

Concrete surfaces are finished using a hand or power trowel. This process is typically taken off and estimated by the SF. The SF units more closely define the scope of work than do CY, as slabs generally have a larger SF area than CY volume.

The contractor must calculate the SF area of the slab to be finished based on areas shown on plan view drawings. These are simple length x width calculations of the surface area.

Other Takeoff Considerations

The cost placement of concrete is affected by one main question: How much is the concrete handled during the pour? Concrete that is poured directly from the chute of the ready-mix truck involves the lowest placement cost. To direct-chute concrete, the truck must have sufficient access to the pour and the formwork must be lower than the highest point on the chute, as this method relies on gravity. Some specifications require the use of a vibrator to evenly distribute the mix within the form and to fill voids caused by air pockets. This is included as part of the placement cost.

A second method of placement, mainly for ground-level flatwork, is the use of a power buggy or wheelbarrow. This method is less productive than the direct chute method but may be the most economical for inaccessible areas and small quantities. This situation may warrant the separation of CY placement units from SF finishing units to allow the contractor the chance to analyze different options.

For any application of concrete where access is limited or height above the ground makes direct-chute or wheeled placement impractical, concrete pumping equipment may be necessary. Concrete pumping involves depositing fresh concrete into a truck-mounted pump that uses a piston-type action to push the mix. The cost of placement per unit by such means is typically expensive. The contractor must determine whether the quantity of concrete required warrants the use of pumping equipment. Most concrete pumping companies charge by the day or a minimum of

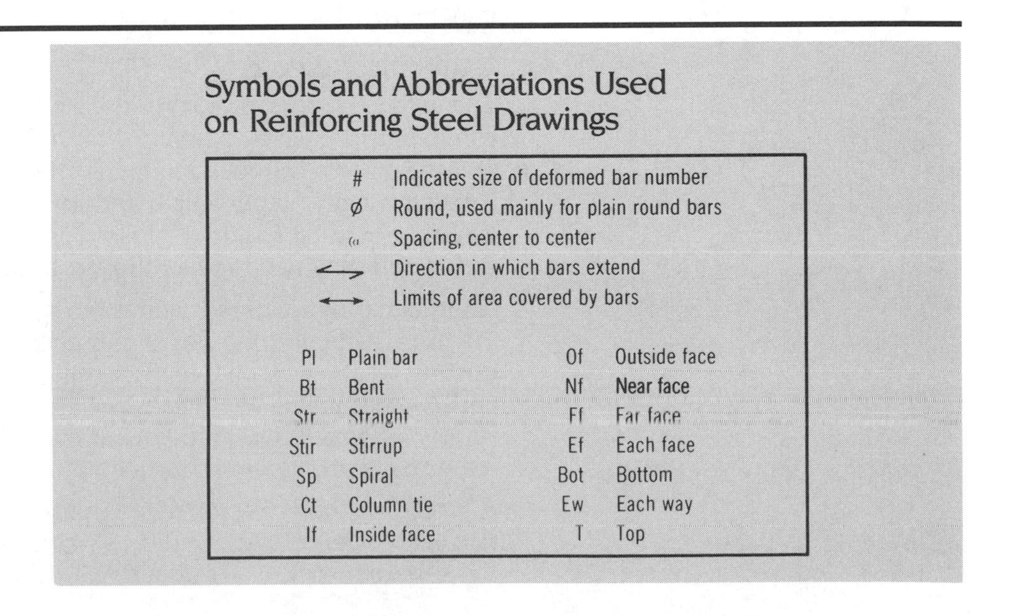

Symbols and Abbreviations Used on Reinforcing Steel Drawings

#	Indicates size of deformed bar number
ø	Round, used mainly for plain round bars
@	Spacing, center to center
↔ (half arrows)	Direction in which bars extend
↔	Limits of area covered by bars

Abbr.	Meaning	Abbr.	Meaning
Pl	Plain bar	Of	Outside face
Bt	Bent	Nf	Near face
Str	Straight	Ff	Far face
Stir	Stirrup	Ef	Each face
Sp	Spiral	Bot	Bottom
Ct	Column tie	Ew	Each way
If	Inside face	T	Top

Figure 6.13

one-half day regardless of the quantity pumped. Concrete pumping is often used for difficult access situations; for example, when the formwork takes up the entire excavated area and the access, making it difficult to place by chute, power buggy, or crane and bucket. The cost of erecting, maintaining, and dismantling riser piping (temporarily attached vertically to the structure) should be listed and estimated separately.

Another placement method involves the use of a bucket and crane. Concrete is deposited directly from the mixer to a bucket hoisted by a crane to the forms, thereby minimizing the amount of handling. This method is often more economical than the concrete pump, but it does require that the pour be free from overhead obstructions. Again, the contractor must analyze the pour to evaluate whether there is sufficient quantity to justify the use of a bucket and crane.

The contractor should sequence the pour—that is, schedule the pouring of concrete—in such a way as to maximize cost effectiveness. This means pouring only when sufficient forms are available, or using concrete pumps or the bucket and crane method when there is a sufficient quantity to keep the equipment busy for the entire day.

Exposed concrete walls such as retaining walls may, in some cases, require a rubbed finish to make the work more visually appealing. This is achieved by stripping the forms earlier than normal while the concrete is still "green" and rubbing the exposed surface with an abrasive carborundum stone to roughen the surface and remove any irregularities caused by the formwork. This process is taken off by the SF of exposed surface area to be rubbed. The work is done by hand and may require the mixing of small quantities of mortar to patch holes or voids in the rubbed surface area.

Concrete Curing

Concrete curing involves maintaining the proper moisture and temperature in the environment where concrete has been placed to ensure the proper hydration (chemical reaction) and hardening. Curing can also be accomplished by the use of chemicals sprayed on the freshly-finished concrete to create a membrane on the slab surface. The membrane prevents the premature evaporation of water in the mix before the concrete has had adequate time to cure properly. Repeated applications may be required in dry or hot weather conditions.

Units for Takeoff

The area to be sprayed for curing is calculated in SF. This figure is converted to gallons, based on the individual product's coverage per manufacturer recommendations. For example, if the product has a coverage of 400 SF per gallon and the area to be covered is 4,000 SF, 10 gallons would be required. Quantities should be rounded to the nearest gallon and may need to be rounded to the nearest common sales unit.

Another curing method requires the spraying of water with a hose after the initial setting of the surface. Applications are as-required based on the existing weather conditions. The unit of takeoff for this method is also SF, and is seen mainly as a labor cost since most projects have an ample supply of water. The duration of curing may be as long as 28 hours, requiring overtime or shift work.

A similar method calls for covering the moistened surface of the finished slab with a vapor barrier, such as polyethelene, to prevent evaporation of the surface moisture. This is also calculated and listed by the SF. The area is then converted to the required size roll. Sufficient allowances for overlap and waste are necessary for full coverage. A maximum of 10% waste and overlap is usually adequate.

During cold weather, additional protection, such as temporary portable heaters and insulating materials, may be required to ensure proper curing. Sometimes the heat produced during the chemical reaction of hydration is sufficient to maintain a temperature above freezing, and covering the work with straw and polyethelene or insulating blankets is sufficient.

Portable kerosene heaters, called *salamanders*, or similar heaters fueled by propane are also used. Both require supervision while in use, because of the danger of fire. Constant manning and the consumption of fuel can be costly; therefore, portable heaters are employed only when no other means will suffice. The units of takeoff for temporary heat are computed based on the time required, usually by the day, and the fuel consumption of the individual heating apparatus. The contractor should allow for the set-up and dismantling of any necessary enclosures to contain the heat, including both materials and labor hours.

Precast Concrete

Precast concrete includes structural concrete components formed, poured, and finished in a location other than their final position in the structure. Precast components are fabricated off site and must be transported to the site. Items such as bulkhead enclosures, precast steps, beams, wall panels, and concrete planking (also known as hollow-core planking) are common examples of precast concrete members.

Units for Takeoff

Precast items are typically shown and identified on structural drawings in plan and sectional views, but can also be noted on site drawings, such as precast steps or wheel stops. The units of takeoff for precast concrete are typically by the individual piece (EA). Quantities should be listed according to size, length, or any other characteristics.

As most precast items require the use of equipment for handling and installation, the contractor must include sufficient equipment and labor hours based on standard production rates for the individual items. Production rates for the installation of precast items are available in R.S. Means Construction Cost Data books.

On-site cutting of precast plank or caulking of joints between the planks should be included as part of the installation cost. Both are quantified by the LF. Other costs associated with the installation of precast wall panels, such as welding or bolting, are taken off by location and quantified by the piece.

Cementitious Decks

Cementitious decks are lightweight, noncombustible panels used in floor and roof construction with steel framing. In place of concrete, a gypsum core is used.

Units for Takeoff

The units of takeoff for cementitious decks are either the individual piece (EA) or SF. Hoisting these panels to the roof or floor is usually accomplished with a crane, but the actual installation can be done manually. Installation costs should include the welding of clips or metal edging to the steel support framing.

Embedded Items and Miscellaneous Items

Embedded items are encased or poured within the formwork for connecting future work or supporting reinforcing within the formwork.

Sleeves

Sleeves are blockouts within the formwork that hold back the concrete, so that when the formwork is stripped, there is a hole in the concrete wall or slab. Sleeves can be made of PVC pipe, wood, or rigid insulation. A sleeve is fit between each formwork panel and fastened by nails or other means. Sleeves are specifically located to facilitate the passage of piping through the wall at a later date. When the sleeve is removed, it should allow clear passage of the piping.

Units for Takeoff

Sleeves are taken off by the piece and listed according to size. Included in the price of the sleeve should be the cost of patching around the piping with mortar or concrete after the pipe has been installed. Sleeves are not always shown on the drawings, but may be required where utility piping passes through the foundation wall. The contractor should refer to mechanical/electrical drawings for the locations, size, and quantity of sleeves.

Anchor Bolts

An anchor bolt is a threaded rod or bolt with a right-angle bend that is embedded in the poured concrete. Anchor bolts are provided for the future connection of steel columns, beams, or wood sill plates. They anchor the future work to the concrete. The size, quantity, spacing, and location of anchor bolts are noted on the structural drawings.

Units for Takeoff

Anchor bolts are taken off by the piece by counting. In addition to the actual anchor bolt, a template (usually made from wood) may be needed to hold the bolts in position during the hardening. The template is stripped along with the formwork. Figure 6.14 illustrates a template supporting anchor bolts.

Other Items

Steel window frames for basement window sashes in residential construction are installed in a manner similar to sleeves. They are taken off

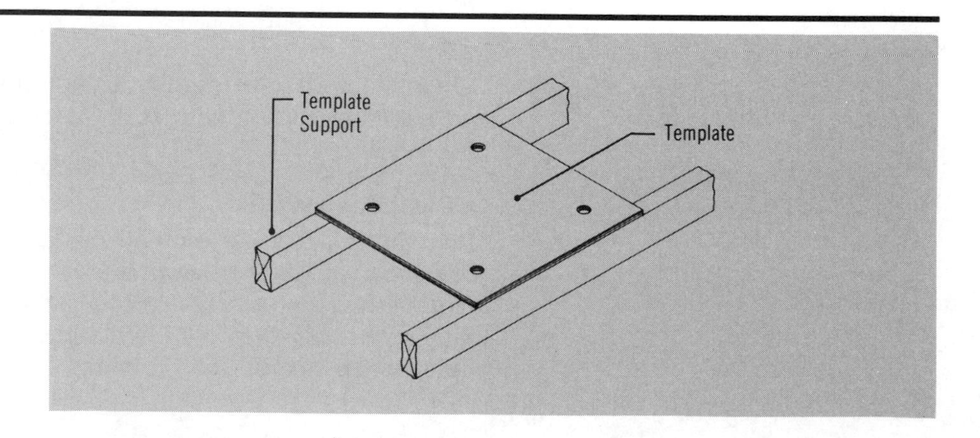

Figure 6.14 Template Supporting Anchor Bolt

by the piece and listed according to size. After the forms are stripped, the steel frames stay and the wood or steel sash is installed within the frame. The locations are noted in both plan and elevation views on the architectural drawings.

Polyethelene is sometimes specified as a vapor barrier between the compacted sub-base and the cast-in-place slab-on-grade. A vapor barrier prevents the transmission of moisture from the ground through the slab. The thickness of the poly vapor barrier is noted on the drawings in mils (with one mil equal to .001 inch). Polyethelene is commonly specified in 4 or 6 mil thicknesses. The vapor barrier is installed by spreading the polyethelene on the sub-base just prior to pouring the concrete. An allowance of 7 – 10% is usually sufficient for the overlapping of seams and waste. Vapor barrier is taken off by the SF and listed separately by thickness. The contractor should allow additional time for setting and finishing concrete slabs with vapor barriers, as water is retained longer as a result of the inability of the water to pass directly to the sub-base.

After the formwork has been stripped, the ends of the ties used to hold the panels in place will be exposed. The protruding portion of each tie must be broken off, and the small indentation filled. This is done by hand and requires mixing small quantities of mortar to patch the indentation. This work is taken off by the SF.

Summary

The takeoff sheets in Figure 6.15 are the quantities for the work of Division 3 – Concrete for the sample project. The author has purposely included the quantities for concrete and formwork for the foundation walls that are shown on the sample project as concrete masonry units. This was done to illustrate an alternative method of foundation construction popular in some areas of the United States. This alternate quantity takeoff might be used in determining the most cost effective method of foundation construction for this particular project.

Division 3 Checklist

The following list can be used to ensure that all items in the takeoff for this division have been accounted for.

Concrete

- ☐ Strength of mix
- ☐ Admixtures
- ☐ Hot Water
- ☐ Size of aggregate

Formwork

- ☐ Footings, strip and spread
- ☐ 4′ walls
- ☐ 8′ walls
- ☐ Walls over 8′
- ☐ Round or odd-shaped walls
- ☐ Piers
- ☐ Grade beams
- ☐ Elevated slabs
- ☐ Stripping and cleaning of forms
- ☐ Ties

- ☐ Release agent
- ☐ Edge forming
- ☐ Keyways

Expansion and Control Joints

- ☐ Saw-cut joints
- ☐ Tooled joints
- ☐ Premolded joint fillers

Concrete Reinforcement

- ☐ Welded wire fabirc
- ☐ Rebar
- ☐ Shop drawings
- ☐ Storage, handling, and protection

Cast-in-Place Concrete

- ☐ Finishing
- ☐ Pumping
- ☐ Crane and bucket
- ☐ "Short loads"
- ☐ Rubbed finishes

Concrete Curing

- ☐ Chemical curing
- ☐ Moisture containment
- ☐ Temporary heat and protection

Precast Concrete Items

- ☐ Precast plank
- ☐ Bulkhead enclosures
- ☐ Precast stairs
- ☐ Precast structural members
- ☐ Miscellaneous precast items

Cementitious Decks

Embedded and Miscellaneous Items

- ☐ Anchor bolts
- ☐ Vapor barriers
- ☐ Breaking ties
- ☐ Sleeves and frames
- ☐ Templates

Means Forms

QUANTITY SHEET

PROJECT	SAMPLE PROJECT
LOCATION	
ARCHITECT	HOME PLANNERS
TAKE OFF BY	WJD
EXTENSIONS BY:	WJD
SHEET NO.	3-1/4
ESTIMATE NO.	1
DATE	
CHECKED BY:	KF

DESCRIPTION	NO.	DIMENSIONS				UNIT		UNIT		UNIT		UNIT
DIVISION 3: CONCRETE												
FORMWORK FOR												
CONTINUOUS STRIP FTG.												
20"x 10" MAIN HOUSE	2	124	.83		206	SF						
" " CRAWL SPACE	2	45.33	.83		76	SF	282	SF			282	SF
NO FOOTINGS @ PORCH,												
ENTRY OR GARAGE												
2"x4" TAPERED KEYWAY	1	124			124	LF						
	1	45.33			46	LF	170	LF			170	LF
FORMWORK FOR 8"												
FOUNDATION WALLS												
@ COVERED PORCH	2	23.67	4		190	SF						
@ GARAGE	2	49.67	4		398	SF						
@ ENTRY	2	17.33	4		139	SF	727	SF			727	SF
FORMWORK FOR 12"												
FOUNDATION WALLS												
@ MAIN HOUSE	2	124	8		1984	SF						
@ CRAWL SPACE	2	45.33	4		363	SF						
@ GARAGE	2	5	4		40	SF	2387	SF			2387	SF
FORMWORK FOR FTG.												
@ FIREPLACE	1	34.67	1		35	SF					35	SF
CONCRETE STOP @												
GARAGE DOOR	1	8	.67		8	LF					8	LF
30"x 30"x 16" ISOLATED												
SPREAD FTG. @ INT.												
COLUMN	1	10	1.33		1	EA					1	EA
3000PSI CONCRETE												
w/3/4" AGGREGATE												
@ STRIP FOOTINGS	1	170	1.67	.83	236	CF						
@ ISOLATED FTG.	1	2.5	2.5	1.33	9	CF						
@ FIREPLACE FTG.	1	12.67	4.67	1.0	60	CF	305	CF			12	CY

Figure 6.15

Means Forms

QUANTITY SHEET

SHEET NO. 3-2/4

PROJECT SAMPLE PROJECT

ESTIMATE NO. 1

LOCATION

ARCHITECT HOME PLANNERS

DATE

TAKE OFF BY WJD

EXTENSIONS BY: WJD

CHECKED BY: KF

DESCRIPTION	NO.	DIMENSIONS				UNIT		UNIT		UNIT		UNIT
3000 PSI CONC. w/ 3/4" AGGREGATE												
@ 8" WALLS 4'-0"	1	90.67	.67	4	243	CF						
@ 12" WALLS 4'-0"	1	50.33	1.0	4	202	CF						
@ 12" WALLS 8'-0"	1	124	1.0	8	1718	CF	2163	CF			81	CY
3% MATERIAL WASTE	1	93	.03								3	CY
CONCRETE FLATWORK												
PLACE & FINISH 4" CONCRETE SLAB												
@ GARAGE	1	19.67	13.33		263	SF						
@ BASEMENT	1	30	28		840	SF						
@ COVERED PORCH	1	13.33	9.0		120	SF						
@ ENTRY	1	6	4.33		26	SF						
DEDUCT FOR FIREPLACE	1	11.67	3.67		(43	SF)						
@ TERRACE	1	14.00	20.75		291	SF					1497	SF
3000 PSI CONC @ SLABS	1	1497	.33		494						19	CY
3% MATERIAL WASTE ON CONCRETE	1	19	.03								.5	CY
4" EDGE FORM @ PERIMETER OF:												
TERRACE	1	43.75			44							
ENTRY	1	17.33			18	LF					62	LF
CONCRETE REINFORCING												
6"x6" – #10/10 WWF												
@ ENTRY	1	5.0	7.33		37	SF						
10% WASTE	1	37	.10		4	SF					41	SF

Figure 6.15 (continued)

Means Forms

QUANTITY SHEET

PROJECT: SAMPLE PROJECT
SHEET NO. 3-3/4
ESTIMATE NO. 1
LOCATION:
ARCHITECT: HOME PLANNERS
DATE:
TAKE OFF BY: WJD
EXTENSIONS BY: WJD
CHECKED BY: KF

DESCRIPTION	NO.	DIMENSIONS				UNIT		UNIT		UNIT		UNIT
CONCRETE REINFORCING (CONT'D)												
6"x6" #8/8 WWF												
@ TERRACE	1	20.75	14.		291							
@ COVERED PORCH	1	13.33	9.0		120							
@ GARAGE	1	13.33	19.67		263						674	SF
10% MATERIAL WASTE	1	674	.10		67						67	SF
EMBEDDED ITEMS												
32"x16"x12" DEEP BASEMENT WINDOW UNITS w/ SCREENS	3	EA			3	EA					3	EA
1/2"x8" ANCHOR BOLT w/ 3" HOOK @ 6' O.C. @ TOP OF WALL	1	220	6		37	EA					37	EA
BEAM POCKET @ BASEMENT FOR STEEL 8"x8"x8"	2	EA			2	EA					2	EA
16"x8" SCREENED VENT FOR CRAWL SPACE	2	EA			2	EA					2	EA
SLEEVES FOR WATER & WASTE LINES	1	6"	12"		1	EA						
	1	4"	12"		1	EA					2	EA
FORMS FOR 32"x24" CRAWL SPACE ACCESS	1	9.34	1.0		10	SF					10	SF
FORMWORK FOR STEPS												
@ TERRACE	4	7.5	1.0	.67	20	CF						
@ ENTRY	1	4.0	1.0	.67	3	CF						
@ COVERED PORCH	1	6.0	1.0	.67	4	CF					6	RIS
CONCRETE MATERIALS					27	CF					1	CY

Figure 6.15 (continued)

Means Forms

QUANTITY SHEET

PROJECT SAMPLE PROJECT

SHEET NO. 3-4/4

ESTIMATE NO. 1

LOCATION

ARCHITECT HOME PLANNER

DATE

TAKE OFF BY WJD

EXTENSIONS BY: WJD

CHECKED BY: KF

DESCRIPTION	NO.	DIMENSIONS			UNIT	UNIT	UNIT	UNIT
Scoring of Conc.								
Slab @ Terrace	5	20.75'			104 LF			
	5	14.0'			70 LF			174 LF

Figure 6.15 (continued)

Chapter Seven

MASONRY

(Division 4)

Chapter Seven

MASONRY
(Division 4)

Masonry construction includes brick; block; glazed block or brick; glass block; fieldstone; cut stone; and the labor, tools, and equipment required to install these materials. Brick and block are typically called *masonry units*, and the process of installing them is *unit masonry*.

The use of masonry in construction is appealing for a number of reasons. It is fireproof and durable, requires little or no maintenance, and can be used structurally as a bearing wall. In addition, masonry units are available in an enormous selection of colors, shapes, textures, and sizes that can be installed for an aesthetically pleasing appearance.

The two most common types of masonry units are brick and block. Brick is modular and adaptable, and can be installed in a variety of pleasing patterns, called *bonds*.

Concrete masonry units, commonly referred to as *CMU*, are concrete block. CMU is primarily used for walls and partitions that will support a structural load, or as the backup for a face brick veneer. It is used extensively in commercial and industrial building applications, and as a foundation material in residential construction.

All masonry work is installed with mortar. Mortar is spread between the joints of individual masonry units and allowed to harden, which bonds the units together.

When taking off masonry work, the following considerations must be addressed:

- Does the work involve:
  - brick unit masonry?
  - concrete unit masonry?
  - stone?
  - glass or glazed block unit masonry?
- What will be the:
  - depth and width of joints?
  - reinforcement requirements (horizontal and vertical)?
  - bonding pattern?
  - allowance for waste?
- Are there miscellaneous items, such as lintels, flashings, and ties?
- Is scaffolding required?
- What are the likely weather conditions during installation?

Mortar

Mortar is a composition of water; fine aggregates, such as sand; Portland, hydraulic, or masonry cement; and lime. The purpose of mortar is to bond the individual units of brick, block, or stone to form a building system.

Types of Mortar

The compressive strength of mortar varies with the proportions of the ingredients. The four basic types of mortar and their uses are based on mixing proportions and strengths.

- *Type M* is a high-strength mortar used primarily in foundation masonry, retaining walls, walks, sewers, and manholes. In general, Type M mortar is used when maximum compressive strength is required.
- *Type S* is a relatively high-strength mortar that develops maximum bonding strength between masonry units. It is recommended for use where lateral and flexural strength are required.
- *Type N* is medium-strength, general-use mortar for above-grade exposed applications.
- *Type O* is a low-strength mortar for interior nonload-bearing applications.

Figure 7.1 lists the mixing proportions and strengths of the four basic types of mortar.

Units for Takeoff

Since the mortar quantity required is directly related to the number of brick or block, the contractor must first calculate this number in order to determine the quantity of mortar needed. The size of the joints is another important factor.

Mortar quantities are typically in cubic feet or, at the discretion of the contractor, can be converted to cubic yards, where 27 CF = 1 CY. For example:

Using the table in Figure 7.2, the quantity of mortar can be determined for laying up 50,000 standard brick with a 1/2" joint thickness.

50,000 brick /1,000 brick = 50

50 is then multiplied by 11.7 CF per 1,000 brick; the result is 585 CF of mortar.

If the 585 CF is increased by 15%, allowing for waste, 585 CF × 1.15 = 672.75 or 673 CF.

This can be converted to CY by dividing by 27; the result would be 673/27 = 24.93 or 25 CY

The quantity of mortar for setting stone varies with the size of the stone and the joint. However, a general rule is that 4–5 CF of mortar will be required per 100 CF of stone.

Quantities for various strengths of mortar should be listed separately for accurate pricing. Mortar that requires special coloring should also be listed separately, as the pigment required to color mortar will affect the cost.

Brick Masonry

Brick is a solid masonry unit made of clay or shale and formed into a rectangular prism while soft, then burned or fired in an oven, called a *kiln*, until hard.

Types of Brick

Most brick can be classified in one of the following groups:

- *Face brick* is used where appearance is important. Its manufacture is closely controlled so that color, size, hardness, strength, and texture are uniform (e.g., veneer walls).
- *Common brick* is used in applications where performance is more important than appearance (e.g., below-grade masonry, as a backup for face brick, in manholes).
- *Glazed brick* is fired with a ceramic or other type of glazing material on the exposed surfaces. It is typically used in applications where durability and cleanliness are essential (e.g., rest rooms, kitchens, hospitals).
- *Fire brick* is used in areas of high temperatures, such as furnaces and fireplaces.

Brick Mortar Mix Proportions

This chart shows some common mortar types, the mixing proportions and their general uses.

Brick Mortar Mixes*					
Type	Portland Cement	Hydrated Lime	Sand (maximum)**	Strength	Use
M	1	1/4	3-3/4	High	General use where high strength is required, especially good compressive strength; work that is below grade and in contact with earth.
S	1	1/2	4-1/2	High	Okay for general use, especially good where high lateral strength is desired.
N	1	1	6	Medium	General use when masonry is exposed above grade; best to use when high compressive and lateral strengths are not required.
0	1	2	9	Low	Do not use when masonry is exposed to severe weathering; acceptable for non-loadbearing walls of solid units and interior non-loadbearing partitions of hollow units.

*The water used should be of the quality of drinking water. Use as much as is needed to bring the mix to a suitably plastic and workable state.

**The sand should be damp and loose. A general rule for sand content is that it should not be less than 2-1/4 or more than 3 times the sum of the cement and lime volumes.

Compressive Strengths of Mortar

This table lists the expected 28 day compressive strengths for some common types of mortar mixes.

Mortar Type	Average Compressive Strength at 28 Days
M	2500 p.s.i.
S	1800 p.s.i.
N	750 p.s.i.
0	350 p.s.i.

Masonry and Concrete Construction, Ken Nolan, Craftsman Book Co.

Figure 7.1

Mortar Quantities for Brick Masonry

This table can be used to determine the amounts of mortar needed for brick masonry walls for various joint thicknesses and types of common brick.

Brickwork	Joint Thickness (in inches)	Actual Requirement (in C.F. per 1,000)	Requirement with 15% Waste (in C.Y. per 1,000)
Standard	3/8	8.6	0.4
Standard	1/2	11.7	0.5
Modular	3/8	7.6	0.3
Modular	1/2	10.4	0.4
Roman	3/8	10.7	0.5
Roman	1/2	14.4	0.6
Norman	3/8	11.2	0.5
Norman	1/2	15.1	0.6
Type	**Thickness (in inches)**	**Actual Requirement (in C.F. per S.F.)**	**Requirement with 15% Waste (in C.Y. per 1,000 S.F.)**
Parging or Backplastering	3/8	0.03	1.3
	1/2	0.04	1.7

Note: Quantities are based on 4" thick brickwork. For each additional 4" brickwork, or where masonry units are used in backup, add parging or backplastering.

Modular dimensions are measured from center to center of masonry joints. To fit into the system, the unit or number of units must be the size of the module, less the thickness of the joint. A standard face brick with a 1/2" joint would measure 8-1/2" long, whereas a modular brick with the same joint would measure 8".

Mortar Quantities for Concrete Block Masonry

This table can be used to determine the amounts of mortar needed for concrete block masonry walls for joint thicknesses for various types of masonry units.

Type	Joint Thickness (in inches)	Actual Requirement (in C.Y. per 1,000)	Requirement with 15% Waste (in C.Y. per 1,000)
Concrete Block			
Shell Bedding			
All thicknesses	3/8	0.6	0.7
Full Bedding			
12 x 8 x 16, 3-Core	3/8	0.9	1.0
12 x 8 x 16, 2-Core	3/8	0.8	0.9
8 x 8 x 16, 3-Core	3/8	0.7	0.8
8 x 8 x 16, 2-Core	3/8	0.7	0.8
6 x 8 x 16, 3-Core	3/8	0.7	0.8
6 x 8 x 16, 2-Core	3/8	0.7	0.8
4 x 8 x 16, 3-Core	3/8	0.7	0.8
4 x 8 x 16, Solid	3/8	0.8	0.9
Type	**Joint Thickness (in inches)**	**Actual Requirement (in C.F. per 1,000)**	**Requirement with 15% Waste (in C.Y. per 1,000)**
4S Glazed Structural Units			
(nominal 2-1/2" x 8")			
4 SA (2" thick)	1/4	2.6	0.1
4 S (4" thick)	1/4	5.6	0.2
4D Glazed Structural Units			
(nominal 5" x 8")			
4 DCA (2" thick)	1/4	3.3	0.1
4 DC (4" thick)	1/4	7.1	0.3
4 DC 60 (6" thick)	1/4	10.9	0.5
4 DC 80 (8" thick)	1/4	13.7	0.6
6T Glazed Structural Units			
(nominal 5" x 12")			
6 TCA (2" thick)	1/4	4.2	0.2
6 TC (4" thick)	1/4	9.3	0.4
6 TC 60 (6" thick)	1/4	14.2	0.6
6 TC 80 (8" thick)	1/4	19.1	0.8
6P Glazed Structural Units –			
1/4" Joints (nominal 4" x 12")			
6 PCA (2" thick)		4.0	0.2
6 PC (4" thick)		8.6	0.4
6 PC 60 (6" thick)		13.1	0.6
6 PC 80 (8" thick)		17.7	0.8
8W Glazed Structural Units			
(nominal 8" x 16")			
8 WCA (2" thick)	1/4	6.0	0.3
8 WCA (2" thick)	3/8	9.1	0.4
8 WC (4" thick)	1/4	12.9	0.6
8 WC (4" thick)	3/8	19.4	0.8

Figure 7.2

Mortar Quantities for Concrete Block Masonry (continued)

This table can be used to determine the amounts of mortar needed for concrete block masonry walls for joint thicknesses for various types of masonry units.

Type	Joint Thickness (in inches)	Actual Requirement (in C.F. per 1,000)	Requirement with 15% Waste (in C.Y. per 1,000)
Spectra-Glaze® Units – 3/8" Joints (nominal 4" x 16")			
44S (4" thick)		16.0	0.7
64S (6" thick)		24.5	1.0
84S (8" thick)		33.0	1.4
Spectra-Glaze® Units – 3/8" Joints (nominal 8" x 16")			
2S (2" thick)		9.0	0.4
4S (4" thick)		19.2	0.8
6S (6" thick)		29.5	1.3
8S (8" thick)		39.7	1.7
10S (10" thick)		50.0	2.1
12S (12" thick)		60.2	2.6
Structural Clay Backup and Wall Tile			
5 x 12 (4" thick)	1/2	20.0	0.9
5 x 12 (6" thick)	1/2	30.4	1.3
5 x 12 (8" thick)	1/2	40.5	1.7
8 x 12 (6" thick)	1/2	35.6	1.5
8 x 12 (8" thick)	1/2	47.5	2.0
8 x 12 (12" thick)	1/2	71.2	3.0
Structural Clay Partition Tile			
12 x 12 (4" thick)	1/2	28.4	1.2
12 x 12 (6" thick)	1/2	42.5	1.8
12 x 12 (8" thick)	1/2	56.7	2.4
12 x 12 (10" thick)	1/2	70.9	3.0
12 x 12 (12" thick)	1/2	85.1	3.6
Gypsum Block (all 12" x 30")*			
2" thick	1/4	12.2	0.5
	3/8	18.4	0.8
	1/2	24.6	1.0
3" thick	1/4	18.3	0.8
	3/8	27.6	1.2
	1/2	36.9	1.6
4" thick	1/4	24.5	1.0
	3/8	36.8	1.6
	1/2	49.2	2.1
6" thick	1/4	36.7	1.6
	3/8	55.2	2.4
	1/2	73.8	3.1
Glass Block (all 3-7/8" thick)			
5-3/4 x 5-3/4	1/4		
7-3/4 x 7-3/4	1/4		
11-3/4 x 11-3/4	1/4		
Firebrick Thin joints, including waste 300 lbs. fireclay per 1,000.			

*Note: Gypsum partition tile cement should be used. 900 lbs. are required per cubic yard, plus 1 cubic yard of sand. 1,000 units equals 2,610 square feet.

Figure 7.2 (continued)

- *Brick pavers* are used as a wearing surface for floors, walks, and patios. Pavers are typically hard and durable, with a high resistance to damage from both freeze-thaw cycles and the corrosive salts used to melt snow.
- *SCR brick* is a patented type of brick developed by the Structural Clay Products Institute. SCR brick has a nominal width of 6″, in contrast to the nominal 4″ associated with modular brick. It can therefore be used structurally in a single width. It is most commonly used in the residential market for single-story applications.

To efficiently interpret the plans and specifications for masonry work and make an accurate and thorough takeoff, the contractor should be familiar with brick nomenclature. Figure 7.3 illustrates brick nomenclature.

There are numerous sizes and types of bricks in use today. Most are modular in design, where either the wythe (width), length, or coursing height is a multiple of 4″. Brick sizes are designated as *nominal*, versus actual size. The difference between the actual and nominal is made up by the thickness of the mortar joint. The table in Figure 7.4 provides the nominal and actual sizes as well as the modular coursing of various modular brick in use today.

Units for Takeoff

Since brickwork is modular, the contractor must determine the wythe of brick for each application. For example, face bricks in veneer walls are one brick wythe, while other construction may be two, three, or four brick in wythe. To accurately determine the number of brick in the takeoff, the contractor must list the work separately by wythe of wall.

Brick masonry units are taken off by the SF of wall in the case of single-wythe walls such as veneers. Multiple-wythe walls are taken off in CF. In either case, the final conversion is to the number of brick. The contractor must determine the number of brick per SF or CF based on the size of the brick, the size of the mortar joint, and the brick bond (pattern).

The calculation of SF for veneer walls is the length multiplied by the height. The contractor should always make deductions *in full* for areas greater than 2 SF. The contractor should show, as closely as possible, the actual number of brick needed to do the work.

When deducting window or door openings from the wall, the contractor must consider the depth of the jambs or returns, commonly called *reveals*. If the reveal is only the width of brick (approximately 4″ for most brick), the entire opening must be deducted. If the depth of the reveal varies, the contractor must subtract the additional brick in the reveal from the opening size to be deducted.

For example, if the opening is 5′-4″ x 6′-8″ with a 16″ reveal on either jamb, the calculation would be:

2 sides × 16″ = 32″, or 2′-8″

Subtract 2′-8″ from 5′-4″, which equals 2′-8″

Therefore, the actual deduction for the opening will be 2′-8″ × 6′-8″, or 17.81 SF

Quantities of waste for brickwork vary with the application, type of brick, bond, and quality of workmanship. Each project must be reviewed and the contractor's judgment used in determining an appropriate waste factor. Figure 7.5 lists the most common brick bonds and the associated wastes.

Brick Bonds

Brick *bonds* refer to the patterns in which bricks are laid. Bonding in general refers to the overlapping of brick, either along the length or wythe of the wall, to create a structurally sound wall. The bricks are laid in such a way as to alternate the vertical mortar joints so that no joints from two consecutive courses align (except in the case of stack bond).

There are two basic arrangements of bricks in a bond. The frequency of these arrangements determines the bond. The first is the *stretcher*, where

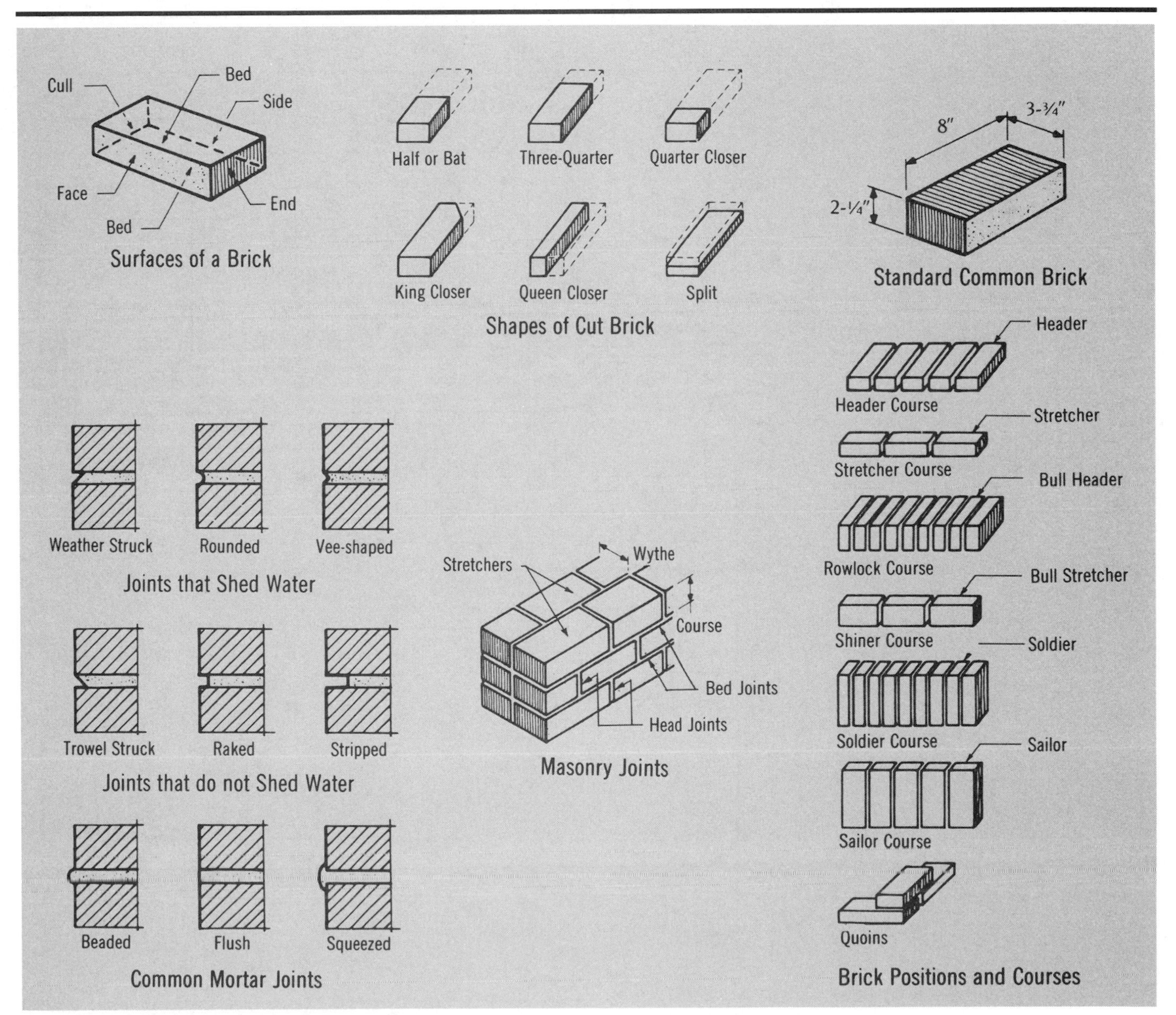

Figure 7.3 Brick Nomenclature

bricks are laid parallel with the face of the wall. The other arrangement is called a *header*, where bricks are laid perpendicular to the face of the wall. In addition, there are ornamental courses such as rowlock, soldier, sailor, and shiner, examples of which are illustrated in Figure 7.3.

The most common types of bonds and their respective coursing are listed below.

- *Running bond* is the continuous use of stretcher courses, where each alternating course overlaps the preceding course by 1/2 brick. Modifications of the standard running bond are also used, such as 1/3 running bond.

Sizes of Modular Brick

Unit Designation	Nominal Dimensions (in inches)			Joint Thickness (in inches)	Manufactured Dimensions (in inches)			Modular Coursing (in inches)
	t	h	l		t	h	l	
Standard Modular	4	2-2/3	8	3/8	3-5/8	2-1/4	7-5/8	3C = 8
				1/2	3-1/2	2-1/4	7-1/2	
Engineer	4	3-1/5	8	3/8	3-5/8	2-13/16	7-5/8	5C = 16
				1/2	3-1/2	2-11/16	7-1/2	
Economy 8 or Jumbo Closure	4	4	8	3/8	3-5/8	3-5/8	7-5/8	1C = 4
				1/2	3-1/2	3-1/2	7-1/2	
Double	4	5-1/3	8	3/8	3-5/8	4-15/16	7-5/8	3C = 16
				1/2	3-1/2	4-13/16	7-1/2	
Roman	4	2	12	3/8	3-5/8	1-5/8	11-5/8	2C = 4
				1/2	3-1/2	1-1/2	11-1/2	
Norman	4	2-2/3	12	3/8	3-5/8	2-1/4	11-5/8	3C = 8
				1/2	3-1/2	2-1/4	11-1/2	
Norwegian	4	3-1/5	12	3/8	3-5/8	2-13/16	11-5/8	5C = 16
				1/2	3-1/2	2-11/16	11-1/2	
Economy 12 or Jumbo Utility	4	4	12	3/8	3-5/8	3-5/8	11-5/8	1C = 4
				1/2	3-1/2	3-1/2	11-1/2	
Triple	4	5-1/3	12	3/8	3-5/8	4-15/16	11-5/8	3C = 16
				1/2	3-1/2	4-13/16	11-1/2	
SCR brick	6	2-2/3	12	3/8	5-5/8	2-1/4	11-5/8	3C = 8
				1/2	5-1/2	2-1/4	11-1/2	
6-in. Norwegian	6	3-1/5	12	3/8	5-5/8	2-13/16	11-5/8	5C = 16
				1/2	5-1/2	2-11/16	11-1/2	
6-in. Jumbo	6	4	12	3/8	5-5/8	3-5/8	11-5/8	1C = 4
				1/2	5-1/2	3-1/2	11-1/2	
8-in. Jumbo	8	4	12	3/8	7-5/8	3-5/8	11-5/8	1C = 4
				1/2	7-1/2	3-1/2	11-1/2	

(courtesy *Masonry and Concrete Construction*, Ken Nolan, Craftsman Book Company)

Figure 7.4

- *Common,* or *American,* bond is similar to running bond, but has a course of headers at every 5th, 6th, or 7th course.
- *Stack bond* is created by stretchers laid up directly over one another, where all vertical joints line up. The bond is a pattern for appearances only, and has relatively little structural value.
- *English bond* consists of alternating courses of stretchers and headers where the vertical joints of alternating courses align.
- *Dutch bond,* or *English cross bond,* is similar to English bond except that the stretcher courses do not align, but alternate by 1/2 brick.
- *Flemish bond* consists of alternating headers and stretchers in each course. In addition, the header is centered over the stretcher in each consecutive course.

There are other types of bonds, but they are variations of the preceding types. Figure 7.6 illustrates the more common types of bonds.

Once the SF area of brick, the size of the mortar joint, and the brick bond are known, the contractor can use the tables in Figures 7.7a & 7.7b to determine the quantity of brick per SF and, finally, the total number of brick. Figure 7.7a refers to brick laid in a running bond pattern with various joints. Figure 7.7b shows factors to use with Figure 7.7a in determining the accurate quantity of brick in a particular bond.

Other Takeoff Considerations

As with all masonry work, the contractor must include allowances for waste when taking off brick.

Masonry work, because of its use of mortar as a main component, is affected by extremes in temperature and weather conditions. Temporary enclosures with heat are sometimes required to provide adequate temperature, both for laying up masonry and during the curing process. The contractor will take off the materials, labor, and equipment to provide

Waste Allowances for Various Brick Bonding Patterns

Type of Pattern	% of Waste
Running or Stretcher Bond	The face brick are all stretchers and are tied to the backing by metal or reinforcing. Waste — 5%
Common or American Bond	Every sixth course of stretcher bond is usually a header course. Waste — 4%.
Flemish Bond	Each course has alternate headers and stretchers with the alternate headers centered over the stretcher. Waste — 3 to 5 %.
English Bond	Consists of alternate headers and stretchers with the vertical joints in the header and stretcher aligning or breaking over each other. Waste — 8 to 15%
Stack Bond	Has no overlapping of units since all vertical joints are aligned. Usually this pattern is bonded to the backing with rigid steel ties. Waste — 3%.
English Cross or Dutch Bond	Built up on interlocking crosses. This wall consists of two headers and a stretcher forming a cross. Waste — 8%.

Figure 7.5

this protection in much the same manner as described for concrete in Chapter 6. Temporary enclosures may need to be dismantled and re-erected in different locations as the work progresses. The contractor must sequence the work to determine the limits of the protection.

The type of building plays an important role in the bricklayer's productivity. Buildings with long, straight, uninterrupted walls, such as a warehouse, tend to go faster than office buildings or schools that have numerous openings. The contractor may want to separate walls on jobs of different building types to estimate labor accurately.

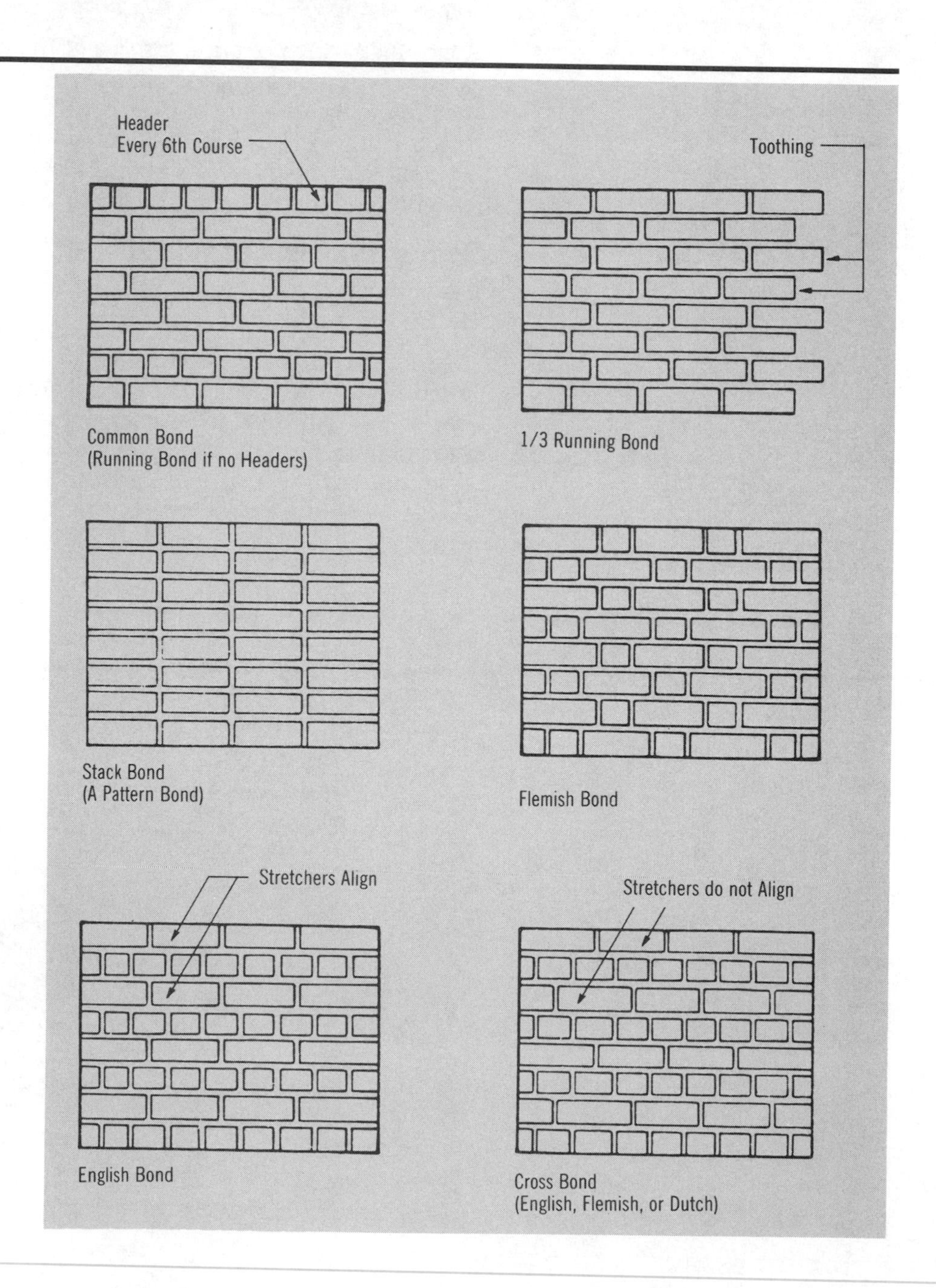

Figure 7.6 Brick Bonding Patterns

Excessive heights also reduce productivity because of the effort and time required for the handling or hoisting of mortar and brick to the elevated work areas. Scaffolding and the cutting of masonry units are other factors in taking off masonry work. They are discussed in detail at the end of this chapter.

Finally, the type of mortar joint also has an effect on production and, ultimately, cost. The *flush joint* is the most economical because it is a one-step process. The bricklayer merely cuts the excess mortar off with the trowel. For rounded, beaded, v-shaped, weather, and trowel-struck, there is a two-step process that starts with cutting the excess mortar off with a trowel and then using either a trowel or a jointing tool to achieve the final joint. For raked and stripped joints, the process is even more labor-intensive in that the excess mortar must be raked out and then tooled.

Concrete Unit Masonry

Concrete masonry units (CMU) have, over the years, come to be known as *concrete block* and *concrete brick*. They are used extensively as interior or exterior load-bearing walls, or as the backup for brick veneer walls. Because of their larger size, fewer units are required. CMU are composed of

Brick Quantities per S.F.

Number of bricks per S.F. based on brick type and size of joints.

Brick Type & Size	Size of Joint (in inches) 1/4	1/3	3/8	1/2	5/8	3/4
Standard face brick (8" x 2-1/4")	6.98	6.70	6.55	6.16	5.81	5.49
Standard common brick (8" x 2-1/4")	6.98	6.70	6.55	6.16	5.81	5.49
Concrete brick (7-5/8" x 2-1/4")	7.31	7.00	6.86	6.45	6.07	5.73
Modular brick (7-1/2" x 2-1/6")	7.68	7.35	7.19	6.73	6.35	5.98
Modular Roman brick (11-5/8" x 1-5/8")	6.47	6.15	6.00	5.59	5.22	4.90
Modular Norman brick (11-5/8" x 2-1/4")	4.85	4.66	4.57	4.32	4.09	3.88

Note: Above constants are net, i.e., no waste is included.

Figure 7.7a

Adjustments to Brick Quantity Factors

This table provides factors for determining the additional quantities of brick needed when specific bonding patterns are used.

For Other Bonds Standard Size Add to S.F. Quantities				
Bond Type	**Description**	**Factor**	**Description**	**Factor**
Common	Full header every fifth course	+20%	Header = W x H exposed	+100%
	Full header every sixth course	+16.7%	Rowlock = H x W exposed	+100%
			Rowlock stretcher = L x W exposed	+33.3%
English	Full header every second course	+50%	Soldier = H x L exposed	—
Flemish	Alternate headers every course	+33.3%	Sailor = W x L exposed	-33.3%
	every sixth course	+5.6%		

(See "Brick Quantities per S.F." table above for basic quantities of brick per S.F.).

Figure 7.7b

Portland cement, water, and a variety of fine aggregates such as sand, crushed stone, and shale; or lightweight aggregates such as perlite, vermiculite, or pumice.

Concrete Brick

Concrete brick are solid modular units of concrete used in much the same way as regular brick. They are typically manufactured to 2-1/4″ x 3-5/8″ x 7-5/8″, although additional sizes may be available in some locations. They are also used as infill material in concrete block walls, where cut block is impractical.

Units for Takeoff

The procedure for taking off concrete brick is the same as for regular brick. The final units of takeoff are *number of brick* derived from the SF area. Refer to the table in Figure 7.7a for the quantity of concrete brick per SF. The contractor should include a 3%–5% waste factor for exposed concrete brick.

Concrete Block

Concrete block are hollow-core, load-bearing masonry units. The standard nominal dimensions of concrete block are 8″ high x 16″ long x 4″, 6″, 8″, 10″, and 12″ in width. The actual size is 3/8″ smaller in the length and height of the block. The actual dimension is reduced so that the addition of a 3/8″ mortar joint will produce an 8″ x 16″ finished unit. Solid versions of the hollow units are available as well as half-block for corners, and specialty block such as bond beam and bullnose block.

Concrete block are manufactured using two basic weights of concrete: heavyweight concrete (weighing approximately 145 lbs. per CF) and lightweight concrete (approximately 100 lbs. per CF).

Concrete block are manufactured and sold in a variety of designs, colors, shapes, and surface textures. The most common are:

- *Scored block,* units scored across the face to give the appearance of smaller units.
- *Split face block,* which are split lengthwise and installed with the split face exposed.
- *Split rib block,* with a corrugated look achieved by molding the block with coarse, vertical ribbing.
- *Deep groove block,* units with deep vertical grooves scored at intervals along the face of the block.
- *Slump block,* units manufactured such that the face of the block sags, giving a unique appearance.
- *Glazed concrete block,* units with factory-applied glazing on one or more faces. These are used for applications that require a durable, washable surface.
- *Ground face block,* units that have been ground on the face to give a coarse, textured surface.

Units for Takeoff

The standard procedure for taking off concrete block is similar to that used for brick. The contractor must calculate the areas of walls and partitions by multiplying the length by the height of the various walls and partitions, and computing a total area in SF. The contractor must separate the different types of block by size, weight, surface textures, shapes, and color, as each has an effect on pricing.

The contractor converts the SF area to the actual number of block by dividing the total area by the area of an individual block, based on its nominal 8″ x 16″ size (when using a standard 3/8″ mortar joint). For example:

> *A standard hollow concrete block with the nominal dimensions of 8″ x 16″ has an individual area of 128 square inches, or .89 SF. If this calculation is applied to 100 SF, the result is approximately 112.5 block per 100 SF, or 1.125 block per SF.*

The contractor should make deductions for all openings over 2 SF to get an accurate count on the block.

Because concrete block are typically installed in a running bond pattern, the contractor should include half-block at vertical terminations of the wall such as control joints and door or window jambs. These are taken off by counting the half-block at alternating courses at each location. For example:

> *Both sides of the control joint are 20 courses high. Therefore, 20 courses divided by 2 (for alternating courses) equals 10 pieces of half-block. Since there are two sides to the control joint, multiply the 10 pieces by 2. The result, 20, is the total half-block needed.*

In addition, specialty blocks, such as bond beam blocks, must be included as part of the takeoff. *Bond beam blocks* are trough-shaped concrete blocks typically installed at the top course (or a story height) to provide horizontal reinforcing. Steel rebar are laid laterally within the trough and grouted solid to provide a continual lateral reinforced beam. (Both grout and reinforcing will be discussed later in this chapter.)

Bond beams are taken off by the LF and converted to the number of blocks by dividing the length by the nominal length of the block (16″). Figure 7.8 illustrates a bond beam and its use.

Other Takeoff Considerations

The contractor must include an allowance for waste. Under normal circumstances, an average of 3% and a maximum of 5% is adequate.

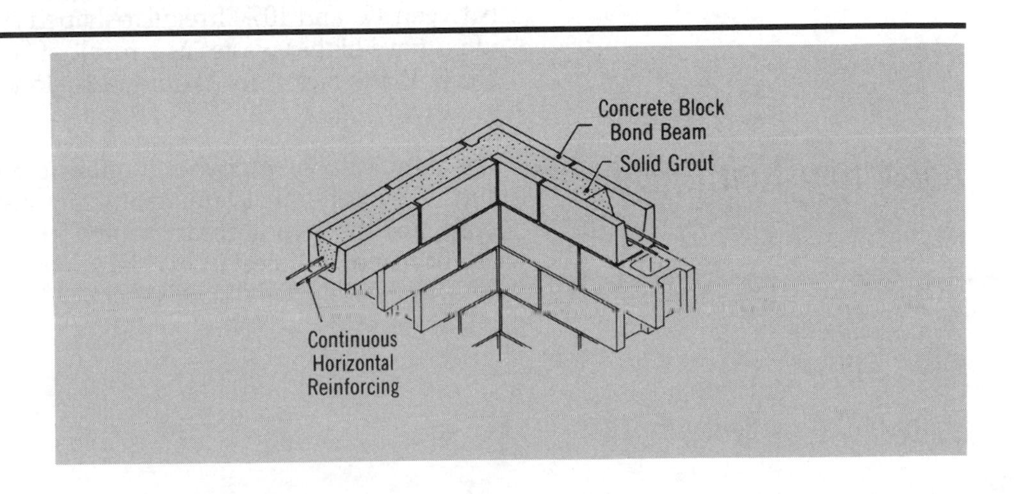

Figure 7.8 Bond Beam

Because of the size and weight of concrete block, some union rules require that two masons set a block. This should be noted as it has a direct effect on productivity.

Other factors that have a direct effect on the cost of the work have been listed in detail under "Other Takeoff Considerations" in the "Brick Masonry" section earlier in this chapter.

Stone

Because it is a product of nature, stone varies dramatically in type, size, shape, and weight with each different species and the geographic location. In general, stone for construction purposes can be classified in one of the following groups:

- *Rubble:* Irregularly-shaped pieces broken from larger masses of rock, installed or "laid up" with little or no cutting or trimming.
- *Fieldstone:* Irregularly-shaped rocks used as they are found in nature. Most commonly used in fireplaces and stone walls in landscaping.
- *Cut stone:* Stone that has been cut to specific shapes and sizes, with a uniform texture. The most common use of cut stone is for veneers.
- *Ashlar:* Characterized by saw-cut beds and joints, usually rectangular in shape, with flat or textured facing.

Units of Takeoff

Units of takeoff may vary with location and species of stone used. In general, the following rules apply:

- Most stone used in landscaping walls is taken off by the CF. This is done by multiplying length × height × width (or thickness) of the wall. Then the CF volume is converted to tons based on the individual stone's volume per ton. Each species of stone will vary depending on the size, shape, and density of the wall. The contractor should consult the supplier of the particular stone.
- Stone in veneer wall and fireplace applications is taken off by the SF and converted to tons in much the same manner.

Other Takeoff Considerations

It is essential that the contractor obtain a direct quote for each species of stone delivered to the job site. Because most stone is extremely heavy, costs for shipping and handling can be high.

Allowances for waste are based on the individual type of stone. A general rule is that for stone, ashlar, or regular-shaped stone, the waste will vary between 4% and 10%. Irregular-shaped stones such as rubble or fieldstone often have higher wastes as a result of "unusable" stone in a delivered batch. Waste factors for lower-grade stone can approach 20%.

Masonry Reinforcement

The term *masonry reinforcement* refers to the use of steel reinforcing bars and wire mesh-type lateral reinforcing installed in the coursing of masonry units, and the vertical rebar grouted into the voids in concrete block. The use of reinforcing steel in masonry work adds tensile strength to the wall or partition in much the same way as reinforced concrete. Masonry reinforcing is classified in two groups: lateral (horizontal) and vertical.

Lateral Masonry Reinforcement

Lateral reinforcing can be further divided into two types: wire mesh, or strip joint reinforcing, and horizontal steel rebar. Both types provide reinforcement against lateral stresses.

Wire Mesh or Strip Joint Reinforcing

Joint reinforcing is accomplished by the use of wire mesh-type strips installed between the courses of the masonry units. The most common types of wire strip joint reinforcement are the truss-type and the cavity-wall ladder type. Figure 7.9 illustrates the major types of joint reinforcement.

Units for Takeoff Horizontal joint reinforcement is taken off by the LF. Most plans and specifications are quite specific as to the location of horizontal joint reinforcement, which is typically specified by the course. To arrive at the total LF of joint reinforcement, the contractor must count the number of courses that require joint reinforcement and multiply by the LF of reinforcement for each course. Special forms of joint reinforcement, such as tees used at the intersection of masonry walls and corners, must be counted and listed separately.

Horizontal Steel Rebar

Horizontal reinforcing in the form of rebar is typically used for bond beam applications. Steel reinforcing bars are laid within the trough of the bond

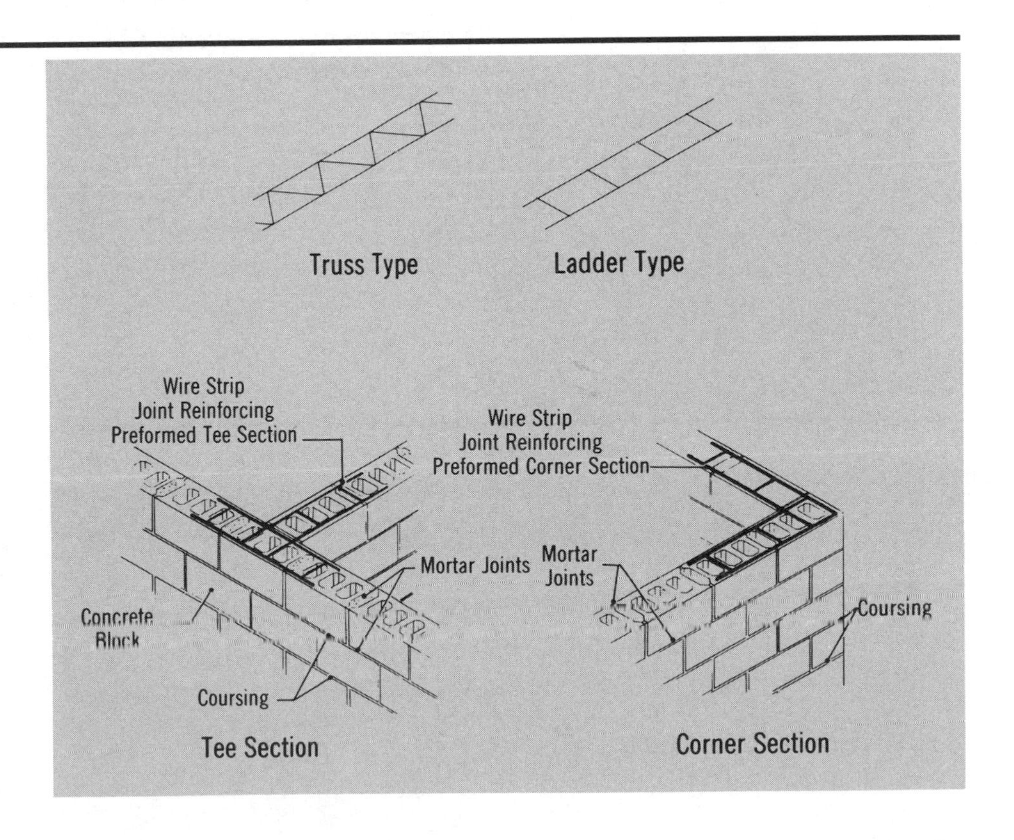

Figure 7.9 Types of Joint Reinforcement

beam and are grouted solid to form a continuous ring of lateral reinforcement at specific locations within the masonry wall or partition.

Units for Takeoff Horizontal rebar in masonry is taken off by the LF. Plans and specifications designate the location of horizontal reinforcing, as well as the size of the bar and the number of bars to be used. Takeoff is done by measuring the total LF of rebar at each location (bond beam).

The contractor must count all bent bars for corners and intersections separately and list them in accordance with bar designation and length for accurate pricing. Allowances must be added for overlap of continuous bars. The procedure for converting the total LF of rebar to its final pricing unit of pounds or tons is the same as discussed in Chapter 6.

Vertical Masonry Reinforcement

Vertical masonry reinforcement refers to the use of steel rebar installed within the cells of hollow concrete block walls, and grouted in place to form a single unit. When installed within an engineered design, the reinforced masonry walls resist stresses exerted by wind, earthquake, and other forces. In the typical application, the vertical bars are spaced at a predetermined "on center" spacing detailed on the drawings. Figure 7.10 illustrates a reinforced concrete block wall.

Units for Takeoff

The takeoff is accomplished with the same method used for vertical rebar in a cast-in-place concrete wall. The total length of the reinforced walls is divided by the on-center spacing to determine the number of bars. Once the quantity has been determined, it is multiplied by the length of the individual bars. Additional bars may need to be added per specification requirements at the corners or jambs of openings.

The contractor should calculate overlap for the specific grouting conditions (covered later in this chapter) and add this to the total length of rebar.

The procedure for converting LF to the final pricing unit of pounds or tons is the same as discussed in Chapter 6.

Grouting

Grout is a composition of Portland cement, sand, lime, and water mixed in similar proportions and strengths to mortar. Additional water is used to bring the consistency to a more fluid or plastic state. The grout is then

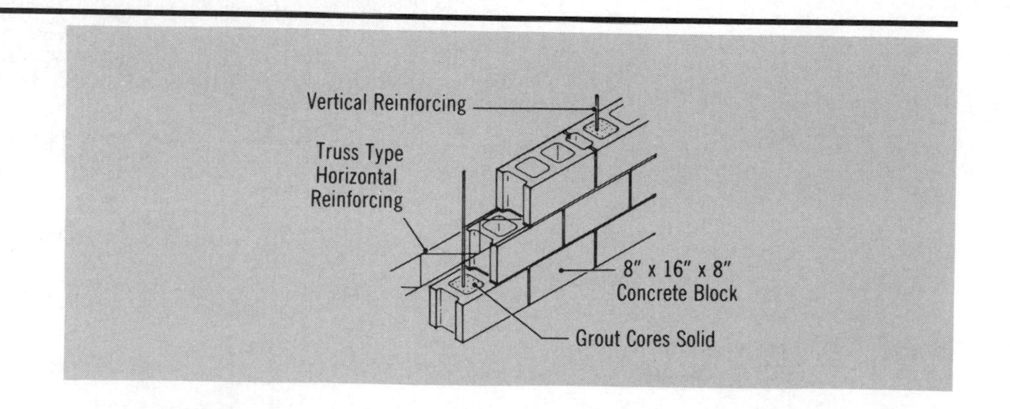

Figure 7.10 Reinforced Concrete Block Wall

pumped or poured into the cells of set concrete block that contain vertical or horizontal reinforcing bars. Vibrating or tamping may be required to ensure complete embedment of the rebar. Once the grout has cured, it forms a single solid unit of grout, block, and rebar.

Units for Takeoff

To determine the quantity of grout needed, the contractor must calculate the volume of the cells to be filled. This is accomplished by determining the location and quantity of vertical columns (cells) to be filled, multiplied by the volume of each column of cells. This figure should be computed in CF, the typical unit of pricing. The contractor may choose to convert the CF quantity to CY, where 1 CY = 27 CF.

To simplify the calculation of grout quantities, the contractor may use the table in Figure 7.11. The quantity of grout specified in the table is listed according to specific spacing and wall thicknesses, and is based on SF of wall area.

Other Takeoff Considerations

In considering grouting applications, the contractor must note the specific requirement within the specifications for the placement of grout. There are two different levels, or "lifts", of accepted grout placement: high-lift and low-lift. There are particular advantages and disadvantages to each.

Low-Lift Grouting

Low-lift grouting calls for the placement of grout within the cells of an erected wall at a maximum of 4'-0". Grout is poured to within 1-1/2" of the top of the top block to allow for a keyway when additional lifts are poured. Low-lift grouting has the advantage of ensuring that cells are filled solid down to the previously poured lift. The disadvantage is the additional cost for delayed production in the erection of the wall.

Volume of Grout Fill for Concrete Block Walls

Center to Center Spacing Grouted Cores	6" C.M.U. Per S.F. Volume in C.F.		8" C.M.U. Per S.F. Volume in C.F.		12" C.M.U. Per S.F. Volume in C.F.	
	40% Solid	75% Solid	40% Solid	75% Solid	40% Solid	75% Solid
All cores grouted solid	.27	.11	.36	.15	.55	.23
cores grouted 16" O.C.	.14	.06	.18	.08	.28	.12
cores grouted 24" O.C.	.09	.04	.12	.05	.18	.08
cores grouted 32" O.C.	.07	.03	.09	.04	.14	.06
cores grouted 40" O.C.	.05	.02	.07	.03	.11	.05
cores grouted 48" O.C.	.04	.02	.06	.03	.09	.04

Note: Costs are based on high-lift grouting method.

Low-lift grouting is used when the wall is built to a maximum height of 5'. The grout is pumped or poured into the cores of the concrete block. The operation is repeated after each five additional feet of wall height has been completed. High-lift grouting is used when the wall has been built to the full story height. Some of the advantages are: the vertical reinforcing steel can be placed after the wall is completed, and the grout can be supplied by a ready-mix concrete supplier so that it may be pumped in a continuous operation.

Figure 7.11

High-Lift Grouting

High-lift grouting is the placement of grout in the cells upon completion of the top course of the masonry work. It requires that cleaning holes be left out at the bottom of each cell. This allows access for cleaning mortar droppings out from within the cell to be grouted. This method requires that the grouting procedure be continuous to the top of each cell. It has the advantage of allowing the grouting operation to be continuous, in contrast to low-lift grouting, which is done in smaller portions. It also allows the vertical reinforcing to be placed after the wall has been completed. The major disadvantage is that it is difficult, if not impossible, to ensure that all rebar has been adequately surrounded by grout through the entire height of the cell. Some structural failures have been attributed to weak spots in the wall where voids have occurred and the rebar has deteriorated. In addition to the specs, the contractor should consult the local building codes for specifics concerning grout placement. Some building codes do not allow the use of high-lift grouting.

Masonry Anchors and Ties

To anchor or tie multiple-wythe masonry walls together, or less stable veneer brick wall to a structural backup wall, ties must be used. Anchors or ties are manufactured of coated metals that will not deteriorate when in contact with the corrosive elements in mortar. They are available in a wide variety of shapes, sizes, and methods for fastening. Some of the more common types of anchors and ties are listed below.

- *Dovetail anchors* are used to tie masonry veneers to cast-in-place concrete backup walls. The dovetail slot is poured within the forms and, when stripped, provides a vertical slot in which to attach the dovetail anchor.
- *Corrugated wall ties* are hot-dip galvanized strips of corrugated metal for tying brick veneers to wood-framed or concrete block backup walls.
- *Box-type cavity wall anchors* are loop-shaped metal wires of various gauges used to tie multiple-wythe masonry walls together.
- *Welded anchors* are used to tie masonry veneers to structural steel columns and beams.

Anchors are embedded in the horizontal courses of masonry work and fastened to the backup wall by screws, welding, or embedment (in the case of dovetail slots). Ties for multiple-wythe masonry walls are embedded in the respective courses of both wythes.

The spacing of anchors and ties is noted in the specifications. They are typically described in terms of both horizontal and vertical spacing. For example, "ties and anchors will be spaced at 24" O.C. both horizontally and vertically."

Units for Takeoff

Anchors and ties are taken off by the piece. This requires that the area of wall to be anchored be calculated by the SF and then divided by the spacing specified. For example:

Calculate the ties required for 1,000 SF of masonry wall, with the ties spaced at 2'-0" O.C. each way.

If the spacing is 2'-0" each way, one tie will be required for every 4 SF of wall;

$$\frac{1000\ SF}{4\ SF\ per\ tie} = 250\ ties$$

The contractor should allow approximately 5–7% for waste resulting from handling.

Masonry Restoration

Masonry restoration refers to restorative work on masonry that is already in place. Most of this kind of work includes the cutting out of old mortar joints and repointing or refilling the existing joints with fresh mortar. This process is sometimes called *tuckpointing*. Masonry restoration may also include the removal and replacement of damaged or deteriorated individual masonry units, most often brick.

The cutting out of old mortar joints is typically done by an electric saw or a grinder with a diamond or carborundum blade. The saw or grinder is set to a specific depth as required in the specifications, and is passed along the existing mortar joint until free of old mortar. The removal of damaged or deteriorated brick is accomplished in much the same manner. Additional chipping by means of a hand mallet and chisel may be required to remove individual bricks.

The newly cut and cleaned joint is then filled with mortar and tooled to achieve the desired joint. Replacement brick are "buttered" with mortar and fitted into place. The replacement of individual brick at areas subject to the most damage, such as outside corners, is most common. The removal of old courses to create a bond for the new work is referred to as "toothing" or "toothing in."

Restoration work may also include sandblast cleaning of the surface area. This is the process of forcing fine sand or slag through a hose at high pressure. The abrasive force removes debris, graffiti, and the old surface. It should be noted that excessive sandblasting can damage the surface and remove the mortar from the joint. Some repointing of damaged joints may be necessary as a result.

Units for Takeoff

The standard unit for takeoff for both cutting out the old joint and repointing is the SF. The contractor may choose to separate the takeoff for each process for pricing purposes. Area is calculated by multiplying the length or width by the height of the surface area to be tuckpointed.

Removal and replacement of old brick can be taken off by the individual piece or by the square foot. Individual brick to be replaced in random locations may be best taken off by the piece. Larger quantities grouped together with definable dimensions should be taken off for pricing by the SF. In either case, the quantities of each should be noted separately on the takeoff.

The takeoff units for sandblasting are SF of surface area.

Other Takeoff Considerations

The cost of restoration work is affected by a variety of factors, including staging required to access the work. Movable staging, such as swing staging or motorized platforms, may best accommodate the frequent moves associated with tuckpointing. Typically, the cost for renting both types of staging exceeds the cost of conventional scaffolding. The contractor should evaluate and price the most efficient method.

Another factor may be the need for temporary bracing when large quantities of damaged or deteriorated brick are removed. The possibility of collapse of old brickwork already weakened by the removal of old mortar is a realistic concern. Bracing should be evaluated by the individual work area. Cost for materials and labor to brace the work should be considered.

Ground labor to support the workers on the staging is usually required and must be accounted for in the takeoff. Ground support provides mortar, materials, tools, and equipment for the crew above.

As cutting out old mortar generates a lot of wear on tools and equipment, an allowance may be necessary for additional blades.

Sandblasting requirements vary with the surface materials and the size of the area being cleaned. Protection of surface features such as windows, doors, or ornaments can slow productivity. Because the use of sandblasting equipment is environmentally regulated, the contractor must check the regulatory requirements for individual areas. Work locations with dense populations generally require the containment of airborne particles. Enclosure of the work should be included in the takeoff, and may require the dismantling and re-erection of the containment enclosure several times as the work progresses. Cleanup of the spent sand or slag should also be included as part of the work.

Masonry Cleaning

Some form of cleaning is required on all new masonry work to remove splatterings of mortar and dust. Masonry cleaning is usually done after the work has set but before the mortar has reached its full strength. Several methods are used, the most common of which involves the use of mild detergents and a coarse hand brush to scrub the surface of the work. Once cleaned, the surface is rinsed with clean, potable water through a hose. Stronger solutions containing chemicals such as muriatic acid may also be required. The basic process is the same, although protective gear may be required.

An alternate method involves high-pressure washing. Water is forced through a nozzle at high pressure to remove the surface debris.

Units for Takeoff

Regardless of the method used, masonry cleaning is taken off and listed by the SF of area to be cleaned. Different methods employed on the same project should be listed separately, as each has a different cost.

Solutions used in hand cleaning and power washing are usually diluted to a specified strength. The contractor should base the quantity of solution on the area to be covered. The quantity of solution often varies with the product used and most often is a matter of judgment or may be stated in the specifications.

Other Takeoff Considerations

Like masonry restoration, cleaning often requires staging. On new work with standard pipe scaffolding, the work is usually cleaned before the staging is dismantled and moved. A ground crew may be needed for larger cleaning jobs to support the personnel doing the cleaning.

The use of toxic chemicals or solutions may also be governed by environmental regulations, and may require protection or cleanup. Protective gear for workers can be costly and may slow productivity.

The contractor should consult the specifications for the approved method of cleaning, and price the work accordingly. Substitutions of less expensive methods may not be acceptable.

Miscellaneous

Masonry Insulation

Masonry insulation consists of the installation of rigid insulation boards between the veneer brick and the CMU backup wall. This item is taken off based on the SF to be insulated.

The more common method of insulating concrete block is to use masonry fill insulation, a granular material composed of water-repellent vermiculite or silicon-treated perlite. Both materials are poured from bags or blown in from trucks into the voids in cavity walls or the empty cells in concrete block.

Units for Takeoff

The procedure for determining the quantity of masonry fill insulation is the same as that used for quantifying grout for reinforced cells. (See "Grout" earlier in this chapter.) The takeoff units are CF.

Flashings

Flashings are impervious sheets of material commonly installed at the base of the exterior masonry walls to deflect water from going into the structure.

Units for Takeoff

Flashings are taken off and listed by the LF. Flashings of different compositions, thicknesses, and widths should be listed separately.

Cutting Masonry Units

The contractor must include the cost of cutting masonry units such as block and brick. Some block and brick are cut with a brick hammer and chisel, but when exposed surfaces require cuts, it is typically done by a gas or electric masonry saw with a diamond or carborundum blade. The cost for minor cutting may be negligible in the overall estimate. Projects that require extensive cutting must be evaluated accordingly.

Units for Takeoff

Cuts are taken off by the piece and can be listed as a total quantity for each type of masonry unit. An alternate method provides for the labor and equipment costs as a function of time. For example, a contractor may carry five days' worth of labor and equipment time to do all the cutting on a particular project.

Forms

Some kinds of masonry work require the use of forms or braces to support installed work until it has cured and can support itself. Classic examples are the forms needed to hold brick or stone over half-round windows, or formwork to brace the bottom of a bond beam poured as a lintel for an opening.

Units for Takeoff

The contractor may elect to quantify such work by the opening or piece. Different types and applications should be listed separately.

Freight of Masonry Units

Because most masonry units and stone are quite heavy, the cost for freight and delivery to the job site can be substantial. The contractor should secure the price from the supplier to include all shipping and handling costs.

Units for Takeoff

Freight costs can be taken off and listed either per ton or for the entire load as a lump sum (LS). Quantities listed by weight will require the approximate weight per 1,000 units (brick, block). Weight of stone may be by the pallet or shipping container. Both should be available from the supplier.

Items Furnished by Other Trades

It is common for masonry contractors to install items furnished by other trades as part of the masonry scope of work. Typical examples are the installation of lintels (structural steel members that support the weight of masonry over openings in masonry walls); joist bearing plates (steel plates with anchors embedded in bond beam courses to tie bar joists to masonry walls); hollow metal door frames; and electrical boxes and conduits enclosed in masonry walls for receptacles and switches. These items must be installed as the various courses of masonry are laid. The specifications should clearly enumerate the items furnished by other trades and installed under the masonry scope of work.

Units for Takeoff

Each individual item should be taken off by the piece (EA) and listed according to size (e.g., length and weight of lintel). The weight of a steel lintel is a factor to consider in the takeoff. The contractor must calculate the weight of the individual lintel (to be discussed in detail in Chapter 8), as equipment may be required for placement. The contractor must separate the lintels that can be installed by hand from those that will require hoisting equipment.

Control Joints

Masonry walls of considerable straight length require the installation of a vertical control joint to allow for expansion and contraction of the wall. Control joints require the use of a compressible premolded joint filler (similar to the type used in concrete construction, discussed in Chapter 6) to break the bond between adjacent units in courses of brick or block masonry. The location of control joints should be clearly shown on both plan view and elevations of masonry walls on the drawings.

Units for Takeoff

Control joints are taken off by the vertical linear foot (VLF). Control joints of varying thicknesses (for different wall thicknesses) should be listed separately.

Incidentals for Fireplace and Chimney Construction

The construction of a fireplace requires the masonry contractor to furnish and install some special masonry items such as dampers, flues, clean-out doors, and fire brick. Dampers are cast-iron operable traps that regulate the draft of the fireplace.

Clean-out doors are access doors and frames in varying sizes to allow the removal of ashes and debris. They are installed within the courses of masonry as the work proceeds, and are located at the lowest point in the fireplace.

Masonry flues are noncombustible, heat-resistant, rectangular or round-shaped tubes made of fireclay to allow the passage of smoke from the fireplace, boiler, furnace, or solid fuel stove through the chimney. They vary in size as required by the design of the system.

Units for Takeoff

Typically there is only one damper per fire box. It should be listed by the piece (EA). Any special characteristics (such as manual or rotary operation) should be noted, as these can affect the cost. Clean-out doors are taken off by the piece (EA). Different sizes should be listed separately. Flue lining is taken off by the LF. Each size and shape should be listed separately for accurate pricing. Figure 7.12 illustrates a flue in perspective and plan view.

Staging and Scaffolding for Masonry Work

Staging and scaffolding, for the purpose of this discussion, will be considered interchangeable terms. Staging consists of a temporary elevated platform and its supporting structure erected against, within, or around the work to support workers, materials, and equipment for the performance of masonry work. The most common type is conventional pipe scaffolding and planking. It is erected from the ground level and built or added to as required to maintain access to the work.

Units for Takeoff

The takeoff unit for staging is SF of surface area of the masonry. The contractor can elect to convert the SF area to actual sections of scaffolding and planks needed. This conversion is based on the actual size of the available staging units. This type of takeoff will consist of frame sections, cross braces, outriggers (brackets attached to the side of vertical sections of staging to provide support for planking), and planks.

Other Takeoff Considerations

As most masonry contractors own some quantity of staging and planking, the cost of materials may be negligible. In cases where the contractor does not have enough staging available, it will be necessary to include rental of the appropriate amount. Because staging is an ongoing process as the work progresses, and requires dismantling, moving, and re-erection, the contractor must consider the number of moves based on productivity of the crew. This is a matter of judgment based on experience.

Other types of staging may prove more efficient for some jobs. (Refer to the "Masonry Restoration" section earlier in this chapter.)

Cleanup

Most masonry contracts provide for cleanup of debris generated by the work. The amount of cleanup varies to some degree based on the contract or specifications. Cleanup typically consists of picking up broken or discarded brick, block, or other such masonry units, as well as ties and forms, and scraping or sweeping mortar droppings. Disposal may be in on-site dumpsters or may involve removal from the site by the masonry contractor's own forces.

Units for Takeoff

Masonry cleanup is most often accomplished by hand and quantified in labor-hours. The contractor's judgment based on experience is important in determining the labor-hours necessary for the level of cleanup required. Included should be the cost for disposal in the form of dumpster rentals or landfill fees.

Summary

The takeoff sheets in Figure 7.13 are the quantities for the work of Division 4—Masonry for the sample project. The quantities of the various masonry units have been separated according to size and specific use. The quantities have been determined from the dimensions available on the drawings and extended to the final units that would be used in pricing.

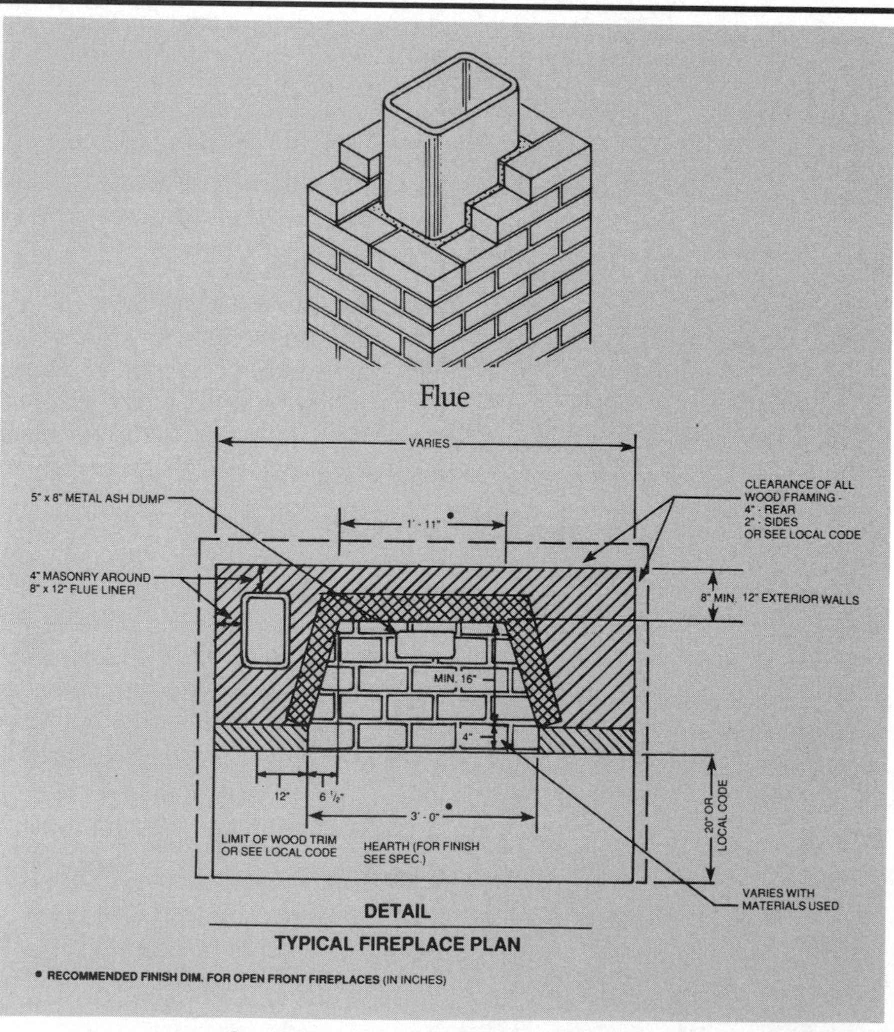

Figure 7.12 Flue in Perspective and Plan View

Division 4 Checklist

The following list can be used to ensure that all items in the takeoff for this division have been accounted for.

Mortar

- ☐ Strength
- ☐ Special additives such as color

Brick Unit Masonry

- ☐ Type and size
- ☐ Bond pattern
- ☐ Waste

Concrete Unit Masonry

- ☐ Weight of units
- ☐ Type and size
- ☐ Bond pattern
- ☐ Bond beams

Stone

- ☐ Type and size
- ☐ Method of setting

Masonry Reinforcement

- ☐ Lateral joint reinforcement
- ☐ Lateral rebar reinforcement
- ☐ Vertical reinforcement

Grout

- ☐ Strength
- ☐ High- or low-lift placement
- ☐ Waste

Anchors and Ties

- ☐ Anchors
- ☐ Ties
- ☐ Spacing

Masonry Restoration

- ☐ Cutting joints and repointing
- ☐ Sandblasting
- ☐ Cleaning
- ☐ Staging

Masonry Cleaning

- ☐ Chemicals
- ☐ Worker protection
- ☐ Staging
- ☐ Water

Miscellaneous

- ☐ Masonry fill insulation
- ☐ Flashings
- ☐ Cutting of masonry units
- ☐ Forms
- ☐ Freight and shipping

☐ Items furnished by other trades
☐ Control joints
☐ Fireplace and chimney construction
☐ Staging and scaffolding
☐ General cleanup

Means Forms

QUANTITY SHEET

PROJECT: SAMPLE PROJECT

SHEET NO. 4-1/3

ESTIMATE NO. 1

LOCATION:

ARCHITECT: HOME PLANNERS

DATE:

TAKE OFF BY: WJD

EXTENSIONS BY: WJD

CHECKED BY: KF

DESCRIPTION	NO.	DIMENSIONS				UNIT		UNIT		UNIT		UNIT
DIVISION 4: MASONRY												
CONCRETE MASONRY UNITS												
12" CMUs @												
FOUNDATION WALL												
@ MAIN HOUSE	1	124	8.0		992	SF	÷ .89 SF/BLK				1115	BLK
@ CRAWL SPACE	1	45.33	4.67		212	SF	÷ .89 SF/BLK				240	BLK
12" CMU's @ FIREPLACE												
FOUNDATION 11'-8"x3'-8"x												
8'-0" HIGH	1	30.67	8.0		246	SF	÷ .89 SF/BLK				276	BLK
6" CONCRETE SLAB												
@ FLUSH HEARTH												
@ COUNTRY KITCH.	1	11.5	5.67	.5	33	CF						
@ LIVING ROOM	1	6.67	4.5	.5	15	CF					48	CF
FORMWORK TO SUPPORT												
POURED HEARTHS	1	11.5	3.0		107	SF						
	1	6.67	1.67		12	SF					119	SF
#4 REBAR @ 4" O.C												
@ HEARTH	35	7.67			269	LF						
	21	6.5			137	LF						
					406	LF	x .668 LB/LF				272	LBS
#3 REBAR @ 10" O.C.												
@ HEARTH	7	11.5			81	LF						
	6	6.67			40	LF						
					121	LF	x .376 LB/LF				46	LBS
4" SOLID CMU @												
FIREPLACE CHIMNEY	1	24	7.5		180	SF						
	1	14.67	8.0		118	SF						
	1	13.34	13.67		183	SF						
					481	SF	÷ .89				540	BLK
4" BRICK VENEER												
@ COUNTRY KITCHEN	1	11.5	8.33		121	SF	x 6.55 BRK/SF				793	BRK
IN RUNNING BOND	1	3.0	8.33									
@ EXPOSED CHIMNEY	1	16	5.33		86	SF	x 6.55 BRK/SF				563	BRK

Figure 7.13

Means Forms

QUANTITY SHEET

PROJECT: SAMPLE PROJECT | SHEET NO. 4-2/3
ESTIMATE NO. 1
LOCATION: | ARCHITECT: HOME PLANNERS | DATE:
TAKE OFF BY: WJD | EXTENSIONS BY: WJD | CHECKED BY: KF

DESCRIPTION	NO.	DIMENSIONS					UNIT
4" STACK BOND PIERS & SOLDIER COURSE HEADER @ ELEV. #16 A-10							
@ PIERS	2	12			24		
@ HEADER	1	21			21		75 BRK
4½" x 9" FIREBRICK @ BOTH FIRE BOXES							
@ COUNTRY KITCHEN	1	2	3		6		
	1	3.33	7		24		
@ LIVING ROOM	1	2	3		6		
	1	3.33	7		24	60 SF x 3.5 BRK/SF	210 BRK
ADD 5% FOR WASTE ON FIREBRICK	1	210	.05				11 BRK
MORTAR FOR BLOCK							
4" CMU —	540 x 1.05				567 BLK ÷ 1000 = .567 x .9 =		.51 cy
12" CMU —	1631 BLK x 1.05 =				1712 BLK ÷ 1000 = 1.71 x .9 cy =		1.54 cy
(FULL BEDDING)							2.05 cy
MORTAR FOR BRICK @ CHIMNEY, VENEER	1431 x 1.05 =				1502 BRK ÷ 1000 = 1.50 x .4 =		.6 cy
MORTAR FOR FIREBRICK @ FIRE BOXES	210 x 1.05 =				221 BRK ÷ 1000 = .221 x 300#/M =		67 #
MORTAR FOR BACK PLASTERING OF 4" CMU @ EXPOSED AREAS	1	8.0	9.0		72 SF		
3/8" THICK	2	16.0	8		256 SF		
					328 SF ÷ 1000 = .328 x 1.3 =		.42 cy
CORRUGATED STRAP TIES @ 24" O.C. E.W. @ BRICK VENEER	1431 ÷ 6.55 =				218 SF x .25 TIES/SF = 55 x 1.10 =		60 TIES

Figure 7.13 (continued)

Means Forms

QUANTITY SHEET

SHEET NO. 4-3/3

PROJECT SAMPLE PROJECT

ESTIMATE NO. 1

LOCATION

ARCHITECT HOME PLANNERS

DATE

TAKE OFF BY WJD

EXTENSIONS BY: WJD

CHECKED BY: KF

DESCRIPTION	NO.	DIMENSIONS				UNIT		UNIT		UNIT		UNIT
12"x8" FLUE FOR FURNACE FROM 4' A.F.F.	1	40'									40	VLF
8"φ THIMBLE FOR FURNACE	1	2.5'									1	EA
12"x12" FLUE FOR COUNTRY KITCHEN FROM 5'-9" ABOVE 1ST FLR.	1	31.33'									32	VLF
12"x16" FLUE FOR LIV. RM. FROM 6'-0" ABOVE 1ST FLOOR	1	31.08									32	VLF
CONCRETE CAP @ TOP OF CHIMNEY INCL. REINF. & FORMWK.	1	5.33	2.5	1.33	18	CF					18	CF
LINTELS FOR OPNGS. ∠ 3½ x 3½ x 5/16"	4	3.67'									4	EA
∠ 3½ x 3½ x 5/16"	2	2.5'									2	EA
9"x5" ASH DUMP	2	EA									2	EA
30" DAMPER	2	EA									2	EA
8"x8" CLEAN OUTS @ BASEMENT	2	EA									2	EA
WASH EXPOSED BRICK VENEER	1	218	SF								218	SF
ROOF STAGING FOR CHIMNEY	1	LS									1	LS
2¼"x5" FLAGSTONE SILL @ BASEMENT WINDOWS	3	EA			3	EA					3	EA

Figure 7.13 (continued)

Chapter Eight

METALS

(Division 5)

Chapter Eight

METALS
(Division 5)

Structural steel has the capacity to support large loads with a relatively compact size and shape. It can be used to support loads in both compression and tension, therefore making it an ideal material for flexural components. The variety of steel shapes and sizes available provides engineers with an economical solution to many structural design problems. Structural steel is also available in different strengths or grades for particular loading or stress conditions. When taking off Division 5 work in general, the following categories must be considered:

- Structural steel shapes
- Open-web bar joists
- Metal decking
- Erection and crane
- Field welding
- Items installed by others
- Shop and field priming
- Shop drawings

This chapter is limited to a discussion of the more common shapes of structural steel, open-web bar joists, and metal decking.

Structural Steel Shapes

Structural steel can best be defined as the members that make up the structural frame of the building that will transmit the load to the foundation. Figure 8.1 lists eight of the more common shapes.

Each shape is prefixed by a letter and numbers. These designations are more than simple identifications; they provide important information about the individual piece. The letter is the classification of the piece by shape. The first number refers to the actual or nominal depth in inches of the section. The second number refers to the weight per linear foot of the section, in pounds per LF.

The exceptions to this rule are the angle shapes. The first two numbers in an angle designation are the lengths of the "legs" of the particular angle, in inches. The third number is the thickness of each leg in inches. Figure 8.2 lists the designations and their corresponding depths and weights.

Structural steel is also available in different strengths, expressed as *yield stress*. Simply put, the yield stress, typically defined as *KSI* or *kips per square inch*, is the maximum allowable stress that can be exerted on the material before it fails. A kip is a unit of measure equal to 1,000 pounds. The different grades of steel are named according to the number of the test conducted by the ASTM (American Society of Testing and Materials) to determine the characteristics of the species. Not all of the previously mentioned shapes are available in some of the more specialized grades of steel. Figure 8.3 lists some common structural steel grades.

Structural steel work is shown on structural drawings in plan views. Elevations, sections, and details are often added for clarity. Using schedules to list columns or lintels saves time during the takeoff. The details show the connections to be used for the individual pieces. Figure 8.4 is a structural steel drawing shown in plan view, and the corresponding details.

Note: The term "DO" indicates duplication, similar to the expression "ditto", and is used to show more than one of the same designation beam, girder, joist, etc.

Units for Takeoff

Structural steel is taken off by the length of the piece, then converted to weight in pounds, and finally tons (tns).

Individual components should be listed according to shape and designation. Pieces can be further classified by use (for example, beams,

Common Steel Sections

The upper portion of this table shows the name, shape, common designation and basic characteristics of commonly used steel sections. The lower portion explains how to read the designations used for the above illustrated common sections.

Shape & Designation	Name & Characteristics	Shape & Designation	Name & Characteristics
W	Wide Flange Parallel flange surfaces	M C	Miscellaneous Channel Infrequently rolled by some producers
S	American Standard Beam (I Beam) Sloped inner flange	L	Angle Equal or unequal legs, constant thickness
M	Miscellaneous Beams Cannot be classified as W, HP or S; infrequently rolled by some producers	T	Structural Tee Cut from W, M or S on center of web
C	American Standard Channel Sloped inner flange	H P	Bearing Pile Parallel flanges and equal flange and web thickness

Figure 8.1

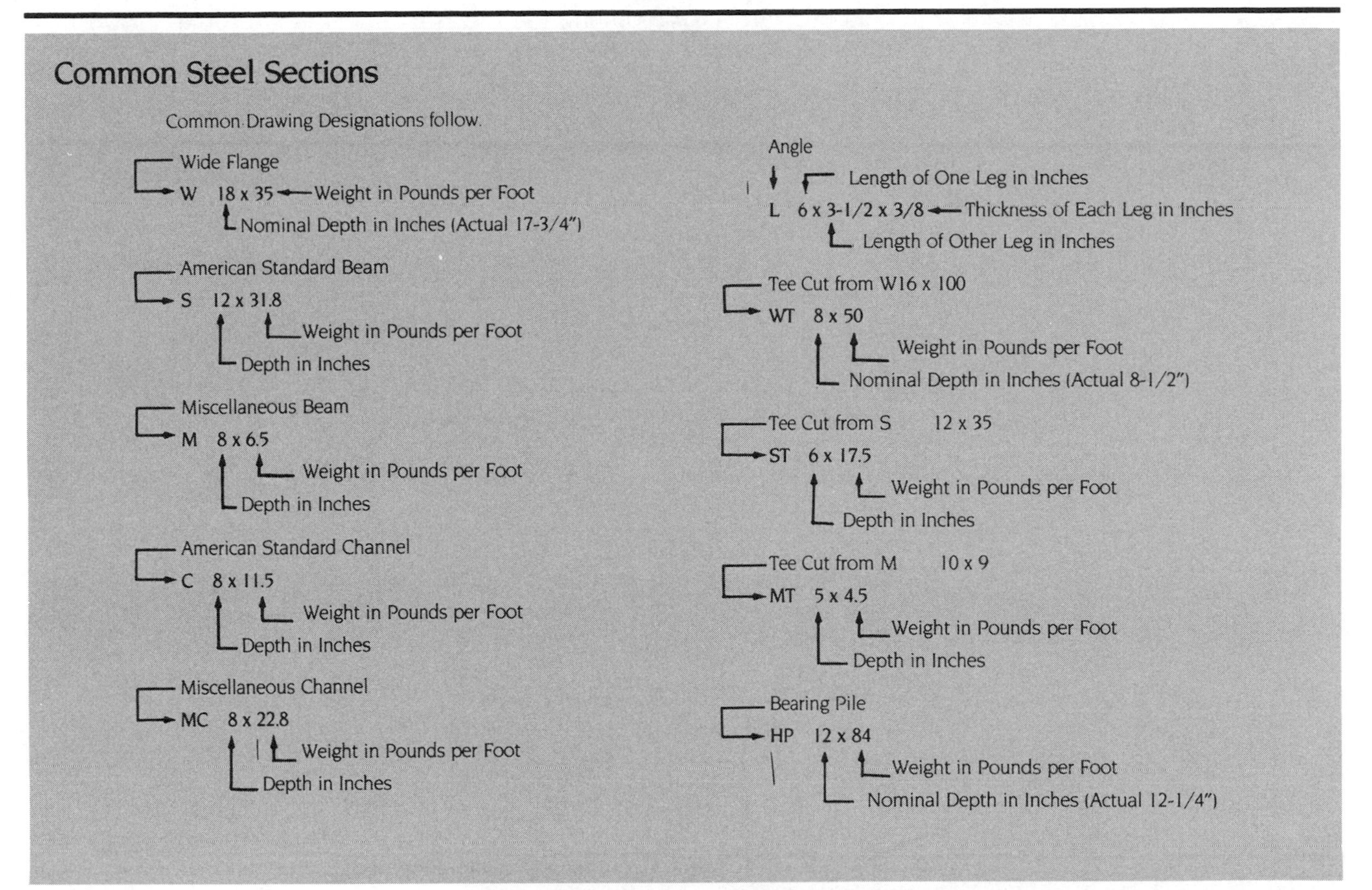

Figure 8.2

Common Structural Steel Specifications

ASTM Designation	Yield Stress in KSI	Description
A36	36	Most common carbon steel, all shape groups and plates and bars up to 8"
A529	42	All shape groups, as A36, but plates and bars up to 1/2" thick
A441	40–50	High-strength, low-alloy steel, shapes and plates, but in lesser variety
A572	42–65	High-strength, low-alloy steel, all shapes and plates
A242 & A588	42–50	High-strength, corrosion-resistant; drop in strength as sizes increase
A514	90–100	Quenched and tempered alloy; plates and bars only, with special care so as not to impair the heat treatment

Figure 8.3

columns, girders). Separating the pieces by application is done for the purpose of pricing the erection rather than the materials.

Referring to Figure 8.4, the weight of the steel can be calculated for each individual member. For example:

Calculate the weight of the W12 x 22 beams between column lines 1 and 2 shown in Figure 8.4.

From the designation number it can be determined that the W12 x 22 beam weighs 22 lbs. per LF, and the plan shows 7 beams 20′-0″ long. Therefore, the following calculation can be made.

7 beams × 20′ each × 22 lbs./LF = 3080 lbs. = 1.54 tns

Many structural steel drawings provide the column lengths. The highest elevation on a column or beam is called the *top of steel* and is shown on the drawings as T.O.S. The proposed elevation for the top of the leveling plate is also given; the difference between the two is the length of the column.

Some shapes, such as angles, do not include a weight designation. Determining the weight of angles, plate steel of various thicknesses, and tube steel requires the use of a table, such as the "Manual of Steel Construction" available from the American Institute of Steel Construction.

Other Takeoff Considerations

When taking off steel quantities, the contractor should not bother to subtract for the length of the connection at each end of the beam. The overall length of 20′-0″ (as shown in Figure 8.4) is acceptable for estimating purposes.

The costs for the connecting hardware and for the fabrication of connections can be taken off separately on small projects. On larger projects, an additional 10% of the weight of the steel is added to cover the cost of items such as plates, clips, and hardware.

Metal Joists

Steel joists are manufactured by welding hot-rolled or cold-formed sections to angle web or round bars to form a truss. Standard open-web and long-span steel joists were developed as an alternative to wood frame construction. The steel joist's capacity to carry loads spanning greater distances has made it popular for all types of light-occupancy construction. In addition, steel joists can be used in fire-rated construction. The joists' open webbing allows the passage of mechanical piping and electrical conduits without the drilling or coring of holes.

Types of Steel Joists

Steel joists can be classified according to one of the following three categories or series:

- **K-Series** are open-web, parallel-chord steel joists manufactured in standard depths of 8″, 10″, 12″, 14″, 16″, 18″, 20″, 22″, 24″, 26″, 28″, and 30″, and lengths up to 60′.
- **LH-Series** refers to long-span, open-web steel joists manufactured in depths of 18″, 20″, 24″, 28″, 32″, 36″, 40″, 44″ and 48″, and lengths up to 96′.
- **DLH-Series** refers to deep long-span, open-web steel joists manufactured in depths of 52″, 56″, 60″, 64″, 68″ and 72″ and lengths up to 144′.

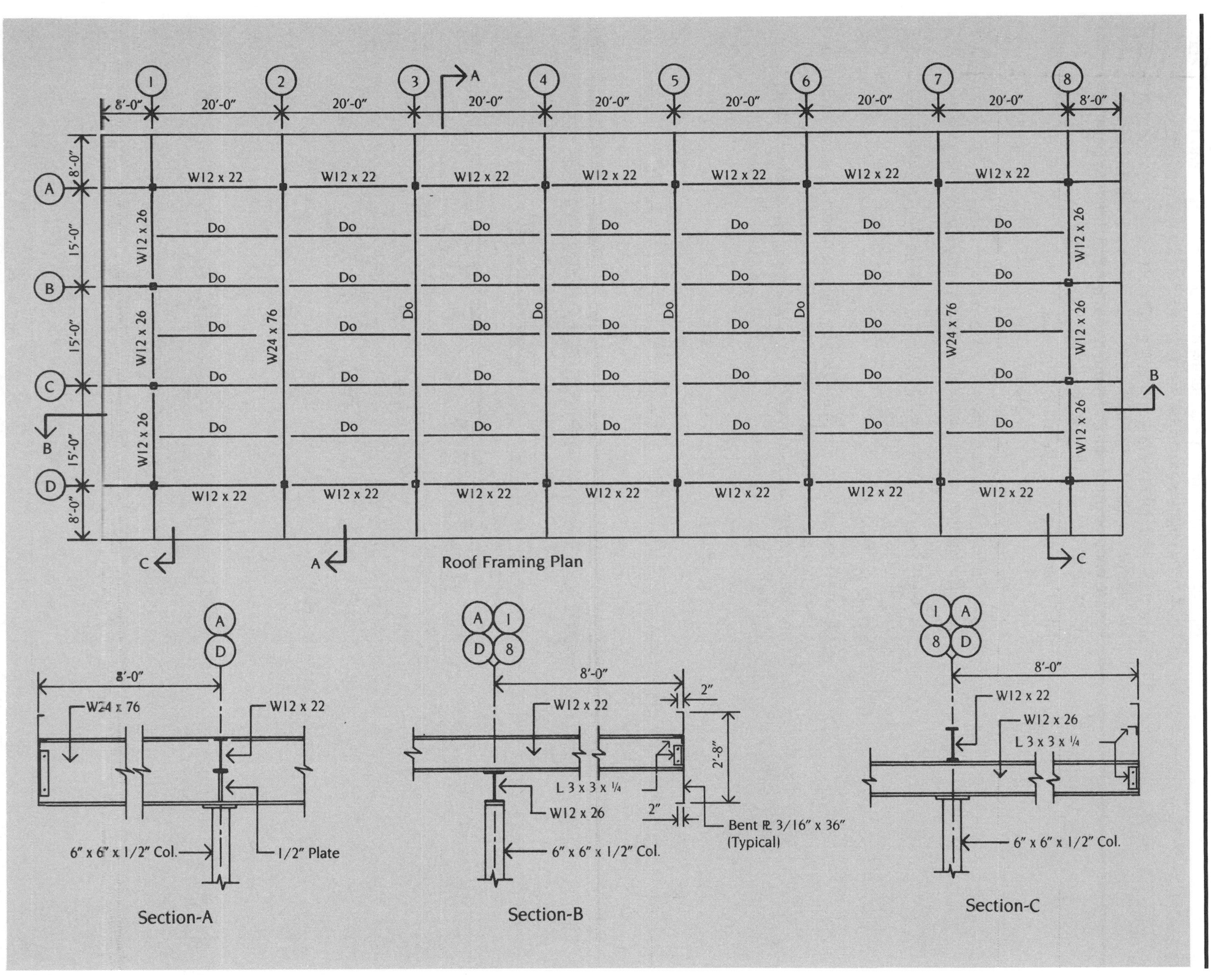

Figure 8.4 Structural Steel Plan

Figure 8.5 shows the typical details for both the K-Series and the LH- and DLH-Series open-web joists.

The LH- and DLH-Series are also available with top chords that are pitched or parallel to the bottom chords. The ends of the joists can be square ends or underslung. Figure 8.6 shows the different types of open-web steel joist designs available.

Joist designations are defined as follows:

24K10

The first number, **24**, *is the depth in inches of the joist.*
The letter **K** *indicates that it is a K-Series joist.*
The last number, **10**, *indicates the load capacity/size of the chords.*

In Figure 8.7, open-web steel joists are shown as they would appear in a typical structural drawing.

Units for Takeoff

Open-web steel joists are quantified by the pound and then extended to tons. They are typically taken off by the LF and quantity of each series and designation, then converted to weight using tables as shown in Figure 8.8.

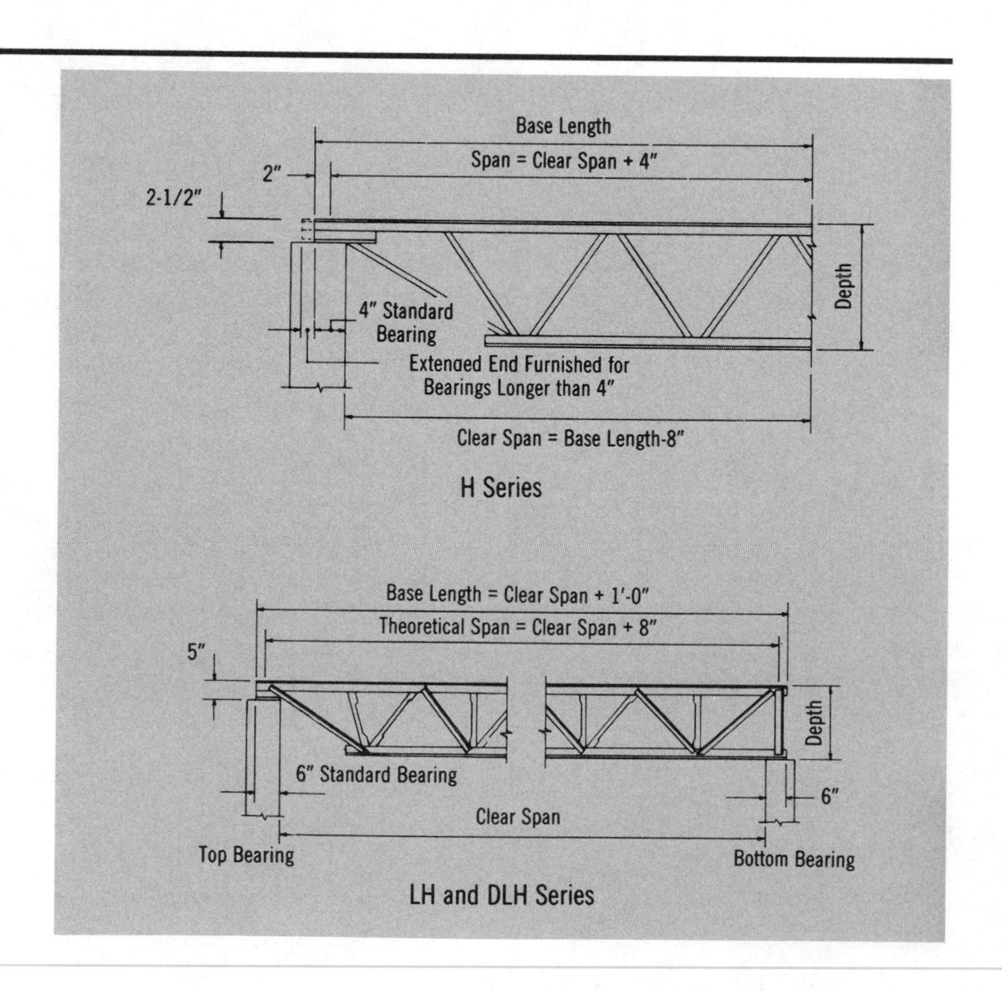

Figure 8.5 Standard Joist Details

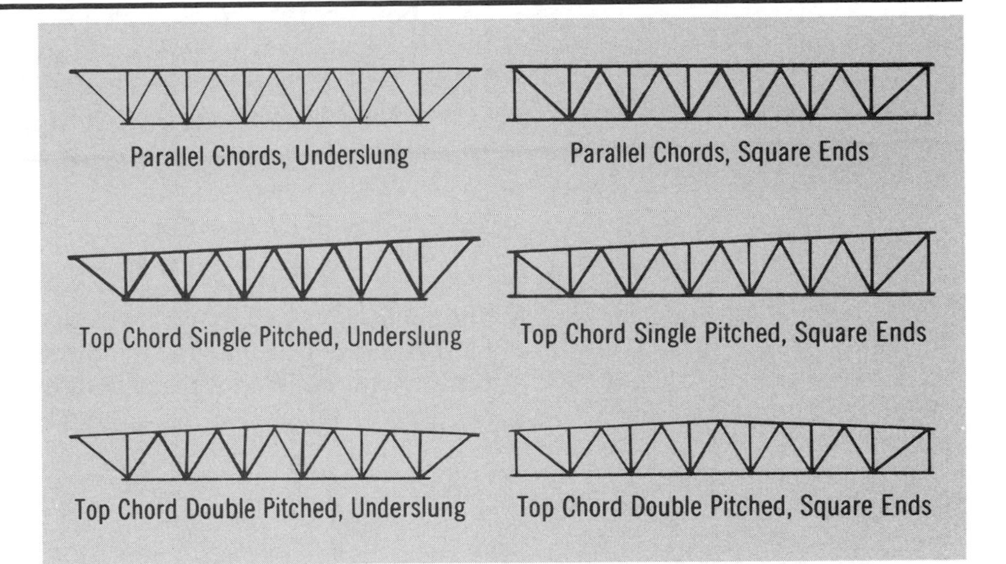

Figure 8.6 Open-Web Steel Joist Types

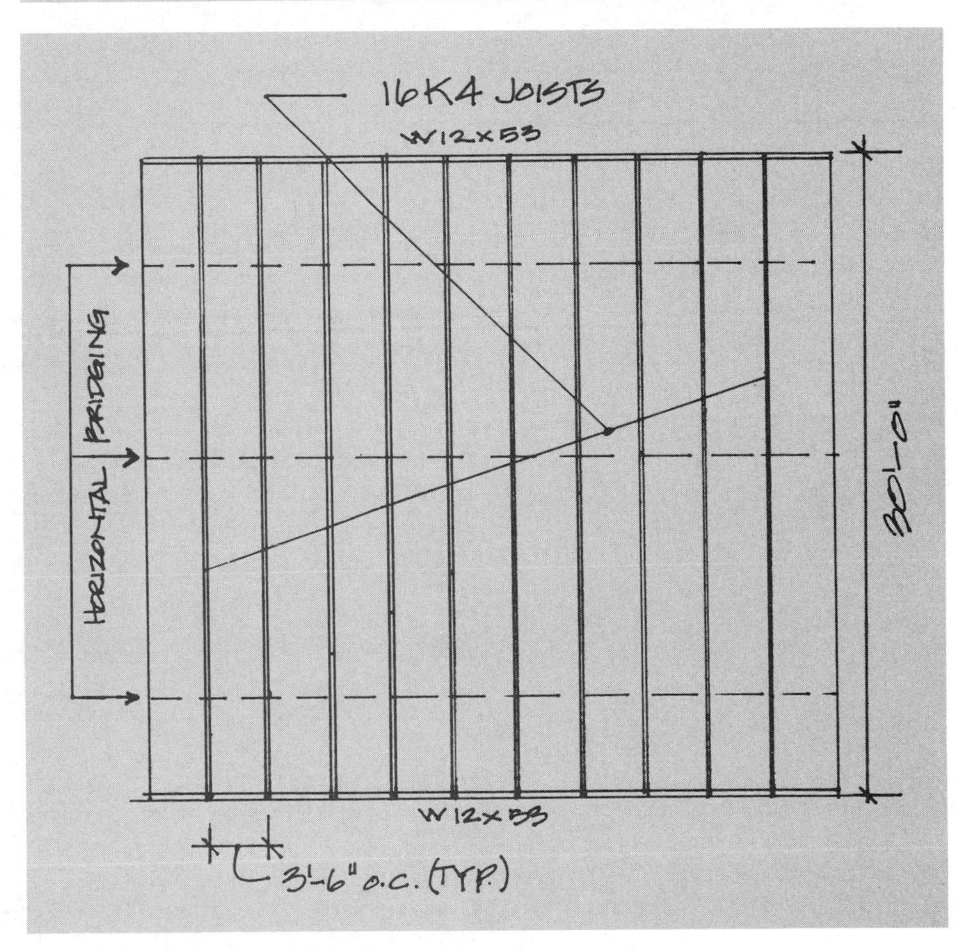

Figure 8.7 Open-Web Joist Plan

By referring to the plan view of a typical open-web, K-Series steel joist (Figure 8.7) and using the table in Figure 8.8, it is possible to calculate the total weight of the K-Series steel joists for pricing. For example:

Calculate the weight of the K-Series steel joists in Figure 8.7.

Joists shown are 16K4. A takeoff shows a quantity of 10 with a span of 30′-0″. Locating a 16K4 joist in the table in Figure 8.8, one can see that it has a weight of 7.0 lbs. per LF.

10 EA × 30′ LF = 300 LF × 7.0 lbs./LF = 2100 lbs. = 1.05 tns

The contractor should separate bar joists by designation number and series for accurate pricing. For pricing purposes, steel joists that are welded in place should be listed separately from those that are connected by nuts and bolts.

Weights of Open Web Steel Joists (K Series)

The approximate weights per linear foot as shown in the following table do not include any accessories or connection plates.

Joist Designation	8K1	10K1	12K1	12K3	12K5	14K1	14K3	14K4	14K6	16K2	16K3
Depth (in.)	8	10	12	12	12	14	14	14	14	16	16
Approx. Wt. (Lbs./Ft.)	5.1	5.0	5.0	5.7	7.1	5.2	6.0	6.7	7.7	5.5	6.3
Joist Designation	16K4	16K5	16K6	16K7	16K9	18K3	18K4	18K5	18K6	18K7	18K9
Depth (in.)	16	16	16	16	16	18	18	18	18	18	18
Approx. Wt. (Lbs./Ft.)	7.0	7.5	8.1	8.6	10.0	6.6	7.2	7.7	8.5	9.0	10.2
Joist Designation	18K10	20K3	20K4	20K5	20K6	20K7	20K9	20K10	22K4	22K5	22K6
Depth (in.)	18	20	20	20	20	20	20	20	22	22	22
Approx. Wt. (Lbs./Ft.)	11.7	6.7	7.6	8.2	8.9	9.3	10.8	12.2	8.0	8.8	9.2
Joist Designation	22K7	22K9	22K10	22K11	24K4	24K5	24K6	24K7	24K8	24K9	24K10
Depth (in.)	22	22	22	22	24	24	24	24	24	24	24
Approx. Wt. (Lbs./Ft.)	9.7	11.3	12.6	13.8	8.4	9.3	9.7	10.1	11.5	12.0	13.1
Joist Designation	24K12	26K5	26K6	26K7	26K8	26K9	26K10	26K12	28K6	28K7	28K8
Depth (in.)	24	26	26	26	26	26	26	26	28	28	28
Approx. Wt. (Lbs./Ft.)	16.0	9.8	10.6	10.9	12.1	12.2	13.8	16.6	11.4	11.8	12.7
Joist Designation	28K9	28K10	28K12	30K7	30K8	30K9	30K10	30K11	30K12		
Depth (in.)	28	28	28	30	30	30	30	30	30		
Approx. Wt. (Lbs./Ft.)	13.0	14.3	17.1	12.3	13.2	13.4	15.0	16.4	17.6		

Figure 8.8

Bridging

Bridging is the lateral bracing of open-web steel joists to provide stiffness and stability and to prevent wracking of the joists. Bridging is most often accomplished with small lightweight steel angles bolted or welded perpendicular to the span of the joist. The locations of the courses of bridging are shown on the structural plans and are determined in accordance with the joist manufacturer's recommendation.

There are two main types of bridging:

- Horizontal bridging, which consists of two continuous steel members, one fastened to the top chord and one fastened to the bottom chord.
- Diagonal bridging, which consists of bracing that runs diagonally from the top of one chord to the bottom chord of the adjacent joist.

It is important that the terminations of bridging are securely fastened to the wall or beam, regardless of the bridging installation method.

Units for Takeoff

Both horizontal and diagonal bracing are taken off and listed by the LF. Bridging mechanically fastened with nuts and bolts should be listed separately from welded bridging.

Metal Decking

Metal decking is accomplished with specially formed sheets of steel applied perpendicular to the span of the joists. They serve as a substrate for the installation of roofing materials, such as rigid insulation and membrane, or as permanent forms for concrete floor slabs.

Types of Metal Decking

Metal decking can be classified in two main categories:

- Corrugated, undulated, or "corruform" concrete-fill permanent forms; also used for roofing and siding of industrial-type buildings.
- Cellular-type with well-defined bent contours in trapezoidal or rectangular pitches or depths; used as permanent concrete forms for larger-span spacings or heavier live load applications.

Metal decking is available in 18-, 20-, or 22-gauge thicknesses, in widths of 30", and in standard lengths ranging from 14′ to 31′. The depth of the decking section can vary with type and application, but is usually 1-1/2" to 2-1/2" deep. Special coatings and colors are available, but the most common finish is galvanized.

Units for Takeoff

Metal decking is taken off by the SF of area to be covered and extended to the square (SQ), where one SQ is equal to 100 SF. Allowances should be made for overlap at the sides and ends of the individual sheets. Overlap requirements are typically specified by the manufacturer or in the specification section for metal decking.

Metal decking of different gauges, widths, lengths, or finishes should be listed separately in the takeoff for pricing purposes.

Other Takeoff Considerations

In addition to the previously mentioned takeoff units, other considerations, such as the specified method of installation, should be noted.

For example, thinner-gauge decking that is welded may require the use of thickening washers placed at the point of fusion, to avoid burning through the decking. This practice is common for metal of 20 gauge or higher (thinner metal).

The contractor should also analyze the specifications for field touch-up requirements for the galvanized finish at the welds. This can be a time-consuming process and should be listed separately in the takeoff. It can be quantified by the square foot area, or by the man-hours to perform the work.

Metal decking that serves as a permanent concrete form will require the use of sheet metal angles fastened to the perimeter of the decking and openings to act as an edge form (see "Edge Forms," Chapter 6) for the placement of the concrete. These angles are taken off by the LF and should be listed by size and method of installation for proper pricing.

Light-Gauge Metal Framing

Light-gauge metal framing refers to the method of construction that uses a high-tensile-strength, cold-rolled steel formed in the shape of joists, studs, track, and channel. All components are fabricated of structural grade steel in 12-, 14-, 16-, and 18-gauge thicknesses. The various sections are designed to provide the load-bearing characteristics of steel or wood framing at a reduced weight and cost.

Light-gauge framing is also noncombustible and does not warp, shrink, or swell—in contrast to wood. Sections are available with a galvanized coating or red zinc chromate paint that resists rusting. Slots or channels are factory-punched within the web of the section to allow the passage of wiring or piping.

On-site cutting of the sections is done with a "chop" saw outfitted with a high-speed metal-cutting blade. Layout and erection are similar to the procedures used for wood framing components.

Fastening of components can be done by bolting, screwing with self-tapping screws, or welding.

Joists

Joist sections are available in 6", 8", 10" and 12" depths, and flanges in 1-5/8" or 2-1/2" widths. The most common lengths are from 8' to 30' in 2' increments.

Joists are used in floor and roof construction with the typical 12", 16", and 24" on-center spacings used in wood frame construction.

Units for Takeoff

The unit of takeoff for light-gauge steel joists is the LF. Special channels for use as box joists are also taken off by the LF. Sections of different shapes, sizes, gauges, and coatings (as well as the method of fastening) should be listed separately for accurate pricing.

Other Takeoff Considerations

The contractor should consult the specifications for the use of special sections, such as "V" or "C" channels used as bridging at mid or third points of the spans. They are slid into the factory-punched slots perpendicular to the span of the joists. Special sections are taken off by the LF and listed by the same criteria as the joists.

Studs and Track

Steel studs and track are available in 1-5/8", 2-1/2", 3-5/8", 4", and 6" depths. Standard gauges are 14, 16, and 18. They are also available in 20- and 25-gauge thicknesses, but these are not considered to be of load-bearing capacity, and will be discussed in Chapter 12, Finishes (Division 9).

Studs are C-shaped with folded flanges and are used as the vertical component of the wall system. The specifications determine the on-center spacing requirement. Steel track is the horizontal component to which the steel stud is fastened. It is typically located at the top and bottom perimeters of the wall. Figure 8.9 illustrates a load-bearing wall.

Units for Takeoff

There are three basic methods for taking off metal studs and track:

- Taking off the stud and track separately by the LF. Studs and track totals are then priced separately.
- Taking off the wall by LF, where one LF of wall is equal to one LF of length by the height. For example, 76 LF of wall that is 8′-0″ high is quantified and listed as 76 LF.
- Calculate the SF area of the wall (length times height). For example, the wall cited in Method #2, 76 LF × 8′-0″ high, would be listed in the takeoff as 608 SF.

Regardless of the method selected, the contractor must separate walls of different stud width, gauge, height, method of fastening, on-center spacing, and application.

Other Takeoff Considerations

Added materials for openings in the wall must also be calculated. Similar to wood construction, openings in light-gauge load-bearing metal stud walls

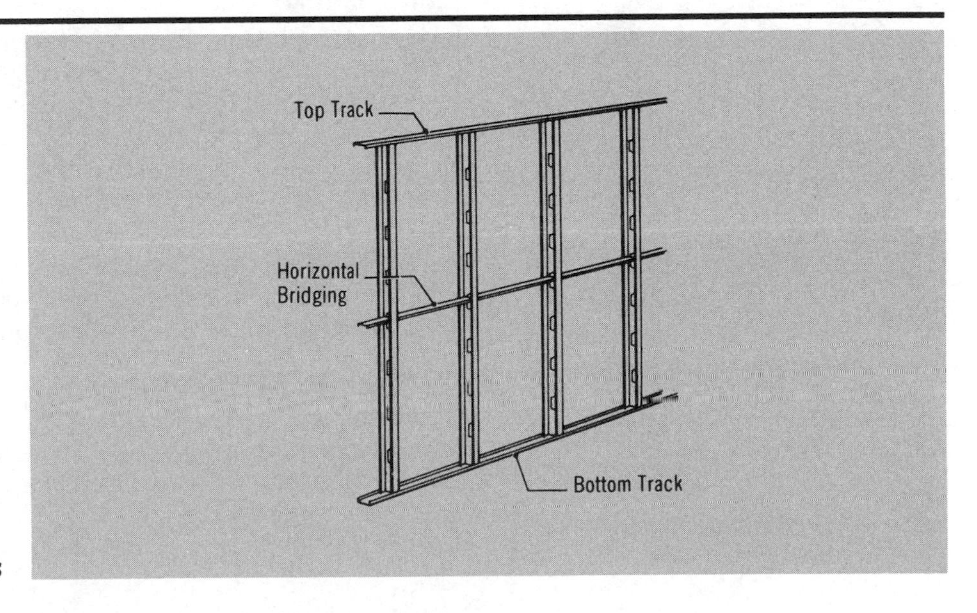

Figure 8.9 Load-Bearing Steel Studs

must be framed to support the load transmitted from above. This involves the use of headers, cripple studs, sills, and doubled jamb studs. Openings and their components may be taken off by the opening (EA) based on size, or by the individual component (e.g., doubled studs per LF, headers per LF, sills per LF). The choice of takeoff method should be based on the type of openings. For example, openings for numerous doors and windows of the same size may best be quantified per opening (EA).

Specialized equipment or staging for the installation of the work may also be needed. The equipment or staging should be listed separately, based on length of time required. Any erection or dismantle charges for moving and setups should also be included as part of this item.

Miscellaneous and Ornamental Metals

Miscellaneous and ornamental metals are typically separated into two categories for takeoff purposes.

Miscellaneous refers to the various items that do not fit into one specific classification.

Ornamental metals refer to railings and special metal fabrications for stairs.

It is not uncommon for projects with limited structural steel to still have a considerable number of miscellaneous and ornamental metal items. This category encompasses a diverse range of items that are furnished and delivered only, or furnished, delivered, and installed. The following is a list of the more common items and their respective units of takeoff.

Lintels

Lintels are structural members, typically installed in masonry walls over windows or doors to support the load of the masonry above the opening. The most common type of lintel is the *steel angle*. These can be cut from stock lengths of various-sized steel angle stock. They are longer than the width of the opening in order to span the opening and provide bearing on either side of the opening. The amount of bearing is dictated by the specifications and/or building code, contingent upon the size of the opening and the load above the opening. Angles can be supplied as either individual units or composite units of multiple angles welded together for walls of multiple masonry wythes.

Units for Takeoff

Like most steel items, lintels are taken off by the individual piece and converted to weight. This is done by adding the bearing requirement for each side to the width of the opening in LF, multiplying the LF by the number of pieces, and then multiplying by the weight per LF to obtain a weight in pounds. For buildings with a substantial masonry scope of work, the engineer will list the lintels in a lintel schedule. The lintel schedule refers to the lintel by name, such as L1, L2, L3, and the species of angle. It also specifies the size in length, and whether the unit is an individual or a composite of angles welded together. Figure 8.10 illustrates a typical lintel schedule as it would appear on a drawing.

Other Takeoff Considerations

Lintels should be listed separately from structural steel items on a takeoff because the unit price per pound for smaller sections is considerably more than for structural sections. Lintels are a furnished-and-delivered item only. Installation is typically included as part of the masonry scope of work because the lintels must be installed as the masonry courses progress.

Pipe Railings

Pipe railings are commonly found at exterior stairs or ramps for pedestrian traffic. They are constructed of welded-steel pipe in configurations to match the profile of the stair or ramp. Diameters of the pipe can range from 1-1/4" to 2". The welds are ground and the railing is often primed or finish-painted ready for installation. They are installed by the use of sleeves, which are slightly bigger than the diameter of the pipe, embedded in the wet concrete. They can also be installed by drilling a hole in the concrete and grouting the rail in place with the use of hydraulic cement. For applications in materials other than concrete, the railings may incorporate the use of a flange welded to the point of attachment that has holes punched for bolted fastening.

Units for Takeoff

The most common unit of takeoff for railings is the LF of completed rail. Railings of various diameters and different methods of installation and finishes should be listed separately.

Ladders

Ladders are typically constructed of flat steel bar stock, with rungs of round steel bars. The ladder is fabricated by welding the rungs perpendicular to the flat stock at the specified spacing. Angles or brackets are welded to the flat stock for fastening to the masonry or concrete wall. Ladders of this nature are meant to be permanently installed for access to high roofs from low roofs. They can be supplied prefinished or primed.

LINTEL SCHEDULE				
QTY.	MARK	SIZE	LENGTH	REMARKS
25	L1	∠4"x3½"x5/16"	5'-0"	GALV.
8	L2	∠4"x3½"x5/16"	4'-4"	"
2	L3	2∠6"x3"x3/8"	8'-4"	"
3	L4	∠4"x3½"x7/16"	3'-4"	"

Figure 8.10 Lintel Schedule

Units for Takeoff

Ladders can be taken off by the vertical linear foot (VLF) or by the individual piece.

Grates

Grates are fabricated items composed of parallel flat bar stock with intermediate round bars welded between. They are used to allow the flow of air or water while supporting vehicular or pedestrian traffic. Grates typically include angles for support that are bolted to the adjacent masonry or concrete surfaces.

Units for Takeoff

The standard unit of takeoff for grating is the SF. The LF measurement of support angles should be included as part of the takeoff.

Joist-Bearing Plates

Joist-bearing plates are small pieces of steel, typically 1/2″ in thickness, with approximate dimensions of 4″ x 6″. Welded to the bottom of the flat plate are hooked anchor bolts. The anchor bolts of the joist-bearing plate are embedded in the bond beam course of masonry wall. Once the grout has cured, the joist-bearing plate is held secure for the attachment of the bar joist. The bar joists are welded to the exposed portion of the joist-bearing plate.

Units for Takeoff

Joist-bearing plates are taken off by the piece. This is accomplished by counting the number of joist ends that bear on masonry walls.

Other Takeoff Considerations

In most circumstances, the joist-bearing plates are a furnished-and-delivered item only. Their installation is typically part of the masonry scope of work.

Lally Columns

Lally columns for residential construction are typically 3-1/2″ diameter concrete-filled steel columns. Separate cap and base plates are provided. The column is used to transmit live and dead loads from the structural members above to the spread footing (see Chapter 6 for a discussion of concrete footings). Lally columns are available in a variety of stock lengths from 7′ to 12′ in even foot increments. Columns can be cut on site using a heavy-duty pipe cutter.

Units for Takeoff

Lally columns are taken off by the individual piece based on the even foot length. To obtain the length of the column, the contractor must calculate the distance from the bottom of the load-bearing beam to the top of the column, and round up to the nearest foot. The cost of cutting should be included as part of the installation.

Stairs

Steel stairs can be of the metal pan type, which require the pouring of concrete into the treads and landings as the wearing surface. They can also be constructed of diamond plate steel or metal grating. Typically the stair fabricator builds as much as possible off-site to reduce the amount of installation time required.

Units for Takeoff

Steel stair takeoff and its supporting structural members should be kept separate from the other miscellaneous metals categories. The quantity and size of the tread should be noted and quantified by the stair tread (EA). Sizes of landings should be noted separately in SF area. Other parts, such as columns, stair stringers, or railings, should be taken off and listed by the LF.

Miscellaneous

The structural steel and miscellaneous metals takeoff often requires items that are not easily classified. Among them are shop drawings, shop painting, and erection.

Shop Drawings

Prior to the actual fabrication of any of the structural steel components, the steel fabricator is required to produce a set of working drawings or *shop drawings* that show the actual connection details, heights, and lengths of the various structural members such as beams, columns, and joists. In addition, the shop drawings provide lintel schedules and show base and cap plate details. Shop drawings are reviewed by the architect or structural engineer for conformance and are approved for fabrication. The fabricator then uses them as a guideline for fabrication and erection.

Units for Takeoff

There are no standard units of takeoff for shop drawings. The cost of producing shop drawings based on the tons of steel and the sophistication of the fabrication and erection.

Shop Painting of the Structural Steel

After the various components have been fabricated, they are generally painted, or primed, prior to leaving the shop, unless the steel will be spray-fireproofed or encased in concrete. The type of primer paint should be listed in the specifications. The contractor should also be aware of any touch-up that may be required in the field where primer has been damaged by welding or handling.

Units for Takeoff

The cost of shop priming is based on the amount of steel to be primed in tons. Field touch-up can be listed as a separate item, and is quantified by labor-hours.

Erection

Structural steel, joists, decking, and so forth are always erected by a crew rather than a single individual. The size of the crew is determined by the complexity and scope of the project. The erection of the structural steel must include the handling and distribution on-site, crane setup and moving, crane rental, and the materials and labor for the on-site welding.

Units for Takeoff

Structural steel erection is taken off and quantified by the ton. The quantity of steel in tons can be obtained from the materials takeoff previously done. Structural steel members should be separated from joist and decking for accurate pricing.

Note: Because steel is an international product, projects that are taxpayer-funded may be bound by the *Buy American Act.* This means that the steel must be of domestic origin. The contractor should account for the steel accordingly.

Summary

The takeoff sheets in Figure 8.11 are the quantities for the work of Division 5—Metals for the sample project. The residential nature of the sample project includes very little structural steel, but other items that are included in Division 5 have been noted. Note the reference to the steel lintels for the fireplace. They have been included as part of the masonry work of Division 4 but have been addressed in Division 5 without a quantity. It reminds the contractor that the lintels have been included in Division 4.

Division 5 Checklist

Structural Steel

☐ Columns
☐ Beams
☐ Channels
☐ Angles
☐ Girders
☐ Cross and wind bracing
☐ Anchor bolts and leveling plates (furnished only)

Metal Joists

☐ K-Series
☐ LH-Series
☐ DLH-Series
☐ Bridging

Metal Decking

☐ Gauge
☐ Floor
☐ Roof
☐ Galvanizing touch-up
☐ Welding
☐ Concrete stops

Light-gauge Metal Framing

☐ Gauge
☐ Stud and track
☐ Joists
☐ Bridging or stiffeners

- ☐ Method of fastening

Miscellaneous Metals

- ☐ Lintels (furnished only)
- ☐ Railings and sleeves
- ☐ Ladders
- ☐ Grates
- ☐ Joist-bearing plates
- ☐ Lally columns
- ☐ Stairs

Miscellaneous

- ☐ Shop drawings
- ☐ Shop and field priming
- ☐ Erection

Means Forms

QUANTITY SHEET

PROJECT: SAMPLE PROJECT | SHEET NO. 5-1/1
ESTIMATE NO. 1
LOCATION | ARCHITECT: HOME PLANNERS | DATE
TAKE OFF BY: WJD | EXTENSIONS BY: WJD | CHECKED BY: KF

DESCRIPTION	NO.	DIMENSIONS				UNIT		UNIT		UNIT		UNIT
DIVISION 5: METALS												
S8 X 18.4 BEAMS @ BASEMENT GIRDER (8" BEARING EA. SIDE.)	1	11.67'										
	1	9.33'				21 LF × 18.4#/LF					387	LBS.
8"x8"x 1/4" BEARING PLATES @ EA. BRG. POINT OF S8 X 18.4	4	.67	.67	1/4"							4	EA
3"ϕ STANDARD CONC. FILLED STEEL COL. w/ BASE & CAP PLATES	1	7.33'									1	EA
METAL AREA WELLS @ BSM'T WINDOWS	3	18" R	3.0								3	EA
FIELD WELD @ 3"ϕ COL. TO S8 X 18.4	1	EA									1	EA
STEEL LINTELS FOR FIREPLACE		INCLUDED IN DIV. 4 — MASONRY										

Figure 8.11

Chapter Nine

WOOD AND PLASTICS

(Division 6)

Chapter Nine

WOOD AND PLASTICS

(Division 6)

Division 6, Wood and Plastics, includes the various facets of wood construction. *Rough carpentry* involves building wood structures, mainly residential or light commercial. It includes the use of framing-grade lumber and sheathings. Wood trusses may also be used. Pre-engineered, prefabricated structural components, used in place of conventional "stick" framing, may also be used, mainly for framing roof and floor systems. *Finish carpentry* involves the materials and techniques for finish-grade wood trims at windows, doors, baseboards, and moldings on the building interior.

In residential construction, most wood framing and finish carpentry work is shown on the architectural drawings, in plan, elevation, section, and detail. *Framing plans* are part of the architectural drawings. The framing plans, a modified version of structural drawings showing the "skeleton" of the structure, illustrate the structural wood members used to construct the frame of the building. The building cross-section and wall section provide essential information for taking off the wall framing and sheathing, roof framing and sheathing, and the ceiling system.

The procedure for takeoff described in this chapter is based on *platform framing*. In platform framing, the load-bearing walls start at the top of the subfloor sheathing and continue to the underside of the platform above. An alternate method, called *balloon framing*, starts the exterior and load-bearing walls at the top of the pressure-treated sills, and is continuous to the underside of the ceiling joists. Platform framing is the more common method among residential/light commercial contractors.

Board Measure

Lumber has its own system of measurement called *board foot measure (BFM)* One board foot is equal to the volume of a piece of wood 1" thick by 1' square. When calculating in BFM, nominal dimensions are used. (Nominal dimensions are explained in the next section.)

The following formula can be used to calculate the board footage of any piece of lumber.

$$BFM = \frac{t \times w \times l}{12} (n)$$

where BFM = board foot measure in feet
t = nominal thickness in inches
w = nominal width in inches

l = length in feet of the individual piece
n = number of pieces
For example: How many board feet are in a 2" x 6" board that is 16' long?

$$BFM = \frac{2 \times 6 \times 16}{12} \times 1 = 16 \text{ board feet (BFM)}$$

Rough Carpentry

Most rough carpentry work can be classified as *frame construction*. Frame construction includes the use of sheathings and *structural grade* lumber, such as 2" x 4", 2" x 6", and 2" x 8". Framing lumber has nominal dimensions from 2" to 6" in thickness and 4" to 14" in width. The length of the lumber differs by the species and availability, but most commonly ranges from 8'-0" to 20'-0" in 2' increments. Larger-dimensioned lumber, such as 2" x 10", 2" x 12", 4" x 6", and 6" x 8", is available in lengths up to 24'-0".

Lumber prepared for frame construction is called *dressed lumber*. The term *dressed* refers to the surfacing of the rough lumber, as accomplished with a thickness plane. This process provides a uniform size and shape with a smooth surface. Lumber that has been surfaced on all four sides is called S4S. Once the lumber has been surfaced on all four sides, it is smaller than its original dimensions. For example, a 2" x 4" piece of rough lumber measures 1-1/2" x 3-1/2" after being dressed. The 2" x 4" dimension is called the *nominal dimension*, and 1-1/2" x 3-1/2" is the *actual size*.

Any lumber takeoff should also begin with a review of the specification for the grade (quality) and species of lumber to be used. The contractor should be careful to separate all materials by species and grade and any other special requirements, such as pressure-treated wood, (defined below) as this will have a definite effect on price.

The framing of a structure can be broken down into three main systems, or assemblies. An *assembly* is a series of components that go together to form a larger system or portion of work. The three main assemblies are the floor framing assembly, the wall/partition framing assembly, and the roof/ceiling framing assembly.

Floor Framing Assembly

Most wood framing quantity takeoffs start with the floor framing, including the girder, sill plates, sill sealer, joists, band or box joist, and the sub-floor sheathing. Figures 9.1a and 9.1b illustrate the floor framing assembly with the floor framing plan and a cross-section through the floor frame.

When the width of the structure is greater than the floor joist can span, a built-up or single-piece member called a *girder* is used. The girder is located at a specific point, frequently the midpoint in the width, to reduce the overall span of the joists. The girder is supported at specified intervals by lally columns (see "Lally Columns" in Chapter 8) that transmit the load to the footings (see "Footings" in Chapter 6).

Sills, or *sill plates*, are typically 2" x 6" or 2" x 8" and on some plans are doubled up. Sills are anchored to the top of the foundation wall by anchor bolts to provide a nailable bearing surface for the joist and box joist. Because they are bolted to the concrete or masonry foundation, they must be *pressure-treated* lumber. Pressure-treating refers to the process of treating wood with preservatives under high pressure to protect against decay.

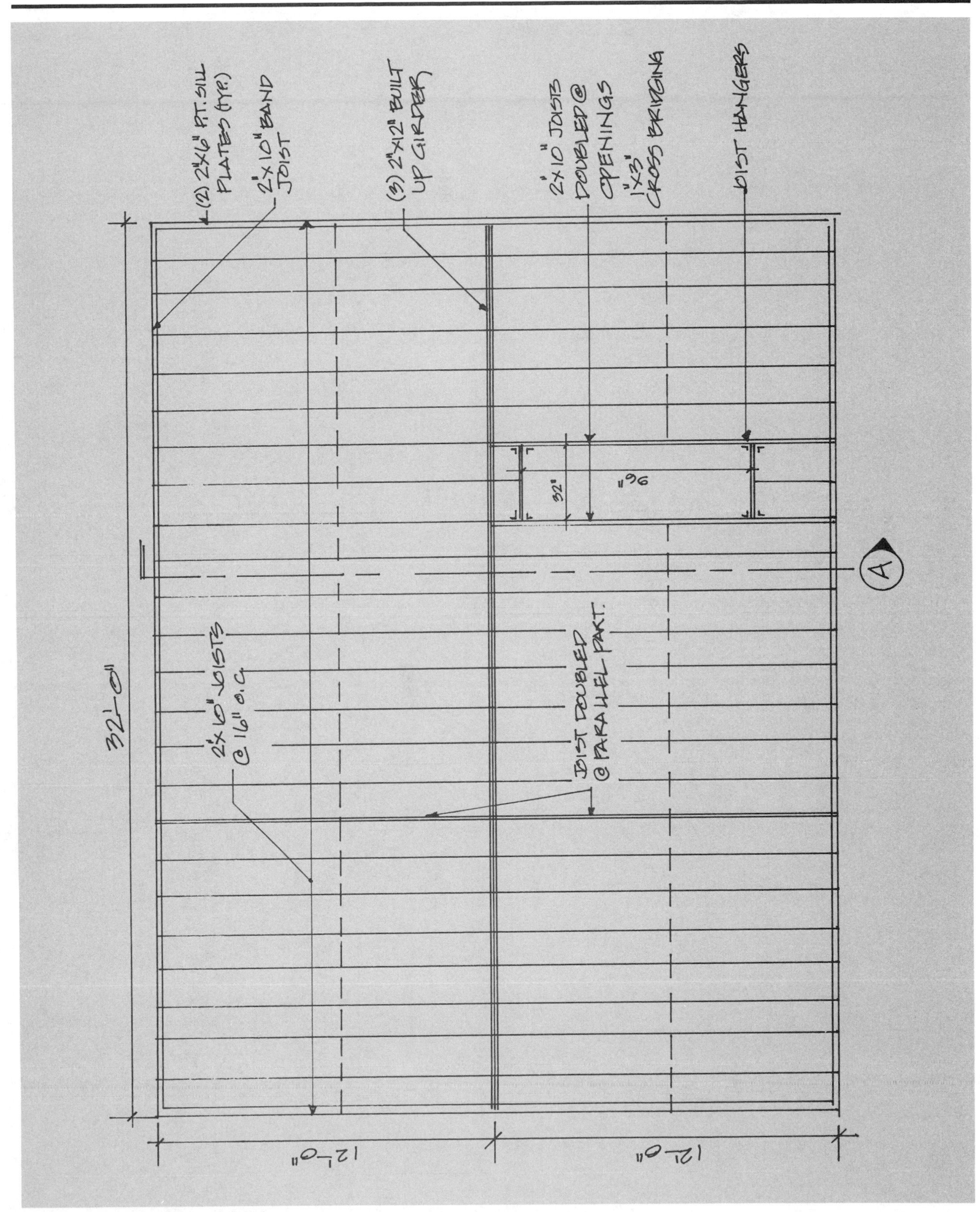

Figure 9.1a Floor Framing Plan

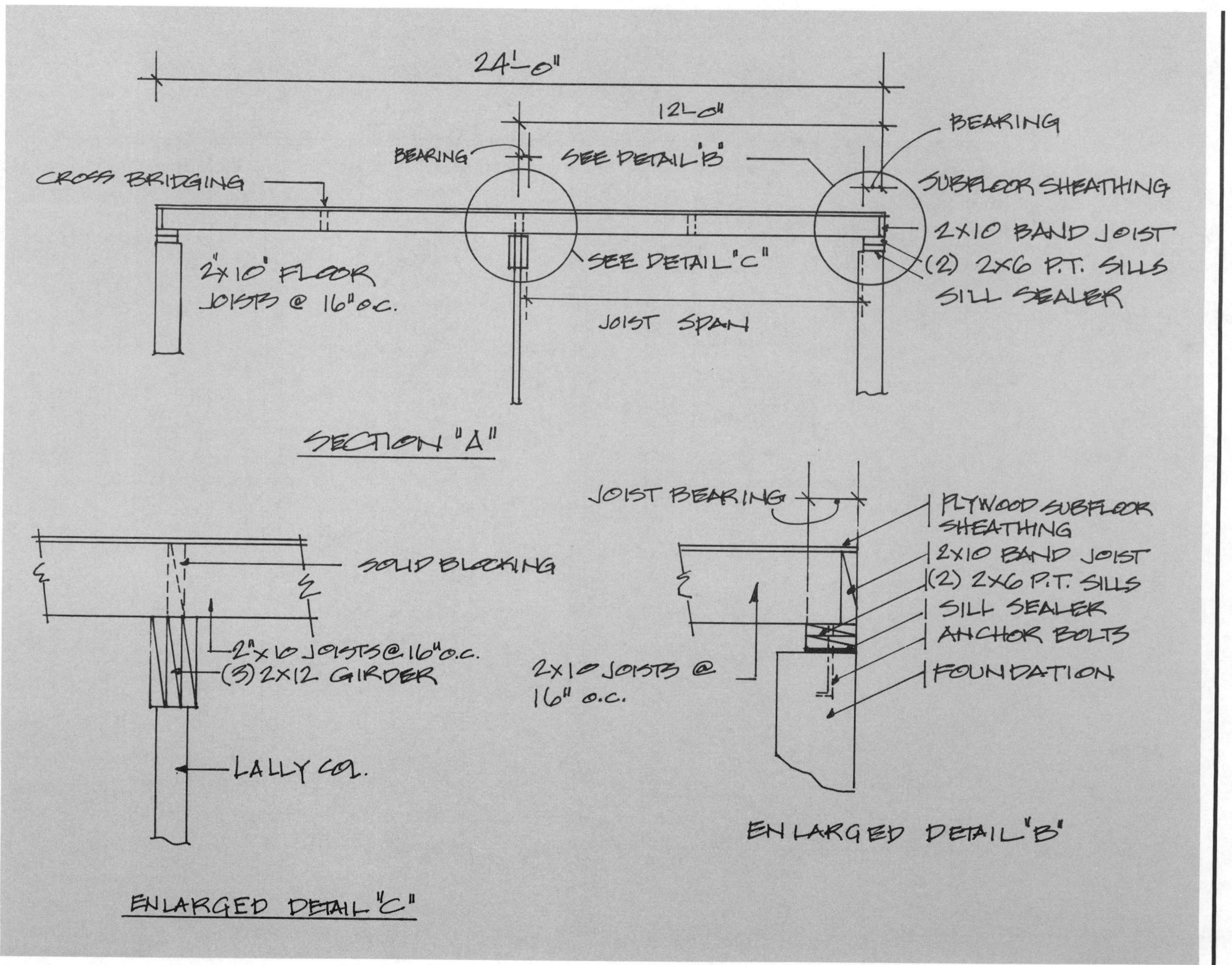

Figure 9.1b Floor Framing Cross-Section

The thin, compressible material installed between the sill and the top of the foundation wall is called the *sill sealer*. It acts as a barrier against insects and air infiltration.

The structural members that provide the support for the floor are called the *joists*. Joists are spaced at regular intervals, most commonly 12", 16", and 24" on center, and span the space between the girder and the sill on the foundation wall. A *box* or *band joist* (also known as a *rim joist*) runs perpendicular to the joists at their ends. It completes the platform or box that the floor framing resembles. To maintain the spacing at the midpoints and to prevent the joists from rolling or buckling, *cross bridging* is installed. Cross bridging is located at approximately 8'-0" on center for joists that span more than 16'-0". Cross bridging pieces are typically 1" x 3" or 1" x 4", cut to fit diagonally between the top of one joist and the bottom of the next. In addition to cross bridging, *solid blocking* can be installed at right angles to the joists, between them. Solid blocking is used as a fire stop between joists at the center of a bearing partition or girder. The flooring laid perpendicular to the joists is called the *subfloor*, and is typically plywood or similar structurally rated sheathing. The specifications may also require that the sheathing be glued and nailed to the joists.

Units for Takeoff

Girders are taken off by the linear foot. Built-up girders consisting of multiple members of the same size should be counted, then multiplied by the length to arrive at the LF. The length should include the clear span plus the required bearing at each end of the girder.

Sill sealer is taken off by the LF of foundation wall that requires sills. Sill pieces are taken off by the LF, and multiplied by two for sills that are doubled.

To determine the quantity of floor joists, the length of the floor frame (in feet) is divided by the spacing (in feet), plus one joist for the end. The takeoff should list the quantity of each length of joist separately, and may be extended to total LF for that length. The length refers to the span of the joist *plus* the required bearing at each end. For example, if we refer to Figure 9.1:

> *The length of the floor frame is 32'-0" divided by 16" O.C. (1.33') = 24 joists + 1 joist at the end = 25 joists.*
>
> *The overall length of the span from the outside foundation wall to the center of the girder is 12'-0". The result is 25 12'-0" joists on each side of the girder, or a total of 50 12'-0" joists. Extended to linear footage: 50 pcs × 12' = 600 LF.*
>
> *To complete the frame, the band joist must be calculated. Band joists run perpendicular to the joists and can be taken off and listed by the LF. Referring to Figure 9.1, the joists are perpendicular to the 32'-0" length of the frame.*
>
> *Therefore, 32'-0" × 2 sides = 64 LF band joist.*

Cross bridging is shown at midspan between the foundation walls and the girder. It is taken off and listed by the LF. To determine the actual LF needed, the length of a typical piece is calculated by using the Pythagorean Theorem (see Chapter 2). Referring to Figure 9.2, the following calculation can be made:

> *The length of a piece of cross bridging "C" is C = 17.2", or 17-1/4" since there are 2 pieces per bay; 2 × 17-1/4" = 34-1/2". There are 32 feet of bridging on each side of the girder, totalling 64 LF.*

64 LF × approximately 3′ per bay, which equals 64 LF divided by 1.33′ spacing, equals 48 bays × 3′, which equals 144 LF of bridging material.

Subfloor sheathing is taken off by the square foot. Since most subflooring used today is a form of sheathing, the contractor can determine the quantity of sheets required by dividing the SF area by the area of an individual sheet. Sheathing is typically manufactured in 4′ x 8′ sheets, with an area of 32 SF. The subfloor sheathing is laid with the 8′ length of the sheet perpendicular to the joists. For the example shown in Figure 9.1, the subfloor required would be determined by a simple calculation of the total area divided by 32 SF per sheet.

The floor area = 32′ × 24′ = 768 SF, divided by 32 SF per sheet = 24 sheets.

Calculating the amount of adhesive required is more a matter of experience than rule. However, a general rule of thumb is that a one-quart tube of adhesive will cover approximately five sheets of sheathing. This rule is subject to change with the weather, handling, and care taken in the application. Colder weather or poor handling will reduce the number of sheets covered by as much as one-third.

Other Takeoff Considerations

To provide an accurate takeoff of the lumber required, the contractor must list separately all framing stock by size, length, grade, and species of wood. Failure to do so could result in inaccurate pricing during estimating.

Depending on the lumberyard, the contractor may want to convert all LF of framing lumber to BFM. With the recent trend of homeowners acting as general contractors, most lumberyards have converted pricing to LF or even to the specific piece of lumber. The contractor should take this into account when soliciting bids for framing lumber.

In determining the lengths required for individual joists, the takeoff should reflect the actual size needed. Since individual lumber is sold in two-foot

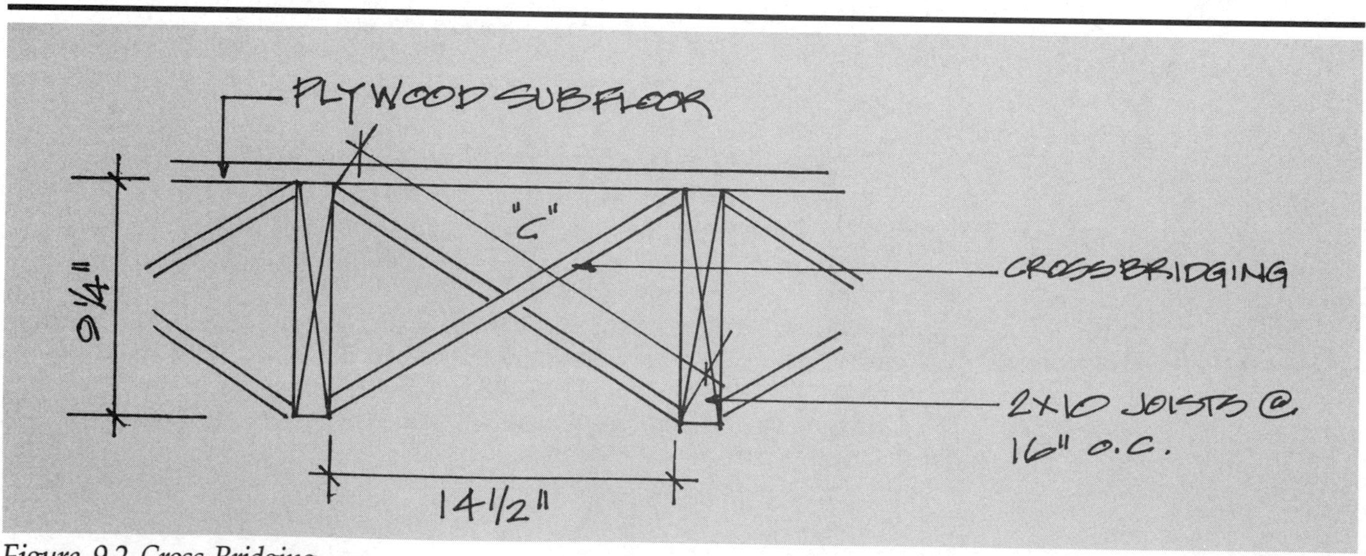

Figure 9.2 Cross Bridging

increments (refer to "Rough Carpentry" earlier in this chapter), the contractor must round up to the nearest incremental length. For example, if the actual size of the lumber needed is 14′-5″, the takeoff should reflect a 16′-0″ length.

When calculating the quantity of joists, it should be noted that most codes require that joists directly adjacent to openings in the floor be doubled up. Similarly, joists must be doubled under partitions that run parallel to the joists. These are often in addition to the regular spacing of the joists.

The contractor must study the drawings for the ends of joists that "hang" or do not bear on anything. Most building codes require the use of joist hangers to support these members.

Since the plywood for all floor frames does not work out to the exact sheet (as in our example), the contractor may need to make adjustments for odd-dimensioned frames. This may include the counting of actual sheets placed over the frame.

Fasteners, including nails and screws, are most often sold by the pound. The contractor should check the specifications and/or manufacturer's recommendations for nailing and fastening schedules that indicate the spacing on-center for fasteners. Many contractors quantify nails as a lump sum (LS) item.

Note: The contractor should use the dimensions shown on the drawing for calculating quantities. Scaling should be a last resort when no other options are available.

Wall/Partition Framing Assembly

The wall and partition framing consists of the exterior walls and sheathing, the interior load-bearing walls, and the interior nonload-bearing partitions. *Load-bearing* walls carry live and dead loads from a part of the structure above, such as a floor, roof, or ceiling. A major component of the load-bearing wall is the *header*. Headers span the openings in load-bearing walls above windows and doors. They are structural members that transmit the load from above the opening to the framing on either side of the opening. Typical header construction consists of 2″ x 6″, 8″, 10″, or 12″ (nominal) framing lumber nailed together with 1/2″ plywood spacers to equal the thickness of the wall. Figure 9.3 illustrates a typical wood header.

The vertical members that support the header are called *trimmers* or *jack studs*. The jack studs are nailed to full studs at each side of the header, sometimes referred to as *king studs*.

All partitions and walls have horizontal members that hold the *studs*, or vertical members, at the desired spacing. These horizontal members are called *plates*. The plate at the top of the wall is referred to as the *top plate*, and the one at the bottom is called the *sill plate* or *sole plate*. Most load-bearing wall construction requires that the top plate be doubled. The horizontal member that runs parallel to the header at the window sill height is called the *sill*. The short studs that fill in under the window sill or above the header are called *cripples*. Figure 9.4 illustrates typical load-bearing wall framing.

To complete the exterior wall system, plywood or similarly rated sheathing is installed over the framed wall. This helps give the wall rigidity and bracing against the wind. Just as in floor framing, wall sheathing is nailed with the 8′ length perpendicular to the studs.

Units for Takeoff

Wall/partition takeoff starts with the exterior walls. The sole plate and double top plate are taken off by the LF by first determining the length of the wall and multiplying by 3. The length of the wall (in feet) is divided by the stud spacing (also in feet), and one stud is added for the end. The height of the stud is calculated by referring to the building section on the architectural drawings for the sole-plate-to-top-plate dimension, and subtracting the width of the three plates. Figure 9.5 shows a partial wall section for determining the stud height.

The example in Figure 9.5 shows that the height of the stud is the plate-to-plate dimension of 8′-0″, less the thickness of three 2″ x 4″ studs: 3 times 1-1/2″, or 4-1/2″. The stud height is 96″ (8′-0″) minus 4-1/2″, or 91-1/2″ (7′-7 1/2″).

Window and door headers are taken off by the LF; headers that are the same size can be taken off by the piece and listed as EA. The size of the stock for the header should be shown on the architectural wall or cross section. To determine the length of the header stock, the contractor should refer to the window or door schedule for the width of the rough opening. The *rough opening* is the clear width and height between the framing into which the window or door will fit. It is expressed as width by height. To the width of the rough opening, the contractor must add the required bearing on either side. This is typically the thickness of a 2″ x 4″ or 2″ x 6″ stud, or 1-1/2″. Figure 9.6 illustrates the rough opening and header size.

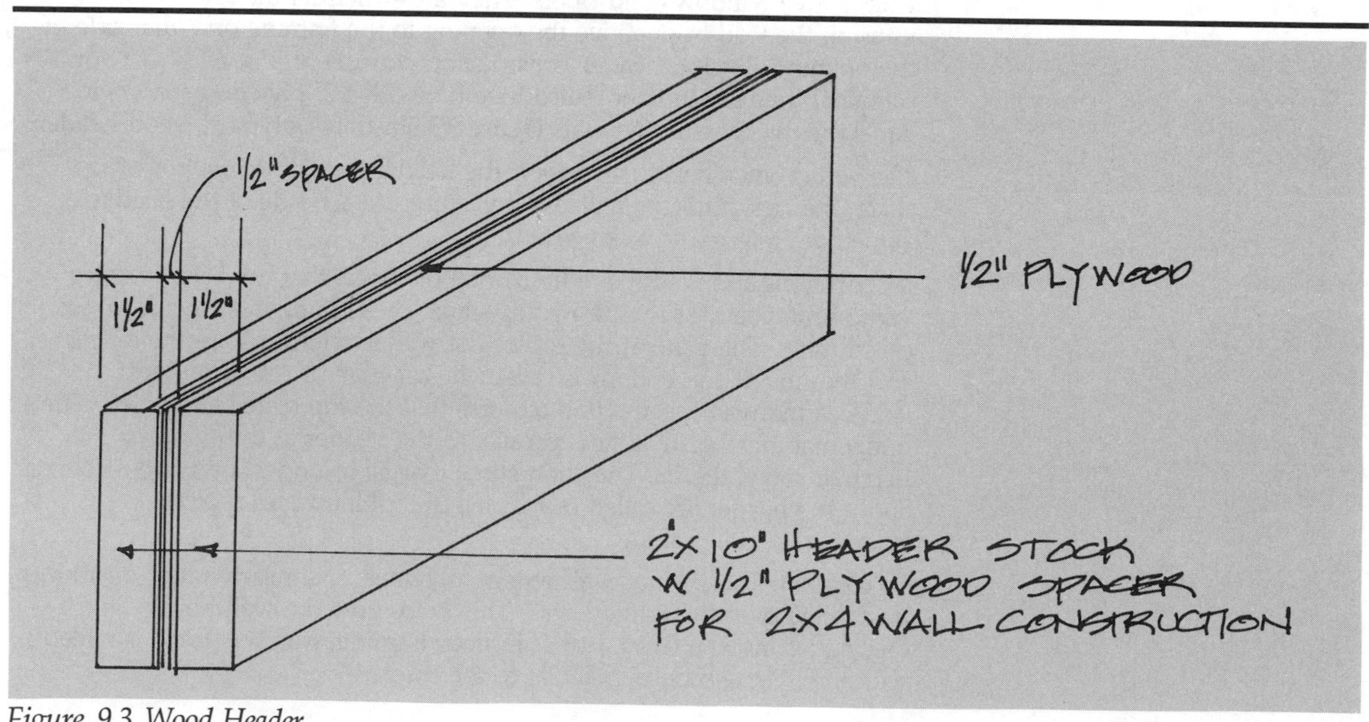

Figure 9.3 Wood Header

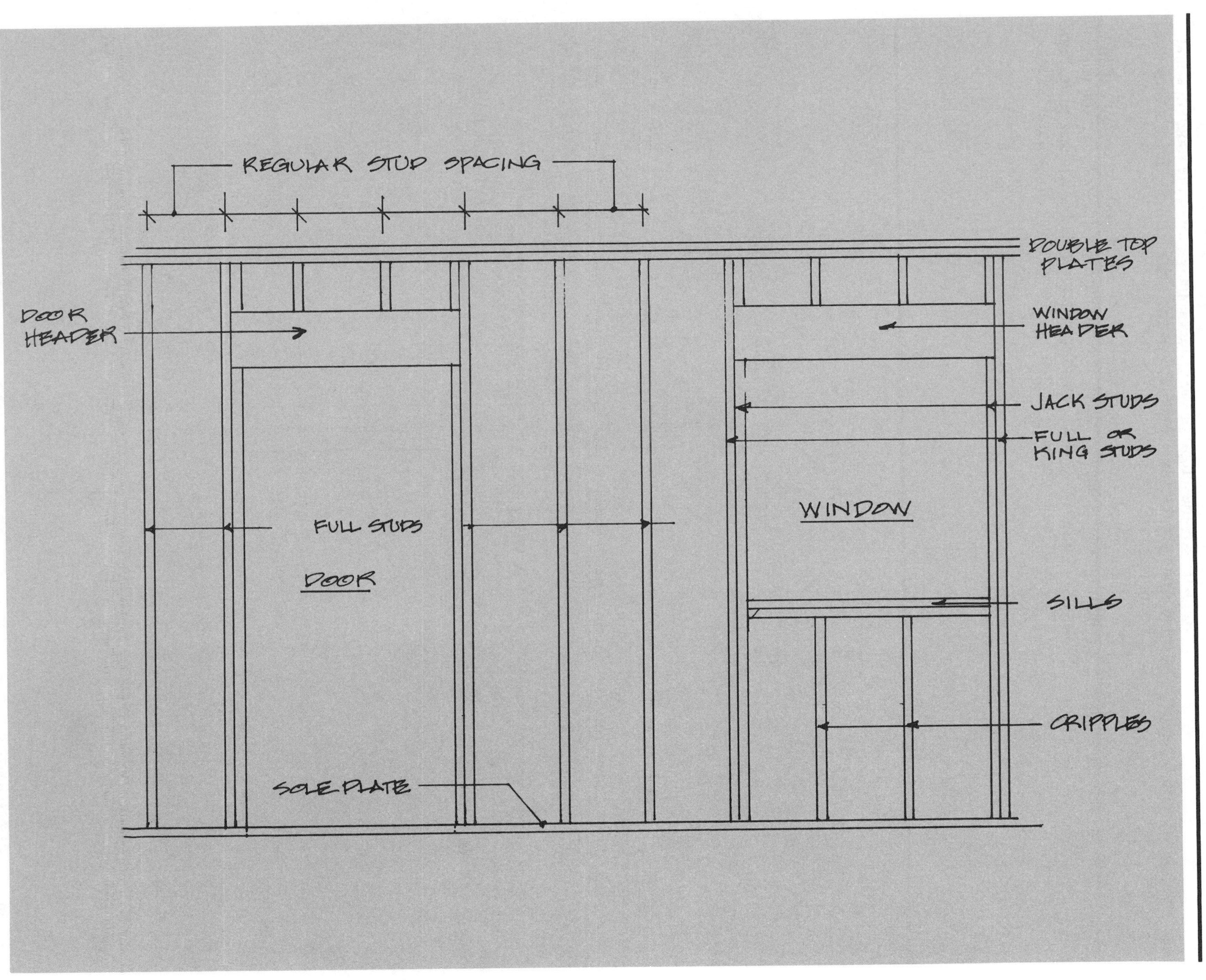

Figure 9.4 Load-Bearing Wall Framing

In Figure 9.6, the length of the header is the width of the rough opening (36″) plus the bearing dimension of the jack studs on either side of the opening, or 2 times 1-1/2″ = 3″, or 39″.

The exterior wall sheathing is taken off by the SF and can be extended to the number of sheets required. The contractor should note that the vertical height of the exterior wall sheathing does not start at the sole plate, but at the bottom of the pressure-treated sills. To determine the vertical height of

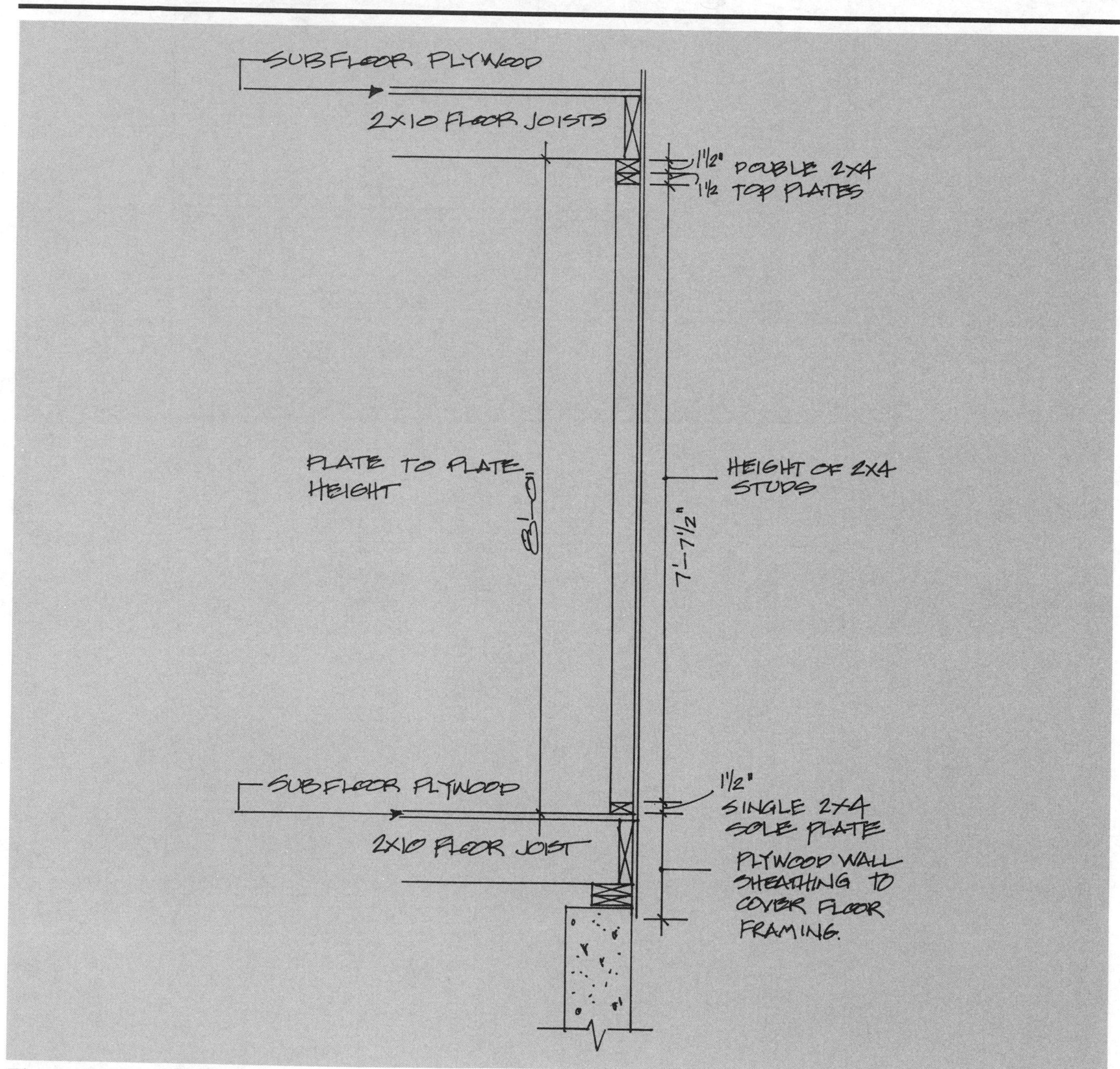

Figure 9.5 Partial Wall Section

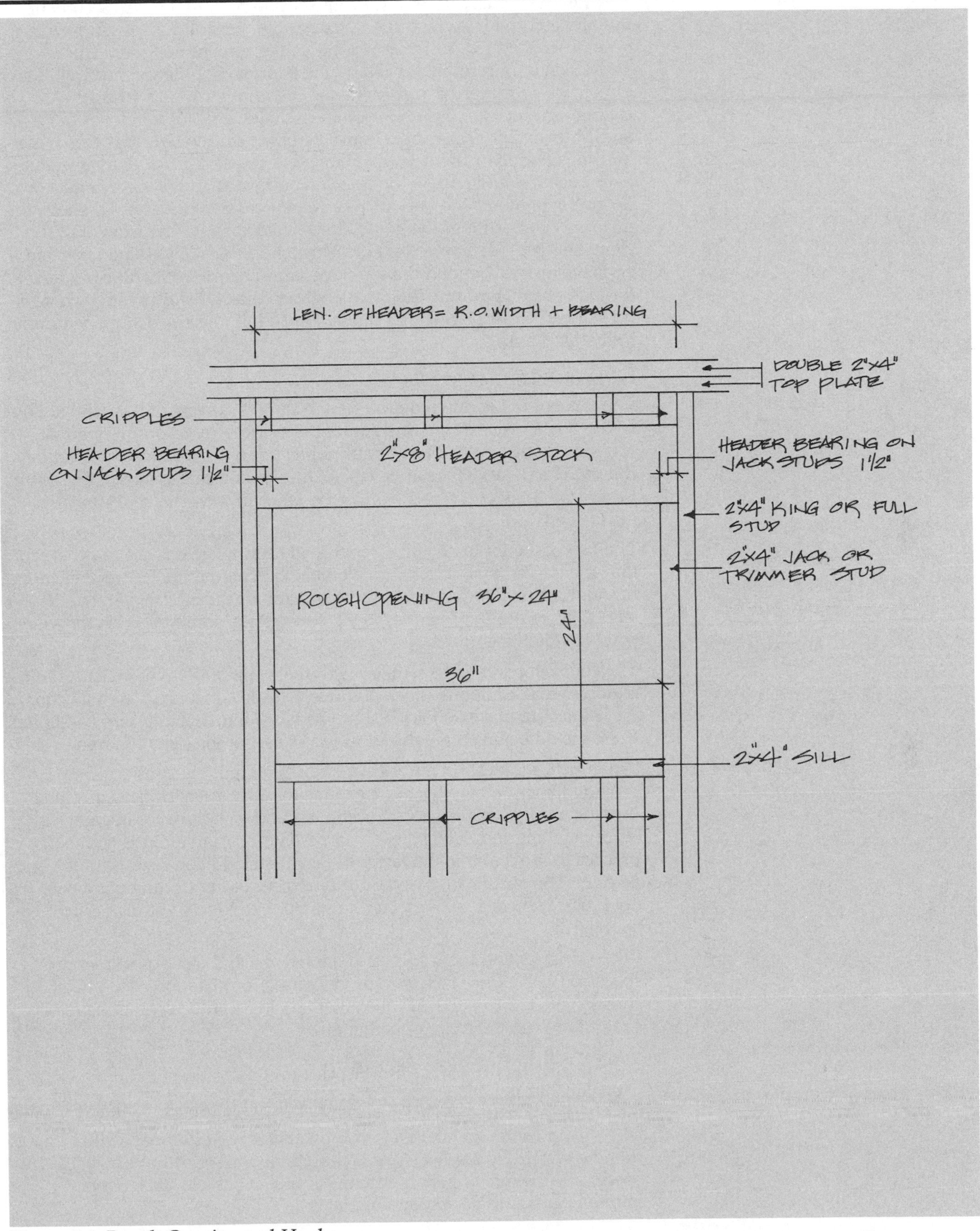

Figure 9.6 Rough Opening and Header

the wall sheathing, the contractor must again refer to the wall section on the architectural drawings and calculate the vertical dimension from the pressure-treated sills to the top plate of the uppermost story. This dimension is, in turn, multiplied by the perimeter of the exterior wall. This dimension may not be consistent for the entire perimeter because of changes in story heights and roof lines. Deductions for windows, doors, and other exterior openings should be based on the size of the opening and the method of framing. Common field practice provides for sheathing over openings and cutting the opening afterward. In this case, deductions do not apply, and the cut-out pieces are considered waste. To check the quantity of sheathing, the contractor should refer to the exterior wall elevations on the architectural drawings. Additional sheathing for gable ends that extend beyond the top plate must be taken off with consideration for waste resulting from cutting at angles to match the roof pitch. There is no standard waste factor that can be applied; it must be determined on an individual basis.

Other Takeoff Considerations

The amount of bearing required for a header is governed by local building codes. Some codes dictate the amount of bearing based on the span, or rough opening, width. For example, openings 6′-0″ wide or more require the use of two jack studs at each side, thereby increasing the length of the header by 3″. The contractor must be familiar with the codes having jurisdiction.

The procedure for taking off interior load-bearing walls is the same, except that there is no wall sheathing. Nonload-bearing partitions are similar, but do not require a double top plate or structural headers. The length of the stud is therefore increased by 1-1/2″ for interior partitions with the same plate-to-plate height.

The contractor should be familiar with the common framing practices in the region so that additional stock used for items like braces or "spring boards" for "plumbing up" exterior walls can be taken into account. This additional stock should be taken off and included as part of the cost of framing.

When determining the number of studs required, the contractor must provide for an additional number for outside corners (*backer studs* that provide nailing for intersecting partitions) and jack studs for openings. Since counting individual studs can be time consuming, it is common practice to use a multipling factor to take these additional studs into account. The factors are based on the number of intersecting partitions, openings, and corners. They can range from 1.10 to 1.35 studs per LF of wall length.

When determining the length of stock for sole and top plates, consider spacing. Twelve-foot stock is commonly used for plates because 144″ (12′) "breaks" on 12″, 16″, and 24″ on-center spacing, and because of its relative straightness compared to longer lengths.

Roof/Ceiling Framing Assembly

Conventional, or "stick," framing is classified as the use of individual members, rather than trusses, to construct the roof. Before starting the takeoff, the contractor should become familiar with the roof and ceiling framing system by studying the architectural and structural drawings, specifically the building cross-sections, wall sections, elevations, roof framing plans, and corresponding details.

The *rafters* are structural members that follow the slope, or pitch, of the roof. *Pitch* is the angle or inclination of the roof. Roof pitch is expressed as a ratio between the horizontal *run* of the roof and the vertical *rise* of the roof. It is typically noted on the drawings as an incremental ratio per 12″, or run. Figure 9.7 illustrates a typical symbol used to designate pitch.

In Figure 9.7, the pitch of the roof is 8″ of rise for every 12″ of run. The rafters are supported at the base by the top plate at the top of the exterior wall. In the same way that floor joists are spaced along the pressure-treated sills, rafters are spaced along the top plate. Common spacing is 12″, 16″, or 24″ on center. The part of the rafter that extends beyond the face of the exterior wall is called the *rafter tail*, or *tail*. It provides the nailing for the fascia and soffit (discussed later in this chapter), and constitutes the roof's overhang. The highest point of the rafter terminates at a perpendicular member in the horizontal plane called the *ridge*, or *ridge board*. To complete the triangular shape of the roof frame, horizontal members called *ceiling joists* provide the floor of the attic space or ceiling of the floor below. Ceiling joists extend from the top plates of bearing walls and span the space, much like floor joists that span from sill to girder or sill. Ceiling joists for gable-end roofs run parallel to rafters. *Strapping* (called furring when applied to walls) is typically comprised of 1″ x 3″ (nominal) boards nailed to the ceiling side of the the ceiling joists. Strapping runs at right angles to the ceiling joists in the same horizontal plane. Strapping is used to maintain the spacing of ceiling joists between bearing points and to provide furring for the ceiling finish. Strapping is commonly spaced at 12″ or 16″ on center.

To increase the rigidity of the roof frame, horizontal members called *collar ties* are installed from rafter to rafter on opposite sides of the ridge. Collar ties are typically located in the top third of the imaginary triangle created by the roof frame.

The *roof sheathing* extends from the rafter tail to the ridge board along the top surface of the rafter and provides a substrate for the application of the roofing. Figure 9.8 shows a typical roof frame as viewed in a building cross-section.

Another type of roof with intersecting roof planes is a *hip roof*. The member that forms the outside intersection of the hip roof is a *hip rafter*. The member at the inside intersection is referred to as the *valley rafter*. In

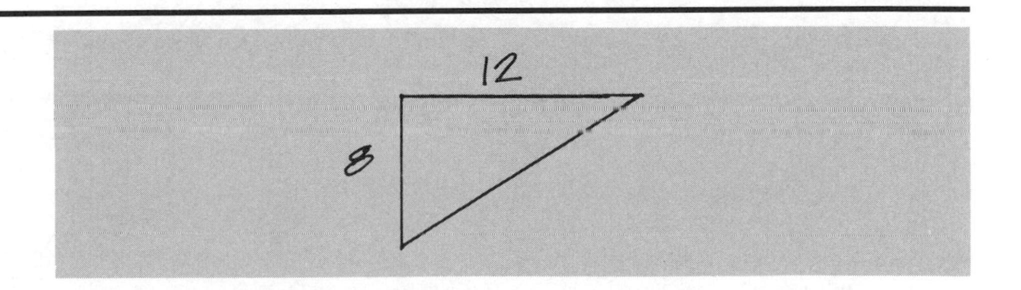

Figure 9.7 Roof Pitch Symbol

addition to the *common rafters*, smaller rafters called *jack rafters* rise from the top plate and terminate at the hip or valley rafters. Figure 9.9 illustrates a hip roof frame.

Units for Takeoff

Rafters are taken off as follows. Count the number of rafters of each length, species, grade, and size. The quantity can be expressed as EA or by LF,

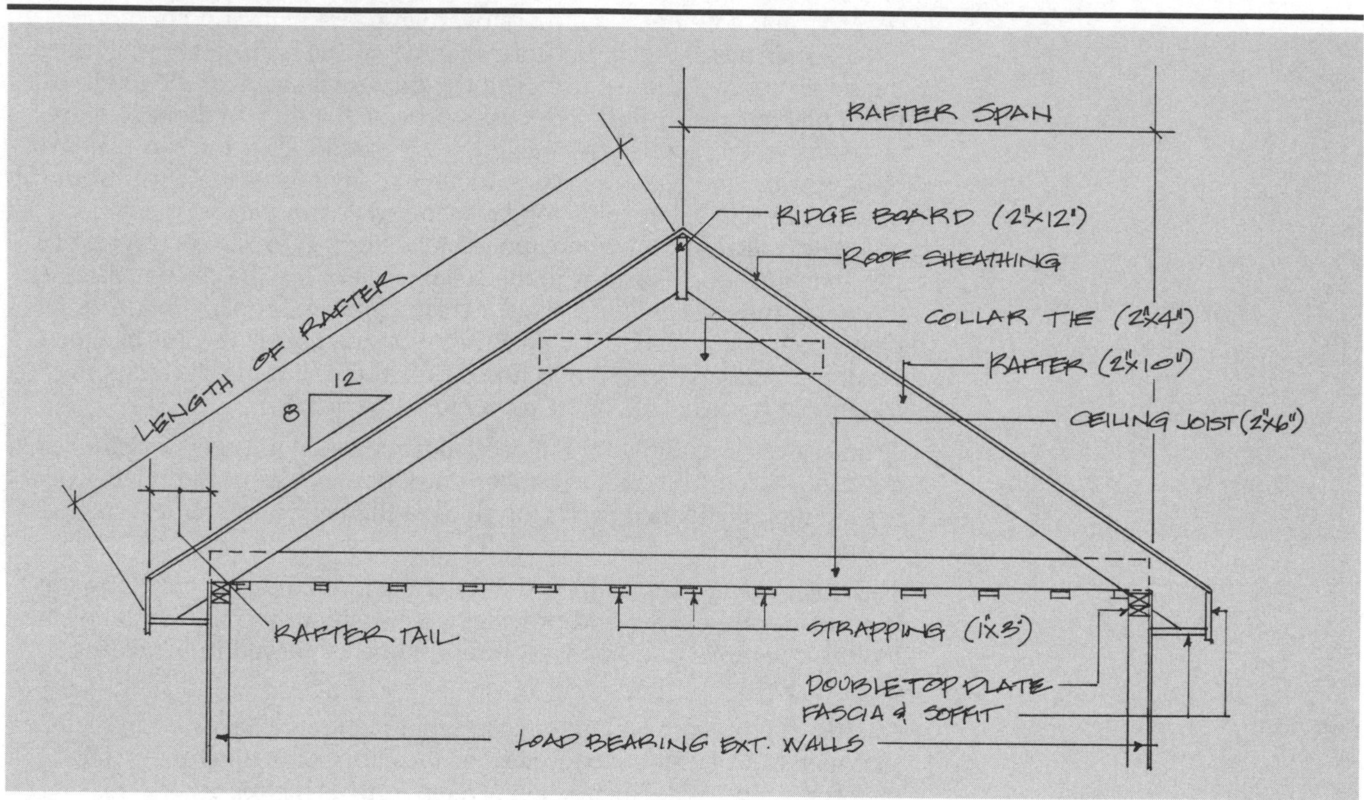

Figure 9.8 Section at Roof Frame

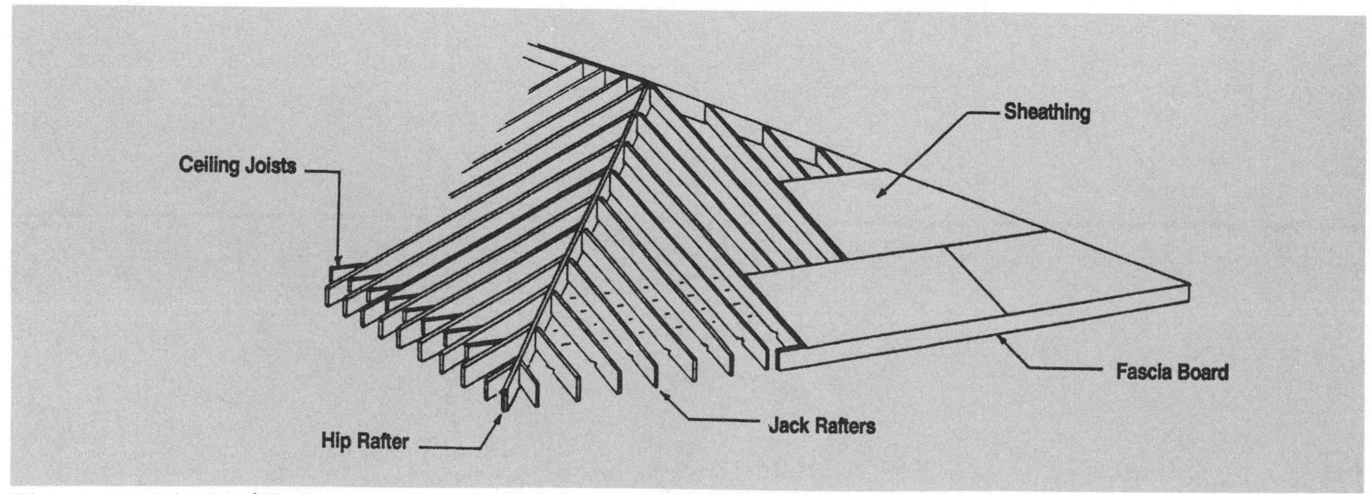

Figure 9.9 Hip Roof Frame

and can be converted to BFM. The quantity of rafters for a gable-end roof is determined in the same manner as floor joists. Divide the overall length (in feet) by the rafter spacing (in feet), and add one rafter for the end. Remember to tally the rafters from both sides of the ridge board.

Determining Common Rafter Length: If a vertical line passing through the center of the ridge were intersected by a horizontal line from the rafter tail to the center of the ridge, the intersection would form a right angle. By adding the rafter to these two imaginary lines, the shape would be a right triangle, subject to the rules of the Pythagorean Theorem (see Chapter 2). Figure 9.10 illustrates the imaginary triangle from the roof frame in Figure 9.8.

To determine the length of the rafter, first calculate the total run ("B") of the rafter. This is done by adding the rafter span to the overhang. The *rafter span* is the horizontal distance from the face of the ridge to the outside face of the exterior wall sheathing. The contractor must determine the distance from the outside of the wall sheathing to the center of the ridge. This is computed from the available dimensions on the architectural

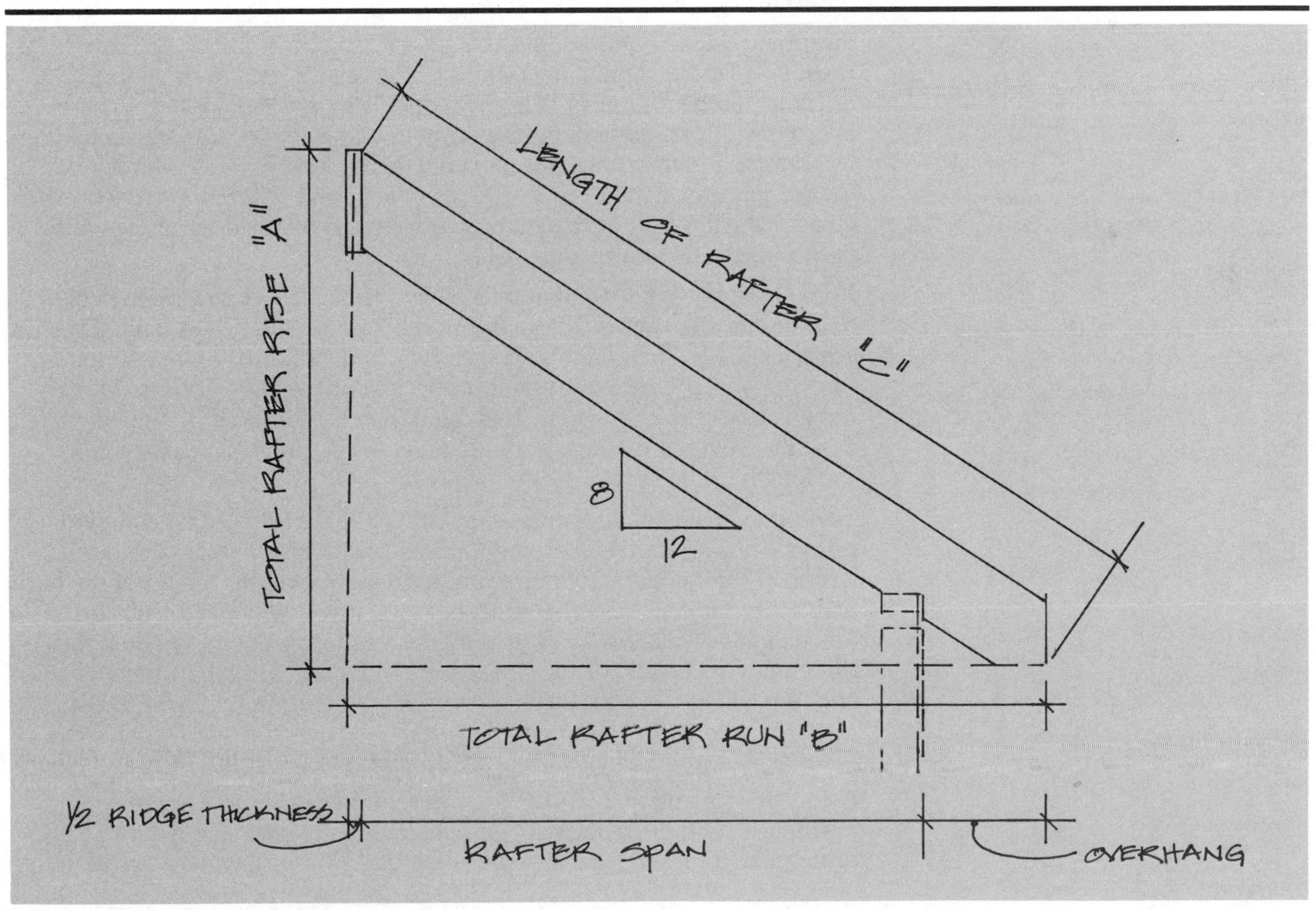

Figure 9.10 Calculating Rafter Length

drawings. For example, in Figure 9.10, if the dimension from the center of the ridge to the outside of the sheathing is 14"-0", the following calculations result:

B = + 14'-0"
+ 1'-0" (overhang)
− 3/4" (half the thickness of the ridge)

B = 14'-11-1/4" = 14.94' = total run of the rafter

Since the ratio of rise to run is 8 to 12, this can be expressed as a decimal: 8 divided by 12 = .67. Once B is known, "A" can be calculated because A = .67 × B = .67 (14.94') = 10.00' With values for both "A" and "B," "C" can be computed using the Pythagorean Theorem: $A^2 + B^2 = C^2$.

$C = \sqrt{(10.00)^2 + (14.94)^2} = 17.97'$ *or 17'-11-11/16"*

Having determined the length of the rafter in the previous example, the contractor can conclude that the stock needed is 18'-0".

Ceiling joists are taken off by dividing the length (in feet) of the building by the spacing (in feet) and adding one joist for the end. The length of the joist should be determined based on dimensions given in the plan. The quantity of ceiling joists should be listed in the takeoff according to length, size, grade, and species. Ceiling joists can be extended to LF or BFM.

Collar ties are taken off using the same process as ceiling joists. They are listed by the piece (EA) or by the LF, or can be quantified in the takeoff as a number of pieces of a specific length.

Strapping is taken off and listed by the LF. To obtain the LF, the contractor must calculate the length and width of the ceiling joists to be strapped, and the number of pieces. This procedure is similar to floor or ceiling joist takeoff, with one exception. Instead of dividing the length of the building by the spacing, the contractor must divide the length of the ceiling joist by the spacing of the strapping (both in feet) and add one piece for the end. This will result in the number of pieces multiplied by the length of each piece to determine the total LF.

Roof sheathing is taken off and listed by the SF. The contractor may elect to translate the unit of SF to the number of sheets, where one sheet is equal to 32 SF. The quantity of sheathing is determined by measuring the length of the roof from the outside face of the starting rafter to the outside face of the last rafter. This dimension is multiplied by the length of the rafter covered by the sheathing. Refer to the previous calculation for the length of the rafter.

Takeoff units for the hip and valley rafters are the same as for common rafters. They are listed according to the quantity, size, species, grade, and length of the lumber. The procedure for determining the length of the hip and valley rafters is also the same, but with one additional calculation. Since hip and valley rafters are at 45 degrees to the common rafter, the dimension of "B" for the hip rafter must be adjusted to reflect the additional length. This is again accomplished by using the Pythagorean Theorem. Figure 9.11 shows the hip rafter first illustrated in Figure 9.9 in plan view as it would be seen on a hip roof framing plan.

If the "B" dimension of 14.94' for the common rafter in the previous example is used, the contractor can determine the total run, or "B" dimension, for the hip rafter in question using the following procedure.

Since the run of the hip rafter is at 45 degrees to the run of the common rafter (in the horizontal plane as measured at the top of the wall), the length of each leg of the triangle would be the same. Using the Pythagorean Theorem, the hypotenuse is the total run or "B" dimension for the hip rafter. For example:

$$B\ hip = \sqrt{(14.94)^2 + (14.94)^2} = 21.12'$$

This represents the total run of the hip rafter and can be used to calculate the total rise dimension "A" by using the relationship of A to B from the roof pitch. If the roof pitch of 8:12 or .67 is used,

$$A = .67 \times 21.12' = 14.15'$$

The length of hip rafter "C" $= \sqrt{(21.12)^2 + (14.15)^2} = 25.42'$

or 25'-5", thereby requiring 26' stock.

The procedure is the same for calculating the length of valley rafters.

The ridge board is taken off by the LF. The length of the ridge board is the same distance as the length of the roof sheathing and can be determined by measuring from the outside face of the starting rafter to the outside face of the end rafter. Ridge boards can be expressed as a quantity of pieces of the same length; for example:

48 LF of ridge = 4 pieces, each 12'-0".

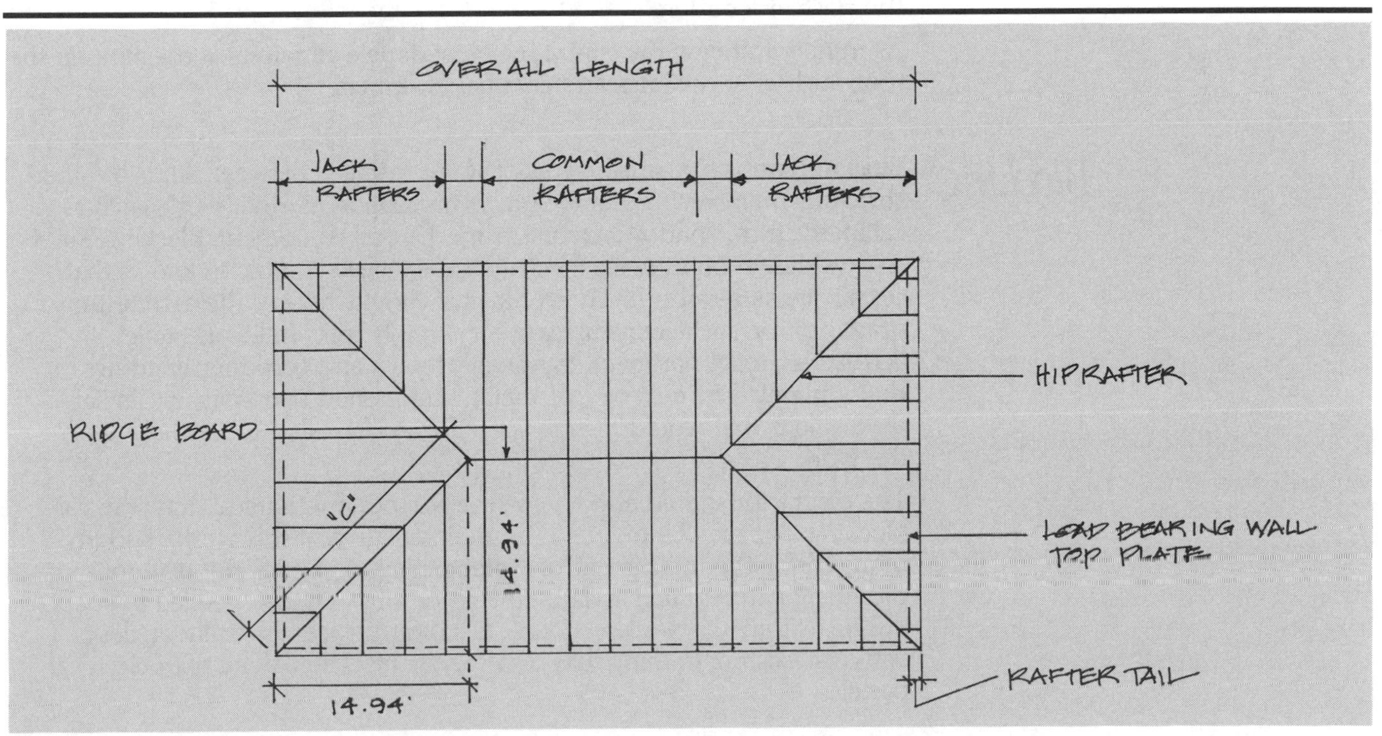

Figure 9.11 Hip Roof Framing Plan

Other Takeoff Considerations

When determining the length of the common rafter, it should be noted that not all roofs have equal pitches for both sides of the ridge. Roofs with varying pitches require separate calculations for each set of rafters.

The procedure for determining the quantity of strapping for cathedral ceilings is identical to that for strapping at ceiling joists. The rafter length can be used (in place of ceiling joist length) to determine the number of pieces required.

Many codes require the use of rafter clips to anchor the rafter to the top plate. The quantity can be determined by counting the number of rafters and multiplying by two for both sides of each rafter.

For most gable-end roofs, the measure used to determine the quantity of rafters and sheathing is the length of the building. Some designs have end rafters that project beyond the plane of the gable end, with an overhang. The contractor should refer to the elevations, sections, and details to determine whether additional rafters, sheathing, or other lumber are needed.

Occasionally hip rafters intersect with common rafters at angles other than 45 degrees; these are irregular, and are sometimes called *bastard hip rafters.* The length of bastard rafters is determined in the same way as conventional hip rafters, except that the length of the triangle leg perpendicular to the rafter (in the horizontal plane) is not the same as the run of the common rafter, and must be determined separately. This is a mathematical calculation based on the angle of the irregular hip rafter.

The length of jack rafters is calculated by the same method as common rafters, but can be time consuming. The contractor should consult framing texts with tables for jack rafter lengths based on the pitch and run of the common rafter. Hand-held calculators designed for contractor use are also helpful in calculating the length of common, hip, valley, and jack rafters at the push of a button.

As in floor framing, the contractor must double all rafters at openings in the roof, such as at skylights and chimney openings.

Blocking

Blocking consists of small pieces of wood installed between studs or other structural members for reinforcing or installation of other work, such as cabinetry, trim, windows, doors, or mechanical equipment. Blocking is not always shown on drawings, but the experienced contractor knows that certain items require it. The contractor should review the architectural drawings for such items as cabinetry, finish trim, millwork, toilet accessories, toilet partitions, handrails at stairs and corridors, windows or doors installed in masonry or metal stud-framed openings, or similar installations that would require special wood blocking for anchoring of the work.

The contractor should also review the electrical/mechanical drawings for blocking that may be required for the installation of this work. Blocking is required for the attachment of water piping as it exits the wall for plumbing fixtures such as sinks, toilets, or tubs. It is also needed to support piping within the walls or joists. Electrical devices in walls or floors between existing framing may also require blocking, along with electrical panels.

In addition to regular blocking, *fire blocking* is required by building codes to slow the transmission of fire through a wall or stair construction, or interior cavity. Again, fire blocking is not always shown on the drawings. In structures with balloon framing, fire blocking is required at the floor level of all walls passing through the floor.

Determining the need for blocking is often a matter of experience or good practice. The contractor should be familiar with the local codes having jurisdiction over the project.

Units for Takeoff

Blocking is taken off and listed by the LF. Quantities of blocking should be separated according to size, species, and grade of material used.

Other Takeoff Considerations

The contractor should study the specifications for specific items that may require blocking. Frequently blocking is required for the installation of work excluded from the contractor's agreement, such as equipment installed by the owner or under separate agreement.

Certain classifications of construction, such as fire-resistant work, may require the use of *fire-treated* wood. Quantities of fire-treated wood should be listed separately, as the cost for such materials is considerably more.

Stair Framing

Stairs are a combination of incremental vertical members called *risers* and incremental horizontal runs called *treads*. The structural members that support the individual treads and risers are called *stringers*. At the framing stage of the work, stringers are installed with a *temporary tread* as a means of access for the workers. Finish treads and risers will be discussed later in this chapter.

Stairs are shown on the architectural drawings. The contractor should refer to the various floor plans to determine the location, size of the stair, number of treads and risers, and the incremental rise and run of the stair. Riser and tread sizes are designated on the plan of the stair itself. Figure 9.12 shows a stair in plan view with the designated riser height.

If the same stair is sectioned, it will be evident that the sum of the risers equals the floor-to-floor dimension of the stairs (see Figure 9.13).

To calculate the length of the stair stringers, the contractor must determine the diagonal measurement between the total rise and total run of the stairs. This is the hypotenuse of the right triangle created by the rise and run of the stair. For stairs that have been designed to a specific riser and tread size, the contractor must multiply the number of risers by the incremental riser height. For example, in Figure 9.12 the following calculations would result:

13 risers at 7-1/2" = 97-1/2" or 8'-1-1/2" total rise
12 treads at 10-1/4" = 123" or 10'-3" total run

Units for Takeoff

Once the total rise and run of the stair have been determined, the contractor must calculate the length of the stair stringers using the Pythagorean Theorem. The vertical leg "A" of the triangle is the total rise of the stair, and the horizontal leg "B" of the triangle is the total run of the stair. Substituting the numbers from the previous example, the following calculations result:

Length of stringer "C" $= \sqrt{(8.125)^2 + (10.25)^2} = 13.07'$ *or approx. 13'-1".*
Therefore, the length of stock needed is 14'-0".

Stair stringers are taken off and listed by the LF, or listed as a multiple of certain length members such as 3 – 14'. Takeoff units for temporary stair treads are by the piece or by the LF. For example, from Figure 9.13:

The stair width is 3'-0", so 12 treads are required at 36" or 3'-0" EA, or a total of 36 LF.

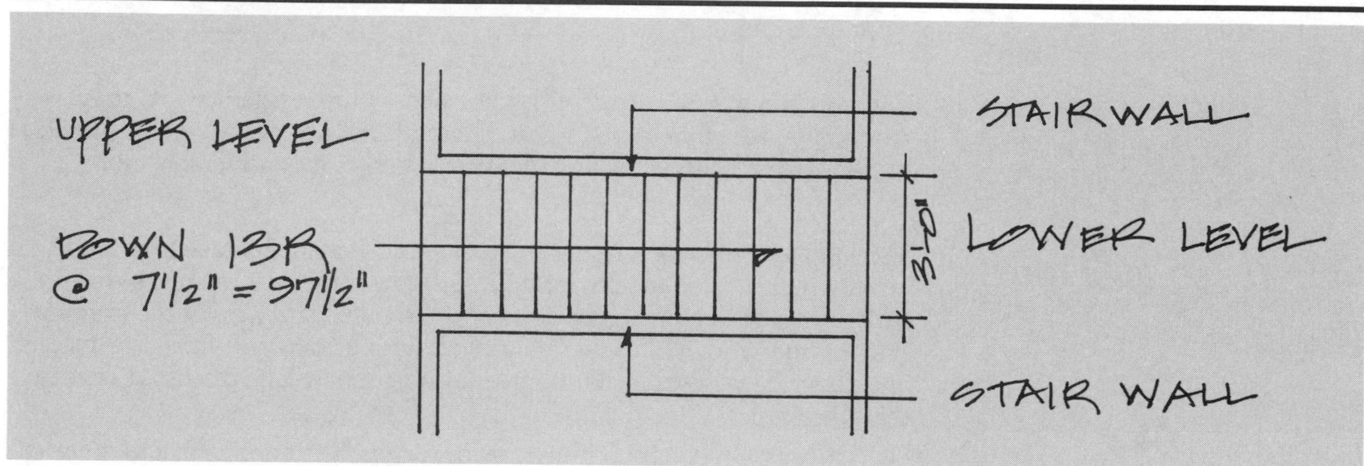

Figure 9.12 Stair in Plan View

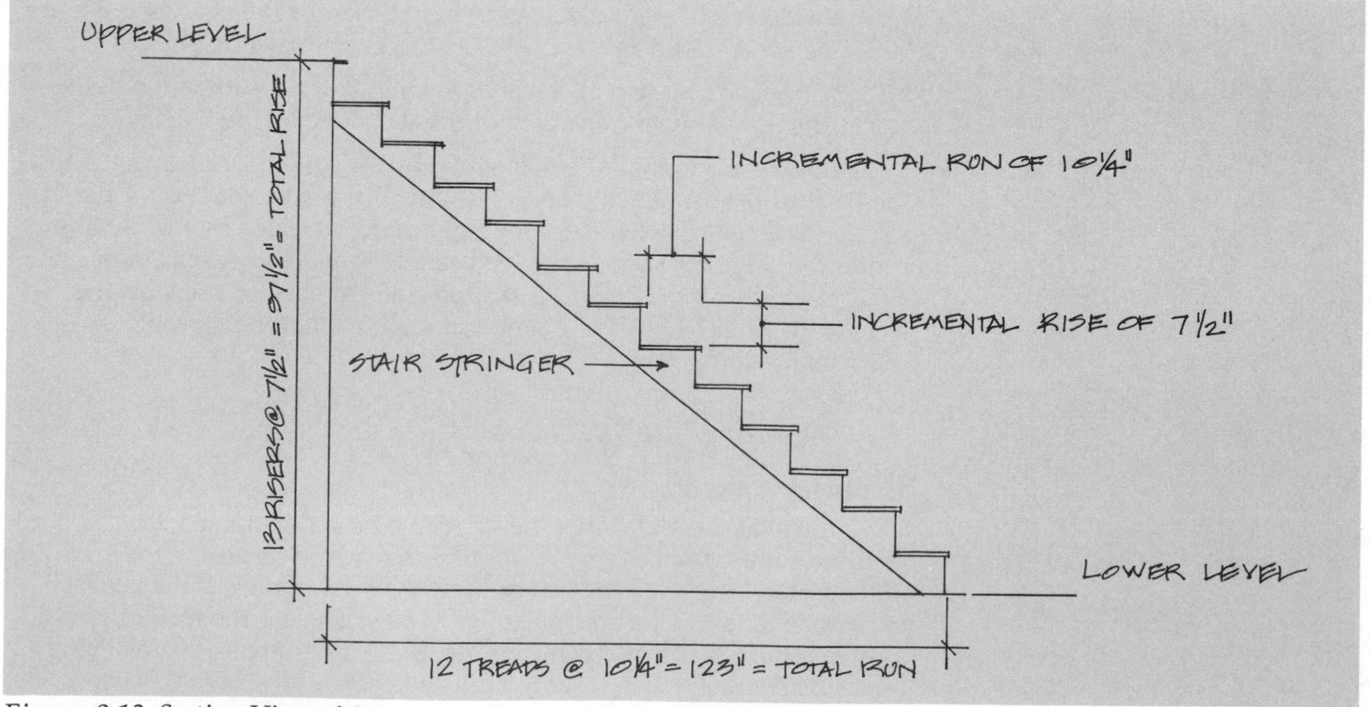

Figure 9.13 Section View of Stairs

Other Takeoff Considerations

The number of stringers required for a stair is based on the width of the stair. In the absence of a specific number of stringers shown on the drawings, local building codes are used. The contractor should be familiar with the building codes that have jurisdiction in the area of the project.

Composite Members

As an alternative to conventional structural grade lumber for the framing of floor and roof systems, special members have been fabricated from structural woods and plywoods, under strict engineering guidelines. These are generally referred to as *composite members* and are used in place of floor and ceiling joists, roof rafters, headers, and other load-bearing members. Composite members used in floor, ceiling, and roof framing look similar to structural steel "I" beams, with flanges constructed of structural grade lumber and webs of plywood or oriented strand board (OSB). Headers and similar load-bearing members that act as beams are generically referred to as *laminated veneer lumber*, or LVL. LVL is comprised of veneers of structural plywood laminated together under pressure to form a structural member.

Composite members offer several advantages over conventional framing lumber. Because of their engineered design and strict manufacturing standards, composite members can span greater unsupported lengths with greater spacing between members. In addition, they are not subject to the shrinking and movement normally associated with conventional lumber, thereby reducing "squeaking" and deflection.

Composite members are available in a variety of cross-sectional sizes and lengths up to 60′. Composite members are handled, cut, and installed using methods similar to conventional framing (see the "Floor and Roof Framing Systems" section in this chapter).

Structural drawings should be reviewed in depth, with special attention to floor and roof framing plans, and sections and details showing load-bearing members and exterior wall openings. Specifications should be reviewed for the required product or manufacturer.

Units for Takeoff

Composite members are taken off by the LF and listed according to depth, width, and length of the member. The contractor should list the various members according to the manufacturer's model or designation number, if available, for accurate pricing. Since composite members are used in place of conventional framing members, the methods and procedures for determining the lengths of common rafters, hip rafters, floor joists, headers, cross bridging, and beams will apply.

Other Takeoff Considerations

Because of the increased spans allowed by composite members, some form of equipment may be required to handle or set in place individual joists or rafters. This should be noted in the takeoff.

Trusses

Trusses are prefabricated structural components composed of a combination of wood members, usually in a triangular arrangement, to form a rigid framework. Trusses are used as an alternative to the roof/ceiling framing in conventional "stick" framing. The architectural roof framing plans or the structural framing plans show the roof truss system. The plan view should show the span, on-center spacing, and number of trusses required to complete the system. Details are often included to show the shape of the truss in elevation view. This view should provide the contractor with the configuration of the truss members, the span and overall length of the trusses, the pitch of the top chord of the trusses, and the bracing requirements.

Installed trusses require *lateral bracing*, or bracing parallel to the length of the building, applied to the underside of the top chords of the trusses. Additional bracing applied to the vertical members closest to the center is called *cross bracing*. It prevents lateral movement in the same way that cross bridging does for floor frame systems. *Diagonal bracing*, or bracing of the bottom and top chords against buckling, may also be required. A combination of all three types of bracing may be needed, depending on the design.

Units for Takeoff

Trusses are taken off by the individual piece. Trusses must be listed separately according to length, span, pitch, and type. To determine the number of trusses, the contractor divides the length of the trussed area by the spacing. Special trusses for end conditions, such as gable-end trusses, should be listed separately.

Bracing for trusses is taken off and listed by the LF. Typical sizes are 1" x 4" or 2" x 4", but the contractor must review the design carefully for location, size, and type of bracing.

Other Takeoff Considerations

Trusses are large and awkward to handle. They may require some type of hoisting equipment to handle and install. The type of equipment is based on the size of the trusses, access to the structure, and number of stories in the building, as well as project coordination.

Exterior Trims

To enclose the the framing system and make the work more visually appealing, exterior wood trim is used. Exterior wood trims are typically 1" x 3", 1" x 4", 1" x 5", and so on, up to 1" x 12". The species of wood depends on the design and may be noted on the drawings or specifications. The contractor must refer to architectural drawings, specifically the elevations, wall, and roof sections, which should show the configuration and sizes of the exterior trims. Vertical trim pieces at the corners of the structure are called *corner boards*. Horizontal trim at the foundation line is called the *water table*. Trim at the top of the exterior wall running around the perimeter of the building is called the *frieze board*. Horizontal trim attached to the vertical portion of the rafter tail is called the *fascia board*, and the trim at the underside of the rafter tail is called the *soffit*. Exterior trim that follows the rafter line on gable-end roofs is called the *rake board*, while smaller trim applied on top of the rake board is called a *rake molding*.

Units for Takeoff

Exterior trim and moldings are taken off by the LF. Exterior trims should be listed according to size (both width and thickness), species of wood, and type or function of the trim piece.

Other Takeoff Considerations

The contractor should review the details of the trim for any special blocking required for the installation of the trim boards.

Finish Carpentry

Finish carpentry can be defined as the use of finish woods in a variety of shapes, sizes, profiles, and species, applied to provide a finished appearance to windows, doors, stairs, and other features of the interior of the building. The trim around window perimeters and the sides and head of doors is called *casing*. The wood trim at the base of the wall that runs along the perimeter of the room is called the *baseboard*. Horizontal trim that runs parallel to the baseboard approximately 3′ above it is called the *chair rail*. A wood surface may be applied between the chair rail and baseboard, in the form of paneling, raised panels, or vertical or horizontal strips; this is referred to as *wainscoting*. Horizontal trim at the intersection of the walls and ceilings is called the *cornice*.

To complete the stairs, finish treads and risers are installed. Additional stair parts such as the *newel posts* that support the railings are also considered finish carpentry, as are the *railings* that run between the newel posts and provide enclosure to the stair, with the *balusters* or individual vertical members attached to the railing. *Skirt boards* are trim pieces at the exterior of an open stair running parallel to the stringer. A similar trim at the interior of a stair is called a *cheek board*.

The contractor should study the architectural drawings for the location, size, and configuration of the interior finish carpentry. Special attention should be paid to the interior elevations and the wall sections that may illustrate the finish carpentry work. Finish carpentry details are often used to clarify or enlarge a particular section or elevation for clarification.

Units for Takeoff

Interior trim such as casings, baseboard, cornice, chair rails, railings, and skirt and cheek boards are taken off by the LF. Quantities can be determined by measuring the perimeter of the room, door, window, or wall.

Other interior finish woodwork, such as wainscoting, is taken off by the SF. Stair parts such as newel posts, balusters, treads and risers are taken off by the individual piece (EA), according to size and spacing.

Other Takeoff Considerations

As in all types of carpentry work, the takeoff for finish carpentry items should be separated according to the size, species, grade, and intended application of the wood.

Cabinetry, Casework, and Plastic Laminate Tops

Discussion of cabinetry in this chapter will be limited to "production line" cabinetry, such as base and wall cabinets for residential kitchens and bathroom vanities, mass-produced and sold as individual units. Custom cabinetry requires special knowledge and experience outside the realm of this text. Most kitchen cabinets and bathroom vanities can be classified as one of three main types for the purpose of takeoff. *Base units* are installed on the floor and terminate at the underside of the countertop. *Wall units* are fastened on the wall above the countertop, terminating below the ceiling line. *Full height units* are continuous cabinets that start at the floor and terminate at the top of the wall cabinets. *Casework* consists of modular or prefabricated units such as bookcases, retail display cases, and storage cabinetry, and is prefinished. One way of prefinishing casework and cabinetry is to cover the exposed surfaces with a thin sheeting of resin-impregnated paper fused together under high pressure and heat. The resulting finish is *plastic laminate.* Plastic laminate, available in a variety of colors and textures, is also used to cover the surfaces of kitchen countertops and bathroom vanity tops.

The contractor reviews the architectural drawings with specific attention to interior elevations, floor plans, sections, and details of the cabinetry. Many drawings include separate plans or elevations for the cabinetry and casework. The specifications should also be studied for the manufacturer, model and series of the cabinetry, casework, or plastic laminate being shown. Figure 9.14 illustrates a plan and elevation view of kitchen cabinetry and countertop.

Units for Takeoff

Cabinetry and casework are taken off by the individual piece (EA) and listed by the quantity of each piece. Individual pieces should be separated according to size, function, and type. For example: 36″ sink base unit, 36″ corner base unit, 24″ 2-door base unit with 1 drawer. Special accessories such as sliding trays, baskets, sliding cutting boards, or adjustable shelving within the unit should also be noted.

Cabinet hardware such as knobs or pulls should be counted and listed separately, as these are not always provided with the cabinetry and may constitute an additional cost for both materials and installation.

Plastic laminate countertops are taken off by the LF. This is typically done by measuring the countertop at the abutting wall. Alternate units of takeoff include the SF of countertop surface. Plastic laminate sheets for field fabrication are taken off by the SF.

Other Takeoff Considerations

Most mass-produced cabinetry is specified according to size and function. Since the majority of base cabinets are 34-1/2″ high and approximately 24″ deep, the size refers to the width of the cabinet. Wall cabinets are designated by width and height. Most are 12″ deep. Specialty cabinets that protrude more than 12″ are also available and should be noted on the takeoff.

Countertops can be manufactured with a *backsplash,* an integral vertical return that abuts the wall. Separate unattached backsplashes are also used and may require a separate listing in the takeoff. They are taken off by the LF.

Plastic laminate taken off for field applications should be converted to the manufacturer's available sheet size. Not all textures and colors are available in every size.

Lumber, Plywood and Nail Specifications

The information in Figures 9.15 – 9.18 is a guide to understanding lumber, plywood, and nail specifications.

Summary

The takeoff sheets in Figure 9.19 are the quantities for the work of Division 6 – Wood and Plastics for the sample project. The takeoff of wood framing and sheathing materials for even a small residence can be quite large and involved. In an effort to avoid omitting segments of work, the

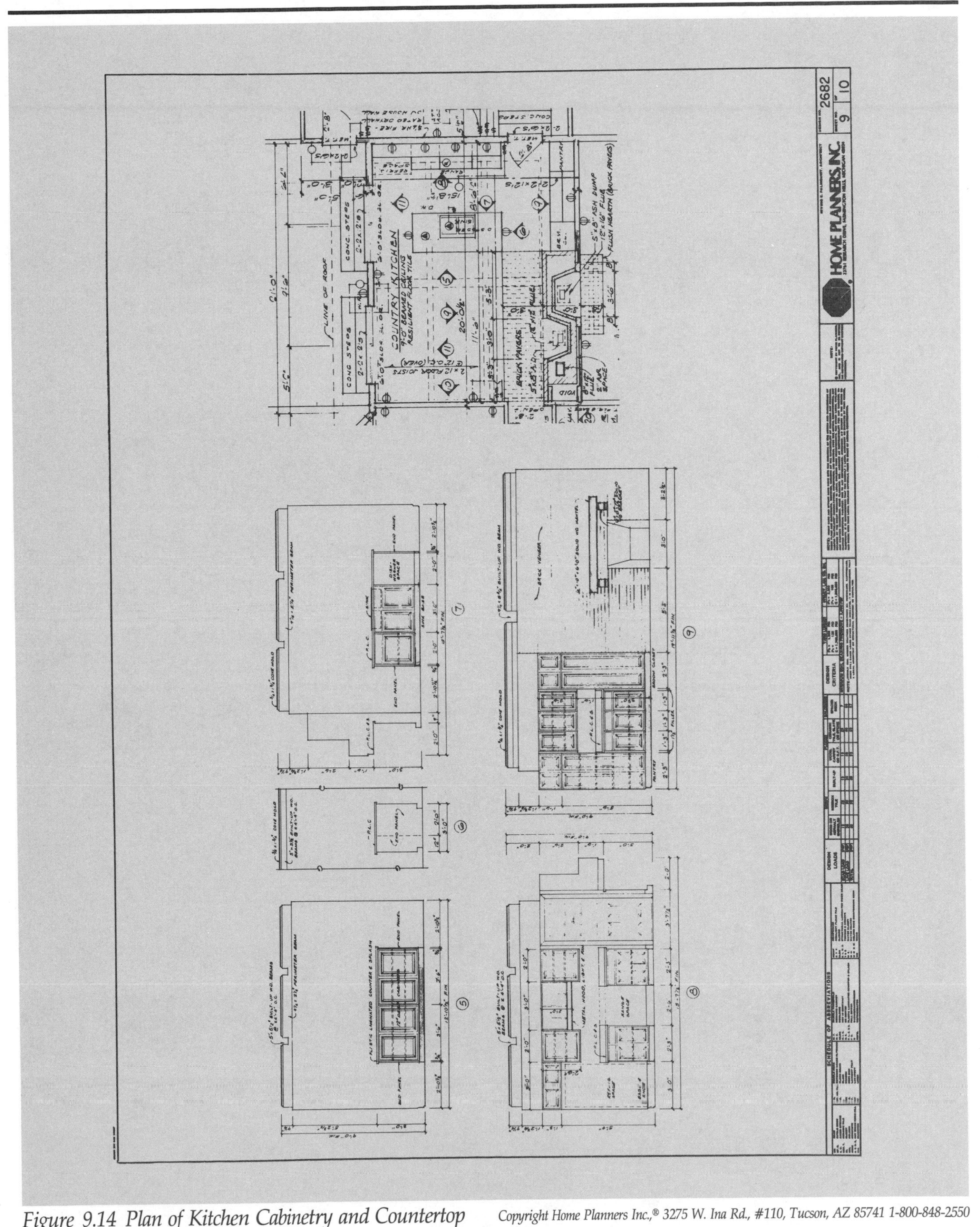

Figure 9.14 Plan of Kitchen Cabinetry and Countertop *Copyright Home Planners Inc.,® 3275 W. Ina Rd., #110, Tucson, AZ 85741 1-800-848-2550*

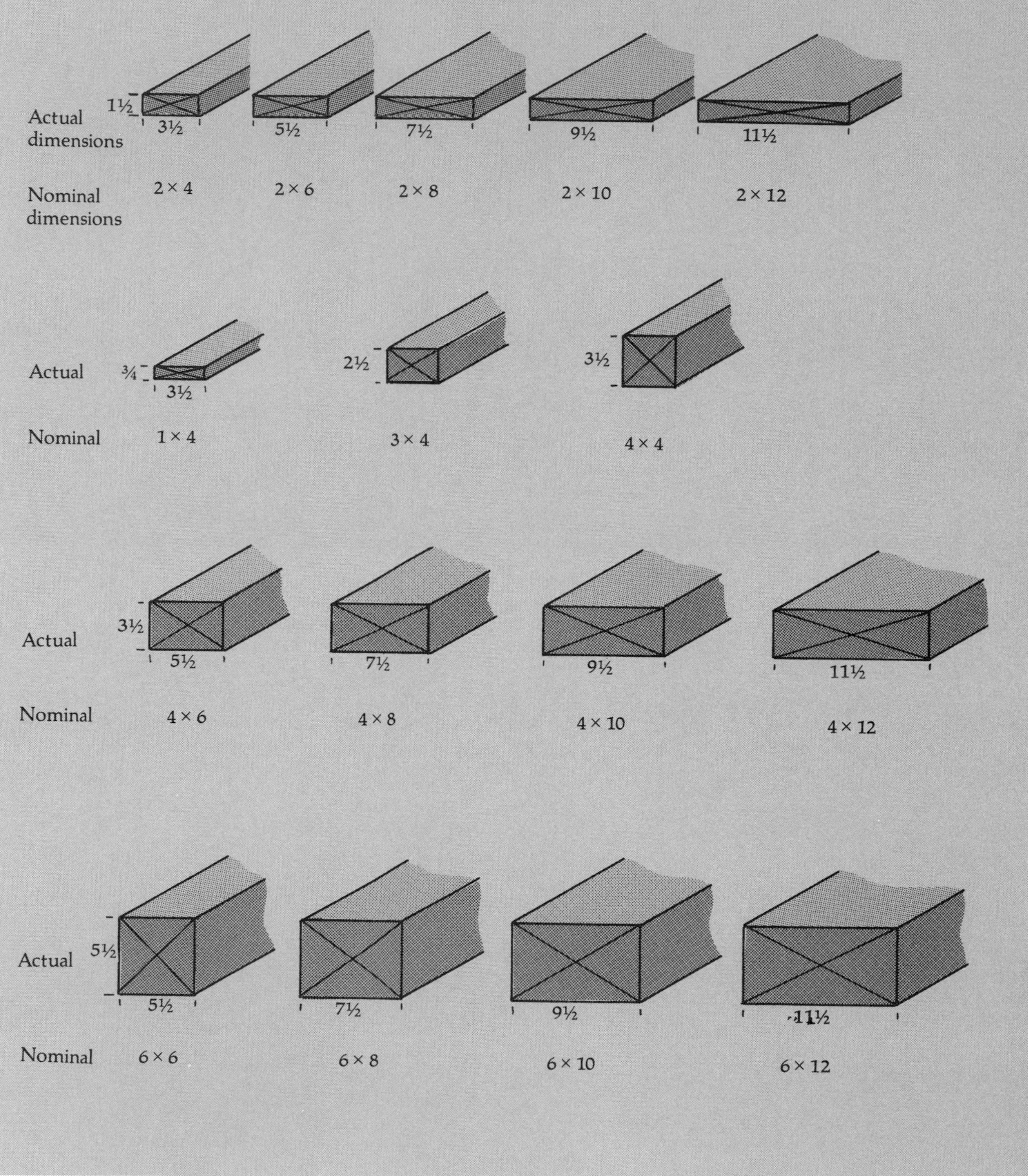

Figure 9.15 Framing Lumber

takeoff proceeds roughly the way the house would be framed. The standard approach to taking off framing materials is to note the actual size of the member as it occurs on the drawings. For example, a rafter may have a calculated length of 13'-10", but the experienced contractor knows that 14'-0" stock would have to be purchased. An alternate method of lumber and sheathing takeoff involves a listing of the materials that would

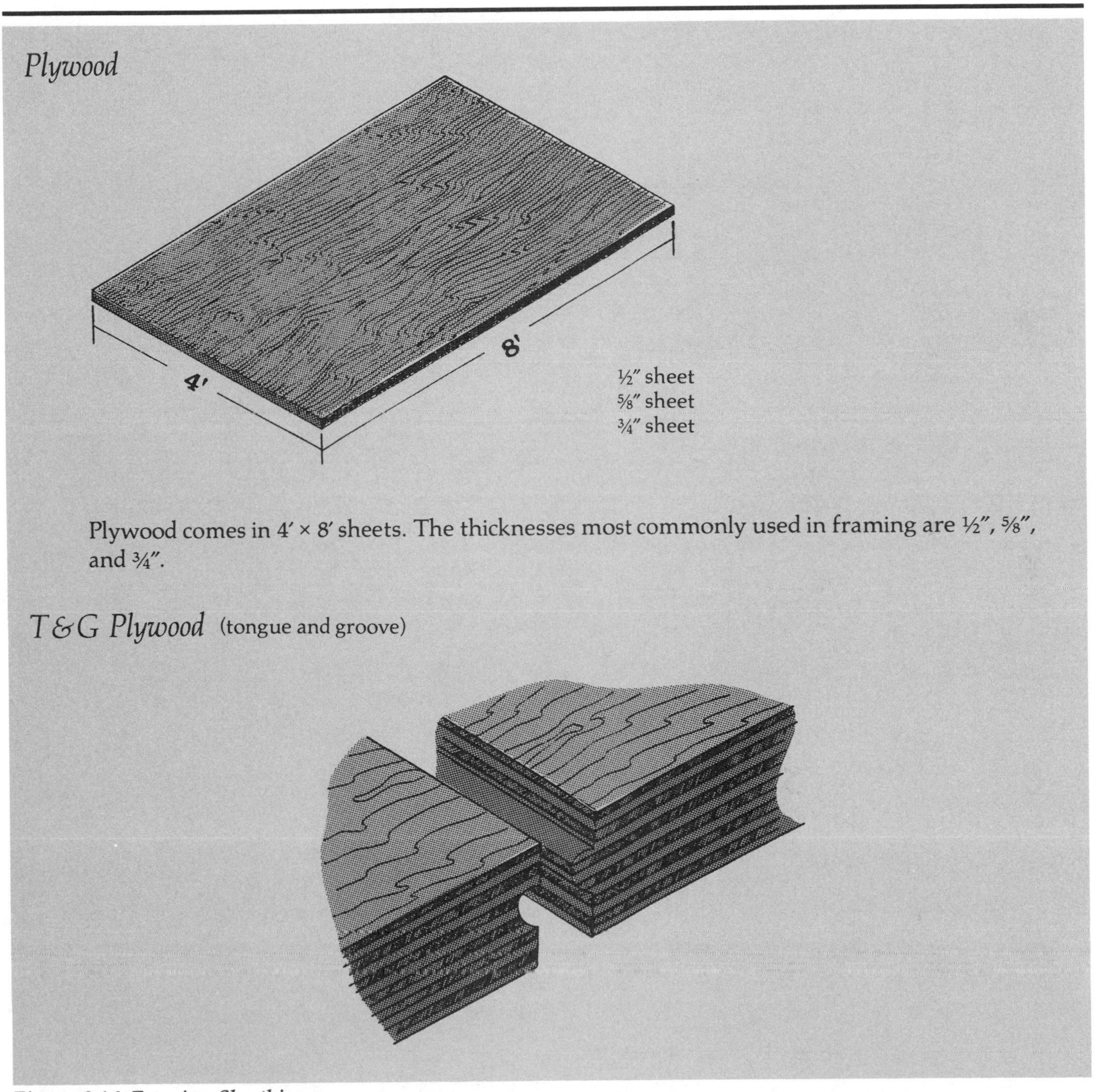

Figure 9.16 Framing Sheathing

Lumber and plywood are graded for strength and different uses. Each piece of lumber is stamped for identification before it is shipped. Architects specify grades of lumber and plywood for various purposes, and framers need to make sure the right wood is used.

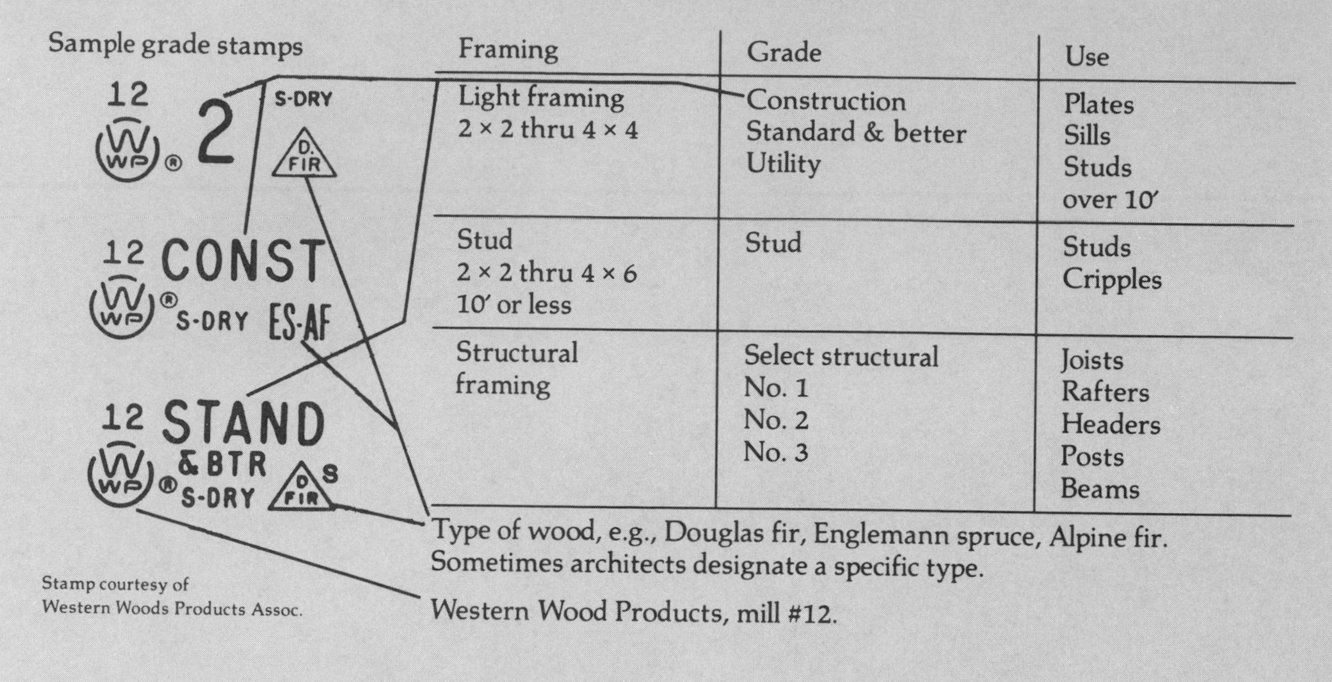

Sample grade stamps

Framing	Grade	Use
Light framing 2 × 2 thru 4 × 4	Construction Standard & better Utility	Plates Sills Studs over 10′
Stud 2 × 2 thru 4 × 6 10′ or less	Stud	Studs Cripples
Structural framing	Select structural No. 1 No. 2 No. 3	Joists Rafters Headers Posts Beams

Stamp courtesy of Western Woods Products Assoc.

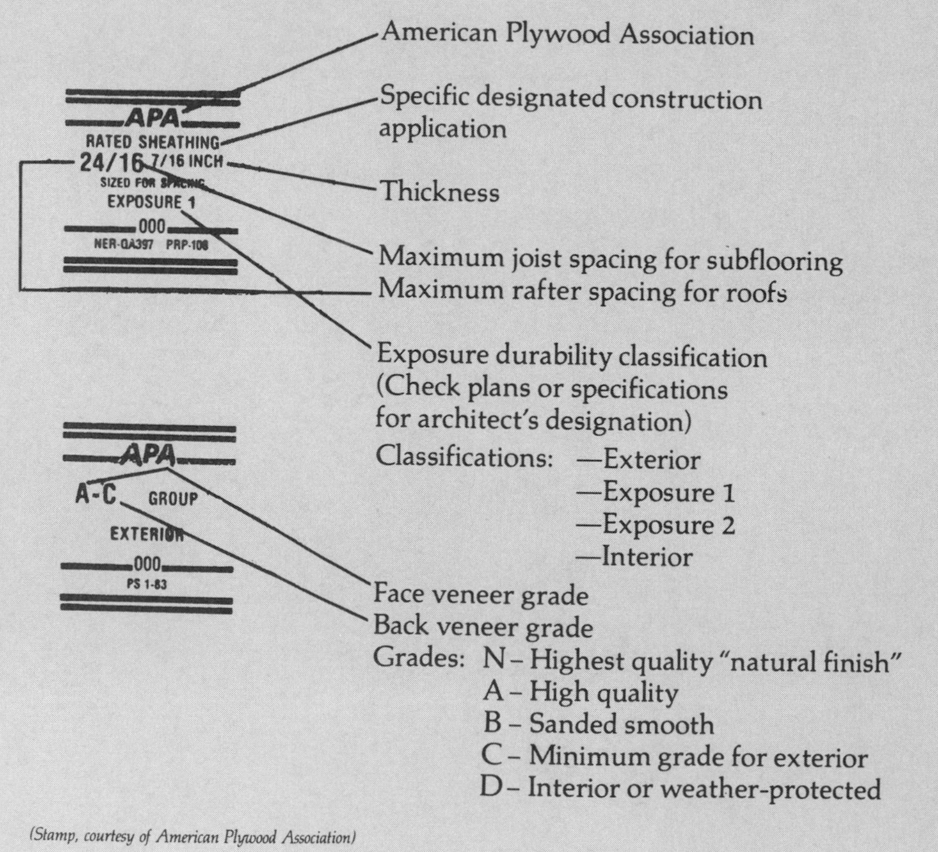

(Stamp, courtesy of American Plywood Association)

Figure 9.17 Lumber and Plywood Grades

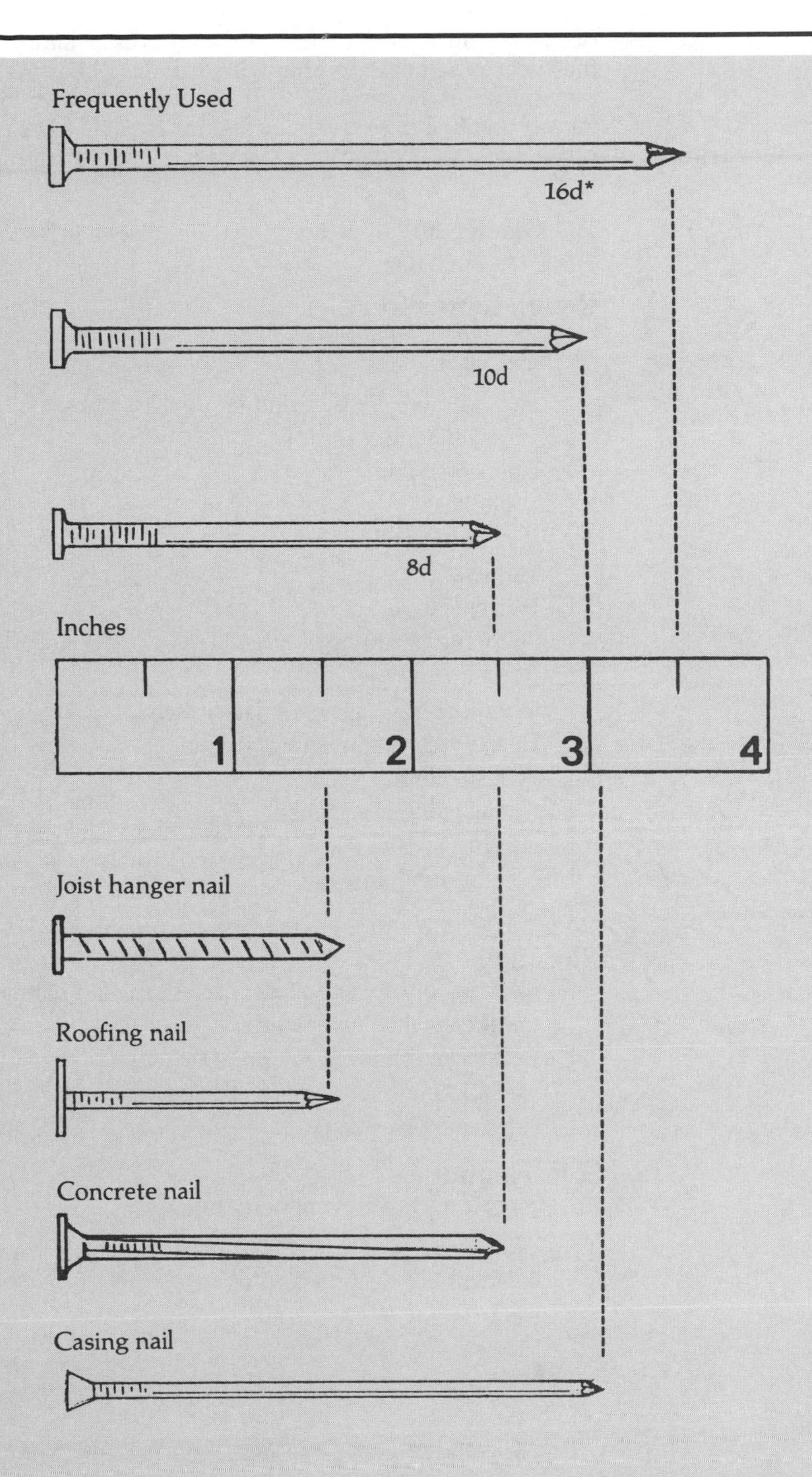

*d = Penny. The abbreviation comes from the Roman word "denarius" meaning coin, which the English adapted to penny. It originally referred to the cost of a specific nail per 100. Today it refers only to nail size.

Figure 9.18 Framing Nails

need to be purchased, typically referred to as a "lumber list". This list is then sent to a lumberyard for a direct quote. The quote is then added into the estimate as a lump sum. Although this may seem more realistic, the takeoff of actual sizes is still needed for accurate labor pricing. The same procedure could be applied to exterior and interior trims.

Division 6 Checklist

The following list can be used to ensure that all items for this division have been accounted for.

Rough Carpentry

- ☐ Pressure-treated sills
- ☐ Sill sealer
- ☐ Floor and box joists, including doubled joists
- ☐ Cross bridging
- ☐ Floor sheathing
- ☐ Wall plates
- ☐ Wall and partition studs
- ☐ Headers
- ☐ Wall sheathing
- ☐ Strapping or furring
- ☐ Ridge board
- ☐ Common, hip, valley and jack rafters
- ☐ Double rafters at openings in roof
- ☐ Roof sheathing
- ☐ Laminated veneer lumber
- ☐ Decking or floor sheathing
- ☐ Nails, rough hardware
- ☐ Composite wood

Blocking

- ☐ Blocking for cabinetry, toilet accessories and partitions, handrails
- ☐ Owner-installed equipment
- ☐ Electrical/mechanical equipment
- ☐ Fire blocking
- ☐ Fire-treated blocking

Stair Framing

- ☐ Stair stringers and temporary treads

Trusses

- ☐ Roof trusses
- ☐ Hip trusses
- ☐ Hip sets
- ☐ Floor trusses
- ☐ Bracing
- ☐ Gable-end or girder trusses

Exterior Trims

- ☐ Water table
- ☐ Corner boards
- ☐ Fascia and soffit

- ☐ Rake boards and rake molding
- ☐ Frieze boards

Finish Carpentry

- ☐ Interior window and door casings
- ☐ Baseboard
- ☐ Chair rail
- ☐ Cornice
- ☐ Wainscotting
- ☐ Stair parts
- ☐ Miscellaneous trims

Cabinetry, Casework, and Plastic Laminates

- ☐ Kitchen cabinetry
- ☐ Bathroom vanities
- ☐ Casework
- ☐ Cabinet hardware
- ☐ Special accessories
- ☐ Plastic laminates
- ☐ Countertop

Means Forms

QUANTITY SHEET

PROJECT	SAMPLE PROJECT	SHEET NO.	6-1/25
		ESTIMATE NO.	1
LOCATION		ARCHITECT HOME PLANNERS	DATE
TAKE OFF BY	WJD	EXTENSIONS BY: WJD	CHECKED BY: KF

DESCRIPTION	NO.	DIMENSIONS		LF	UNIT	BF	UNIT		UNIT		UNIT
DIVISION 6: WOOD + PLASTICS											
FRAMING & SHEATHING											
2"X8" PRES. TREAT SILL											
PLATES @ MASONRY WALL	2	14		28							
@ CRAWL SPACE	1	17.33		18							
@ MAIN HOUSE	2	30		60							
	2	32		64							
				170	LF	227	BF			227	BF
2"X4" P.T. SILL PLATE @											
GARAGE	1	46.67'		47	LF	32	BF			32	BF
2"X6" KD SPRUCE BOLTED	1	11.67									
TO S8X18.4 BEAM	1	9.33		22	LF	22	BF			22	BF
2"X10" KD SPRUCE FLOOR	2	14		28							
JOISTS & BOX JOIST @	7	12		84							
16" O/C @ CRAWL SPACE	7	20		140							
				252	LF	421	BF			421	BF
2"X10" KD SPRUCE FLOOR	2	32'		64							
& BOX JOIST @ MAIN HOUSE	4	14'		56							
	9	10'		90							
	9	16'		144							
	5	12		60							
	10	16'		160							
	6	14		84							
	4	8		32							
	4	12		48							
	7	14		98							
	9	16		144							
				948	LF	1584	BF			1584	BF
3/4" T&G FIR SUBFLOOR											
@ STUDY FLOOR	1	17.33	14	243	SF						
@ MAIN HOUSE	1	30	32	960	SF						
DEDUCT FOR FIREPLACE OPNG.	1	3.67	11.67	(43	SF)						
DEDUCT FOR STAIR OPNG.	1	8.0	9.75	(39	SF)						
				1130	SF					1130	SF

Figure 9.19

Means Forms
QUANTITY SHEET

PROJECT: SAMPLE PROJECT
SHEET NO. 6-2/25
ESTIMATE NO. 1
LOCATION:
ARCHITECT: HOME PLANNERS
DATE:
TAKE OFF BY: WJD
EXTENSIONS BY: WJD
CHECKED BY: KF

DESCRIPTION	NO.	DIMENSIONS			UNIT	UNIT	UNIT	UNIT
5/4"x12" STAIR STRINGERS								
TO BSMT.								
TOTAL RUN = 9'-9"								
TOTAL RISE = 8'-8 1/2"								
LEN = $\sqrt{9.75^2 + 8.71^2}$								
= $\sqrt{95.06 + 75.87}$								
L = 13.07 USE 14'-0"	3	14			42 LF			53 BF
1"x3" CROSS BRIDGING								
@ STUDY FLOOR	2	14			28 LF			
@ MAIN HOUSE	4	32			128 LF			
					156 LF x 3		468 LF	468 LF
CONSTRUCTION ADHESIVE								
FOR SUBFLOOR (1 QT.)	1	1130			1130 SF ÷ 160 SF/TB			7 TUB
2"x4" KD SPRUCE STUDS								
@16" o.c. @ GARAGE								
WALLS STUDS	1	47	10		47 x 1.25 =	59 EA	590 LF	396 BF
TRIPLE PLATES	3	55			55 x 3 =	165 LF		165 BF
2"x6" KD Spruce STUDS								
@16" o.c. @ MAIN HOUSE								
1ST FLOOR EXT. WALLS								
STUDS	126	10			126 LF x 1.25 = 158 x 10 = 1580			1580 BF
TRIPLE PLATE	3	126			126 x 3 = 378 LF			378 BF
2x6 KD Spruce STUDS								
@16" o.c. @ STUDY 1ST FLR.								
STUDS	48	10			48 LF x 1.25 = 60 LF x 10 = 600			600 BF
TRIPLE PLATES	3	48			48 x 3 = 144 LF			144 BF
1/2" CDX FIR SHEATH								
@ EXT. WALLS								
@MAIN HOUSE & STUDY	1	156	11		1716 SF			
@ GARAGE	1	55	9		495 SF			
					2211 SF			2211 SF

Figure 9.19 (continued)

Means Forms

QUANTITY SHEET

SHEET NO. 6-3/25

PROJECT: SAMPLE PROJECT

ESTIMATE NO. 1

LOCATION:

ARCHITECT: HOME PLANNERS

DATE:

TAKE OFF BY: WJD

EXTENSIONS BY: WJD

CHECKED BY: KF

DESCRIPTION	NO.	DIMENSIONS		UNIT		UNIT		UNIT		UNIT
2"x 12" MULTIPLE HEADER @ 1ST FLOOR										
@ COVERED PORCH	2	11.5	23	LF						
@ SLIDER (STUDY)	3	6.5	20	LF						
@ STUDY BAY WINDOW	3	10	30	LF						
@ GARAGE DOOR	2	8.5	17	LF						
@ KIT. SLIDERS	4	6.5	26	LF						
@ DINING ROOM	3	10.25	31	LF						
@ FOYER HALL	2	3.25	7	LF						
@ STAIR TO BASEM'T	2	3.0	6	LF						
			160	LF					320	BF
2"x 10" MULTIPLE HEADERS @ 1ST FLOOR										
@ COVERED PORCH	2	8.5	17	LF					29	BF
2"x8" MULTIPLE HEADERS @ 1ST FLOOR										
@ FRONT DOOR	2	3.5	7	LF					10	BF
2"x6" MULTIPLE HEADERS @ 1ST FLOOR										
@ WINDOWS	8	3.25'	26	LF						
@ DOORS (GARAGE/STUDY)	6	3.08	19	LF						
			45	LF					45	BF
2"x4" INTERIOR LOAD BEARING PARTITIONS STUDS 2x4 @ 16" O.C. TRIPLE PLATES	29	10	29 x 1.25 =		37 x 10' =		370	LF	248	BF
2"x4" INTERIOR NON-LOAD BEARING PARTITION STUDS @ 16" O.C. DOUBLE PLATES	60	10	60 x 1.25 =		75 x 10 =		750	LF	503	BF
1"x3" STRAPPING @ 16" O.C. @ 1ST FLOOR CEILING										
@ STUDY	13	13.5	176	LF						
@ MAIN HOUSE	25	32	800	LF						
			976	LF					976	LF

Figure 9.19 (continued)

Means Forms

QUANTITY SHEET

SHEET NO. 6-4/25

PROJECT SAMPLE PROJECT — ESTIMATE NO. 1

LOCATION — ARCHITECT HOME PLANNERS — DATE

TAKE OFF BY WJD — EXTENSIONS BY: WJD — CHECKED BY: KF

DESCRIPTION	NO.	DIMENSIONS				UNIT		UNIT		UNIT		UNIT
SECOND FLOOR FRAMING												
2"x12" KD SPRUCE @	3	14			42	LF						
12" o.c. @ ATTIC STOR.	15	18			270	LF						
2"x12" KD Spruce @ 16" o.c.	2	14			28	LF						
OVER COVERED PORCH	11	10			110	LF						
@ MAIN HOUSE	4	12			48	LF						
	4	8			32	LF						
	11	14			154	LF						
	9	16			144	LF						
	16	14			224	LF						
2"x12" KD SPRUCE @ 12" o.c.												
OVER COUNTRY KITCH.	24	18			432	LF						
	2	12			24	LF						
					1508	LF					3016	BF
1"x3" CROSS BRIDGING												
@ MID SPAN OF JOISTS	1	74			74	LF	× 3 LF/FT				222	LF
3/4" T&G FIR SUBFLOOR												
@ ATTIC STORAGE	1	27	14		378	SF						
@ MAIN HOUSE	1	32	30		960	SF						
DEDUCT FOR FIREPLACE	1	11	3		(33	SF)						
DEDUCT FOR STAIR OPENING	1	3.5	11		(39	SF)						
					1266	SF					1266	SF
CONSTRUCTION ADHESIVE												
FOR SUBFLOOR (1 QT = TUBE)	1	1266			1266	SF	÷ 160 SF/TB =				8	TUBE
3/4"x12" STAIR STRINGERS												
1ST TO 2ND FLOOR												
TOTAL RUN = 11'-0"												
TOTAL RISE = 10'-1 1/4"												
LEN = $\sqrt{11^2 + 10.10^2}$												
= $\sqrt{121 + 102.01}$												
LEN = 14.93' = 14'-11"	3	15			45	L					57	BF

Figure 9.19 (continued)

Means Forms

QUANTITY SHEET

SHEET NO. 6-5/25

PROJECT SAMPLE PROJECT — ESTIMATE NO. 1

LOCATION — ARCHITECT HOME PLANNERS — DATE

TAKE OFF BY WJD — EXTENSIONS BY: WJD — CHECKED BY: KF

DESCRIPTION	NO.	DIMENSIONS				UNIT		UNIT		UNIT		UNIT
2"x6" KD SPRUCE KNEE												
WALLS @ ATTIC STORAGE												
STUDS @ 16" o.c.	22	1.75'			39	LF						
TRIPLE PLATES	6	14			84	LF						
					123	LF					123	BF
2"x6" KD SPRUCE KNEE												
WALLS @ MAIN HOUSE												
STUDS @ 16" o.c.	31	1.75			55	LF						
TRIPLE PLATES	3	37			111	LF						
					166	LF					166	BF
2"x6" KD SPRUCE EXT												
WALLS @ BEDROOMS												
STUDS @ 16" o.c.	84	7.75'			84	LF × 1.25 = 105 × 7.75					814	BF
TRIPLE PLATES	3	84			252	LF					252	BF
@ ATTIC STORAGE												
STUDS @ 16" o.c.	22	7.75'			171	LF						
TRIPLE PLATES	6	14			84	LF						
					255	LF					255	BF
1/2" CDX FIR SHEATHING												
@ 2ND FLR. EXT. WALLS												
@ ATTIC STORAGE	2	14	2.12		60	SF						
	2	14	8.12		228	SF						
@ MAIN HOUSE	1	84	8.12		683	SF						
	1	37	2.12		79	SF						
	2	10	2		40	SF						
@ ATTIC STORAGE GABLE	2	16	15.33	.5	245	SF						
@ MAIN HOUSE GABLE	2	10	10	.5	100	SF						
	2	15	10	.5	150	SF						
@ GARAGE GABLE	1	21	10.5	.5	111	SF						
					1696	SF					1696	SF
2"x6" KD SPRUCE												
GABLE END FRAMING	1	135			135	LF						
@ ATTIC STORAGE	4	16			64	LF						
@ MAIN HOUSE	2	48			96	LF						
	2	48			96	LF						
					391	LF					391	BF

Figure 9.19 (continued)

Means Forms

QUANTITY SHEET

PROJECT SAMPLE PROJECT	SHEET NO. 6-6/25
LOCATION	ESTIMATE NO. 1
ARCHITECT HOME PLANNERS	DATE
TAKE OFF BY WJD	EXTENSIONS BY: WJD
CHECKED BY: KF	

DESCRIPTION	NO.	DIMENSIONS			UNIT	UNIT	UNIT	UNIT
2"×6" KD SPRUCE								
BLOCKING OVER BAY								
WINDOWS	2	6	12		144 LF			144 BF
2"×6" KD SPRUCE LOAD								
BEARING PARTITIONS @	3	32'			96 LF			96 BF
16" O.C. MAIN HOUSE	32	6.75'			32 LF × 1.25 =	40 × 6.75 =		270 BF
2"×4" KD SPRUCE								
GABLE END FRAMING								
STUDS @ 16" O.C.	2	39			78 LF			
PLATE @ PERIMETER	1	48			48 LF			
					126 LF			126 BF
2"×4" KD SPRUCE LOAD								
BEARING STUDS @ 16" O.C.								
@ 2ND FLOOR HALL								
STUDS @ 16" O.C.	32	7.75			32 LF × 1.25 =	40 × 7.75 × .67 =		208 BF
TRIPLE PLATES	3	32			96 LF			65 BF
2"×4" KD SPRUCE NON-								
LOAD BEARING PARTITIONS								
@ 2ND FLOOR STUDS @ 16" O.C.	74	8			74 × 1.25	93 EA × 8 = 740 LF		496 BF
SINGLE PLATE TOP & BOTTOM	2	74			148 LF			100 BF
1"×3" STRAPPING @ 16" O.C.								
@ MAIN HOUSE	18	32			576 LF			576 LF
2"×12" KD SPRUCE								
MULTIPLE HEADERS @								
2ND FLOOR								
@ BEDROOMS	4	6.5'			26 LF			52 BF
2"×8" KD SPRUCE MULTIPLE								
HEADERS @ 2ND FLOOR	4	6.5'			26 LF			35 BF
2"×6" KD SPRUCE								
MULTIPLE HEADERS @								
2"×4" LOAD BEARING	4	3.0			12 LF			12 BF
PARTITION								

Figure 9.19 (continued)

Means Forms

QUANTITY SHEET

PROJECT: SAMPLE PROJECT — SHEET NO. 6-7/25

ESTIMATE NO. /

LOCATION — ARCHITECT: HOME PLANNERS — DATE

TAKE OFF BY: WJD — EXTENSIONS BY: WJD — CHECKED BY: KF

DESCRIPTION	NO.	DIMENSIONS				UNIT		UNIT		UNIT		UNIT
2"x8" KD SPRUCE												
CEILING JOIST @ 16" O.C.												
@ ATTIC STORAGE	11	16			176	LF					235	BF
@ BED ROOMS	25	12			300	LF					399	BF
@ MSTR. BDRM. & BATHS	25	16			400	LF					532	BF
@ GARAGE	11	22			242	LF					323	BF
1"x3" CROSS BRIDGING												
@ GARAGE CEILING												
JOIST (2 COURSES)	2	14			28	LF	× 3 LF/FT =		84	LF	84	LF
2"x4" KD SPRUCE												
BLOCKING FOR CATWALK												
@ MAIN HOUSE	2	32			64	LF						
@ ATTIC STORAGE	2	14			28	LF						
					92	LF					62	BF
1/2" CDX FIR PLYWOOD												
FOR CATWALK (2'-0")												
@ ATTIC STORAGE	1	14	2		28	SF						
@ MAIN HOUSE	1	32	2		64	SF						
@ GARAGE	1	14	2		28	SF						
					120	SF					120	SF
2"x8" KD SPRUCE												
MULTIPLE HEADER @ STAIR	2	4.5			9	LF					12	BF
2"x4" KD SPRUCE												
BLOCKING OVER STAIR HDR.	4	4.5			18	LF					13	BF
2"x4" KD SPRUCE												
CEILING JOIST @ STAIR												
@ 16" O.C.	4	14			56	LF					38	BF
2"x4" KD SPRUCE CEILING												
JOISTS @ 16" O.C. @ SOFFITS												
@ REAR SOFFIT	1	32			32	LF						
	25	2.6			65							
@ STUDY (FRONT)	1	14			14							
	11	2.6			29							
					140	LF					94	BF

Figure 9.19 (continued)

Means Forms

QUANTITY SHEET

PROJECT SAMPLE PROJECT				SHEET NO. 6-8/25
				ESTIMATE NO. 1
LOCATION		ARCHITECT HOME PLANNER		DATE
TAKE OFF BY WJD		EXTENSIONS BY: WJD		CHECKED BY: KF

DESCRIPTION	NO.	DIMENSIONS				UNIT		UNIT		UNIT		UNIT
2"x4" KD SPRUCE												
COLLAR TIES @ 32" O.C.												
@ STUDY	7	6.			42	LF						
@ MAIN HOUSE	14	10			140	LF						
@ GARAGE	7	10			70	LF						
					252	LF					169	BF
2"x4" KD SPRUCE @ 16"/c												
SUPPORT WALL @ ATTIC												
OF MAIN HOUSE												
STUDS @ 16" O.C.	25	4.14			104	LF						
TRIPLE PLATES	3	32			96	LF						
					200	LF					134	BF
ROOF FRAMING SYSTEM												
2"x6" KD SPRUCE												
RAFTERS @ 16" O.C.												
@ GARAGE												
RAFTER SPAN = "B"												
10'-6"												
0-8 1/4" OVERHANG												
0'-1/2" SHEATHING												
(0'-3/4") DEDUCT 1/2 RIDGE												
11'-2" = 11.16' = "B"												
PITCH 12:12 = 1.0												
"A" = 1.0 x "B" = 11.16'												
LEN = $\sqrt{11.16^2 + 11.16^2}$												
= $\sqrt{124.54 + 124.54}$												
= 15.78 = 15'-9 7/16"												
USE 16'-0"	11	16'			176	LF					176	BF
2"x10" KD SPRUCE												
RIDGE BOARD	1	14			14	LF					24	BF

Figure 9.19 (continued)

Means Forms

QUANTITY SHEET

SHEET NO. 6–9/25

PROJECT SAMPLE PROJECT

ESTIMATE NO. 1

LOCATION

ARCHITECT HOME PLANNERS

DATE

TAKE OFF BY WJD

EXTENSIONS BY: WJD

CHECKED BY: KF

DESCRIPTION	NO.	DIMENSIONS				UNIT		UNIT		UNIT		UNIT
2"x8" KD SPRUCE RAFTER												
@ 16" o.c. @ STUDY												
FRONT												
RAFTER SPAN = "B"												
13'-6"												
2'-8 1/4" OVERHANG												
0'-1/2" SHEATHING												
(0'-3/4") DEDUCT 1/2 RIDGE												
16'-2" = 16.16' = "B"												
PITCH = 12:12 = 1.0												
"A" = 1.0 x "B" = 16.16'												
LEN = $\sqrt{16.16^2 + 16.16^2}$												
= $\sqrt{261.14 + 261.14}$												
LEN = 22.85' = 22'-10 3/8"												
USE 24'-0"	11	24			264	LF					352	BF
REAR												
RAFTER SPAN = "B"												
13'-6"												
1'-0" OVERHANG												
0'-1/2" SHEATHING												
(0'-3/4") DEDUCT 1/2 RIDGE												
14'-5 3/4" = 14.47' = "B"												
PITCH = 12:12 = 1.0												
"A" = 1.0 x "B" = 14.47'												
LEN = $\sqrt{14.47^2 + 14.47^2}$												
= $\sqrt{209.38 + 209.38}$												
LEN = 20.46' = 20'-5 1/2"												
USE 22'-0"	11	22			242	LF					322	BF
2"X 10" RIDGE BOARD	1	14			14	LF					24	BF

Figure 9.19 (continued)

Means Forms

QUANTITY SHEET

PROJECT	SAMPLE PROJECT	SHEET NO.	6-10/25
LOCATION		ESTIMATE NO.	1
ARCHITECT	HOME PLANNERS	DATE	
TAKE OFF BY	WJD	EXTENSIONS BY:	WJD
CHECKED BY:	KF		

DESCRIPTION	NO.	DIMENSIONS				UNIT		UNIT		UNIT		UNIT
2"x8" KD SPRUCE RAFTERS												
@ 16" O.C. @ MAIN ROOF												
FRONT												
RAFTER SPAN = "B"												
15'-0"												
1'-0" OVERHANG												
0'-1/2" SHEATHING												
(0'-3/4") DEDUCT 1/2 RIDGE												
15-11 1/2' = 15.96' = B												
PITCH = 12:12 = 1.0												
"A" = 1.0 × 15.96' = 15.96'												
LEN = $\sqrt{15.96^2 + 15.96^2}$												
= $\sqrt{254.67 + 254.67}$												
LEN = 22.57 = 22'-6 13/16"												
USE 24'-0"	25	24			600	LF					798	BF
PARTIAL REAR (CHEEKS)												
RAFTER SPAN = B												
15'-0"												
2'-8 1/4" OVERHANG												
0'-1/2 SHEATHING												
(0'-3/4") DEDUCT 1/2 RIDGE												
17'-8" = 17.67' = "B"												
PITCH = 12:12 = 1.0.												
"A" = 1.0 × "B" = 17.67'												
LEN = $\sqrt{17.67^2 + 17.67^2}$												
= $\sqrt{312.22 + 312.22}$												
LEN = 24.98' = 24'-11 3/4"												
USE 26'-0"	6	26			156	LF					208	BF
2"x8" KD SPRUCE RAFTERS												
@ 16" O.C. OVER BEDROOMS												
@ REAR												
RAFTER SPAN = "B"												
15'-0"												
0'-8 1/4" OVERHANG												
1'-6" BYPASS RIDGE												
0'-1/2 SHEATHING												
17'-2 3/4" = 17.23' = "B"												
(cont'd)												

Figure 9.19 (continued)

Means Forms

QUANTITY SHEET

PROJECT: SAMPLE PROJECT — SHEET NO. 6-11/25

ESTIMATE NO. 1

LOCATION — ARCHITECT: HOME PLANNERS — DATE

TAKE OFF BY: WJD — EXTENSIONS BY: WJD — CHECKED BY: KF

DESCRIPTION	NO.	DIMENSIONS				UNIT		UNIT		UNIT		UNIT
RAFTERS @ BEDROOM (CONT.)												
PITCH = 6:12 = .5												
"A" = .5 × 17.23' = 8.62'												
LEN = $\sqrt{8.62^2 + 17.23^2}$												
= $\sqrt{74.22 + 296.88}$												
= 19.26' = 19'-3 1/8"												
USE 20'-0"	21	20			420	LF					559	BF
2"×8" KD SPRUCE												
FOR SHORT FALSE												
RAFTERS @ BEDROOM												
ROOF	21	6			126	LF					168	BF
2"×12" KD SPRUCE												
RIDGE BOARD @ MAIN												
ROOF	1	32			32	LF					64	BF
2"×6" KD SPRUCE												
FALSE RAFTERS @ 16" o/c												
REAR OVERHANG												
RAFTER SPAN = "B"												
2'-0"												
0'-8 1/4" OVERHANG												
0'-1/2 SHEATHING												
2'-8 3/4" = 2.73' = "B"												
PITCH = 12:12 = 1.0												
A = 1.0 × 2.73'												
LEN = $\sqrt{2.73^2 + 2.73^2}$												
= $\sqrt{7.45 + 7.45}$												
= 3.85 = 3'-10 5/16"												
USE 4'-0"	21	4			84	LF					84	BF
2"×4" LEDGER (CLEAT)	1	27			27	LF					19	BF
FOR FALSE RAFTERS												

Figure 9.19 (continued)

Means Forms

QUANTITY SHEET

PROJECT SAMPLE PROJECT

SHEET NO. 6-12/25

ESTIMATE NO. 1

LOCATION

ARCHITECT HOME PLANNERS

DATE

TAKE OFF BY NJD

EXTENSIONS BY: WJD

CHECKED BY: KF

DESCRIPTION	NO.	DIMENSIONS				UNIT		UNIT		UNIT		UNIT
5/8" CDX ROOF SHEATHING												
(FIR)												
@ GARAGE	2	14	16		448							
@ ATTIC STORAGE	1	14	23		322							
	1	14	22		308							
@ MAIN ROOF (FRONT)	1	32	23		736							
@ MAIN ROOF (CHEEKS)	2	2.5	25		125							
@ MAIN ROOF OVER BDRM	1	27	19		513							
@ FALSE RAFTERS (OVER HNG)	1	27	4		108							
					2560	SF					2560	S.F.
2x4" KD SPRUCE CLEAT												
FOR SOFFIT SUPPORT												
@ 8'4" & 12" OVERHANGS	25	1			25	LF						
	1	32			32	LF						
	14	1			14	LF						
					71	LF					48	BF
NAILS												
#16d COMMON	2	50#			100	LB					100	LB
#8d COMMON	1	50#			50	LB					50	LB
#6d COMMON	1	50#			50	LB					50	LB

Figure 9.19 (continued)

Means Forms

QUANTITY SHEET

PROJECT: SAMPLE PROJECT
SHEET NO. 6-13/25
ESTIMATE NO. 1
LOCATION:
ARCHITECT: HOME PLANNERS
DATE:
TAKE OFF BY: WJD
EXTENSIONS BY: WJD
CHECKED BY: KF

DESCRIPTION	NO.	DIMENSIONS				UNIT		UNIT		UNIT		UNIT
EXTERIOR TRIM & FINISH												
ALL EXTERIOR TRIMS TO BE NO 2 PINE U.N.O.												
1"x6" FASCIA BOARDS												
@ STUDY/PORCH	2	14			28	LF						
@ BEDROOMS ROOF	1	27			27							
@ FALSE RAFTER OHG	1	32			32							
@ FRONT	1	32			32							
@ GARAGE	1	14			14							
	1	15			15							
					148	LF					148	LF
1"x6" SOFFITS												
@ GARAGE	1	14			14	LF						
	1	15			15	LF						
@ BEDROOM ROOF	1	27			27	LF						
					194	LF					194	LF
1"x10" SOFFIT												
@ COVERED PORCH	1	14			14	LF						
@ FRONT MAIN HSE	1	32			32	LF						
					46	LF					46	LF
3/8" ACX EXT. FIR PLYWOOD SOFFIT												
@ STUDY	1	14	2.7		38	SF						
@ FALSE RAFTERS	1	32	2.7		87	SF						
@ UNDER BAY @ STUDY	1	10	2		20	SF						
@ UNDER BAY @ D.R.	1	10	2.5		25	SF						
@ COVERED PORCH CEILING	1	9.5	13.7		131	SF						
					301	SF					301	SF
5/4"x12" FRIEZE BOARD												
@ STUDY	1	6			6	LF					6	LF
5/4"x10" FRIEZE BOARD												
@ BEDROOMS	1	27			27	LF						
@ GARAGE	1	14			14							
@ FRONT MAIN HOUSE	1	32			32							
@ REAR MAIN HOUSE	1	21			21							
					94	LF					94	LF

Figure 9.19 (continued)

Means Forms

QUANTITY SHEET

PROJECT: SAMPLE PROJECT

SHEET NO. 6-14/25

ESTIMATE NO. 1

LOCATION:

ARCHITECT: HOME PLANNERS

DATE:

TAKE OFF BY: WJD

EXTENSIONS BY: WJD

CHECKED BY: KF

DESCRIPTION	NO.	DIMENSIONS			UNIT	UNIT	UNIT	UNIT
5/4" x 8" FRIEZE BOARD								
@ GARAGE REAR	1	15			15 LF			
@ GARAGE FRONT	1	14			14 LF			
@ INTERIOR OF PORCH	1	48			48 LF			
					77 LF			77 LF
5/4" X 12" WINDOW HEAD								
TRIM	2	5			10 LF			
	5	3			15 LF			
					25 LF			25 LF
3/4" x 4 1/4" CROWN								
MOLDING ON 5/4" x 12"	5	3.5			11 LF			11 LF
5/4" x 6" STANDING &								
RUNNING TRIM								
@ CORNER BOARDS	2	110			220 LF			
@ COVERED PORCH OPNG.	4	28			112 LF			
@ DOORS / WINDOWS	1	144			144 LF			
					476 LF			476 LF
5/4" x 4" WINDOW SILLS	2	5			10 LF			10 LF
1" x 12" @ FRIEZE BOARD								
@ REAR MAIN HOUSE	1	35			35 LF			
@ STUDY	1	12			12 LF			
					47 LF			47 LF
3/4" x 1 1/4" BED MOLDING								
@ REAR MAIN HOUSE	1	35			35 LF			35 LF
3/4" x HALF-ROUND								
MOLDING @ STUDY	1	17			17 LF			17 LF
1" X 12" HEAD TRIM @								
BAY WINDOWS								
@ DIN. RM. & STUDY	2	12			24 LF			24 LF
1" x 6" STANDING & RUNNING								
TRIM								
@ BAY WINDOWS	2	80			160 LF			
@ JAMBS PORCH	2	28			56 LF			
@ JAMBS GARAGE DOOR	3	8			24 LF			
					240 LF			240 LF

Figure 9.19 (continued)

Means Forms

QUANTITY SHEET

SHEET NO.	6-15/25
PROJECT	SAMPLE PROJECT
ESTIMATE NO.	1
LOCATION	
ARCHITECT	HOME PLANNERS
DATE	
TAKE OFF BY	WJD
EXTENSIONS BY:	WJD
CHECKED BY:	MF

DESCRIPTION	NO.	DIMENSIONS				UNIT		UNIT		UNIT		UNIT
3/8" M.D.O. PLYWOOD @												
BASE OF BAY WINDOWS	2	12	2		48	SF					48	SF
1"x8" KICK BOARD @												
COVERED PORCH	1	28			28	LF					28	LF
24"x30" SCREENED												
WOOD LOUVER VENTS	3	EA			3	EA					3	EA
5/4"x8" RAKEBOARDS												
@ GARAGE	1	40			40	LF						
@ MAIN HOUSE	1	96			96	LF						
@ STUDY / PORCH	1	44			44	LF						
					180	LF					180	LF
5/4"x 2 1/4" SHINGLE MOLD												
ON RAKE BOARDS												
@ GARAGE	1	40			40	LF						
@ MAIN HOUSE	1	96			96	LF						
@ STUDY/PORCH	1	44			44	LF						
					180	LF					180	LF
24"x16"x6" SOLID												
WOOD BRACKET	1	EA			1	EA					1	EA
WOOD SHUTTERS												
@ DOORS 18"x80"	2	EA			2	EA					2	EA
@ FRONT WINDOWS 18"x78"	4	EA			4	EA					4	EA
@ SIDE WIND. 18" x 78"	2	EA			2	EA					2	EA
@ " " 18"x 54"	2	EA			2	EA					2	EA

Figure 9.19 (continued)

Means Forms

QUANTITY SHEET

PROJECT: SAMPLE PROJECT	SHEET NO. 6-16/25
	ESTIMATE NO. 1
LOCATION:	ARCHITECT: HOME PLANNERS	DATE:
TAKE OFF BY: WJD	EXTENSIONS BY: WJD	CHECKED BY: KF

DESCRIPTION	NO.	DIMENSIONS			UNIT	UNIT	UNIT	UNIT
INTERIOR FINISH CARPENTRY								
3/8" UNDERLAYMENT								
@ KITCHEN	1	20	15.7		314 SF			
DEDUCT FOR FIREPL/HEARTH	1	(6	11.5)		(69 SF)			
@ HALL & FOYER	1	3.5	11.		39 SF			
" "	1	15.5	3.33		52 SF			
" "	1	4	3.83		16 SF			
@ POWDER ROOM/CLOSET	1	3.25	10		33 SF			
@ SEC FLR. COMMON BATH	1	6	5		30 SF			
@ MASTER BATH	1	5	5.5		28 SF			
					443 SF			443 SF
3/4" X 4 1/4" CROWN MOLDING								
@ CEILING LINE								
@ STUDY	1	60			60 LF			
@ LIVING ROOM	1	66			66			
@ DINING ROOM	1	42			42			
@ FOYER/HALL	1	63			63			
@ UPSTAIRS HALL	1	44			44			
					275 LF			275 LF
3/4" X 5 1/4" BASE MOLDING								
@ CEILING LINE UNDER CROWN								
@ STUDY	1	60			60 LF			
@ LIVING ROOM	1	66			66			
@ DINING ROOM	1	42			42			
@ FOYER/HALL	1	63			63			
@ UPSTAIRS HALL	1	44			44			
					275 LF			275 LF
3/4" X 5 1/4" BASE MOLDING								
@ FLOOR								
@ STUDY	1	53			53 LF			
@ LIVING ROOM	1	57			57			
@ DINING ROOM	1	42			42			
@ FOYER/HALL/KIT.	1	78			78			
@ UPSTAIRS HALL	1	33			33			
@ BEDROOMS	1	161			161			
					424 LF			424 LF

Figure 9.19 (continued)

Means Forms
QUANTITY SHEET

SHEET NO. 6-17/25

PROJECT SAMPLE PROJECT

ESTIMATE NO. 1

LOCATION

ARCHITECT HOME PLANNERS

DATE

TAKE OFF BY WJD

EXTENSIONS BY: WJD

CHECKED BY: KF

DESCRIPTION	NO.	DIMENSIONS				UNIT		UNIT		UNIT		UNIT
3/4"x 1/2" SHOE MOLDING												
@ BASE												
@ STUDY	1	53			53	LF						
@ LIVING ROOM	1	57			57							
@ DINING ROOM	1	42			42							
@ FOYER/HALL/KIT	1	78			78							
@ UPSTAIRS HALL	1	33			33							
@ BEDROOMS	1	161			161							
					424	LF					424	LF
3/4"x 2 1/4" BAND MOLDING												
@ 3'0" AFF (CHAIR RAIL)	1	28			28	LF					28	LF
9/16" x 2" PANEL MOLDING	20	1.75			35							
@ WAINSCOTT DETAIL	20	2.0			40							
					75	LF					75	LF
2"x2" WOOD BLOCKING												
@ PERIMETER OF KITCH.												
CEILING FOR BEAM	2	64			128	LF					128	LF
1"x2" @ PERIMETER												
BEAM (SOFFIT) KITCHEN	1	64			64	LF					64	LF
1"x10" @ PERIMETER BEAM												
@ KITCHEN	1	64			64	LF					64	LF
2"x8" KD SPRUCE BLOCKING												
@ 8 3/4" x 9 1/4" BEAM												
@ KITCHEN CEILING	1	12.5			13	LF					13	LF
2"x4" KD SPRUCE												
BLOCKING @ 5"x 5 1/2" BEAM												
@ KITCHEN CEILING	4	9			36	LF					36	LF
1"x10" @ CENTER BEAM												
@ KITCHEN	2	12.5			25	LF					25	LF
1"x8" @ CENTER BEAM												
@ KITCHEN	1	12.5			13	LF					13	LF

Figure 9.19 (continued)

Means Forms

QUANTITY SHEET

SHEET NO. 6-18/25

PROJECT SAMPLE PROJECT

ESTIMATE NO. 1

LOCATION

ARCHITECT HOME PLANNERS

DATE

TAKE OFF BY WJD

EXTENSIONS BY: WJD

CHECKED BY: KF

DESCRIPTION	NO.	DIMENSIONS			UNIT	UNIT	UNIT	UNIT
1"X4" @ CEILING BEAMS								
@ KITCHEN	4	9			36 LF			36 LF
1"X6" @ CEILING BEAMS								
@ KITCHEN	8	9			72 LF			72 LF
3/4"x 1 3/4" COVE								
MOLDING @ KITCHEN								
CEILING BEAMS	6	25			150 LF			150 LF
2"X3" BLOCKING FOR								
SOFFIT OVER CABINETS								
@ KITCHEN (24" O.C.)	4	21.5			86 LF			
	28	2			56 LF			
					142 LF			142 LF
6"X 10"X 6'-0" SOLID								
WOOD MANTEL @								
KITCHEN FIREPLACE	1	6			6 LF			6 LF
6"X 8"X12" SOLID								
WOOD BRACKET	2	EA			2 EA			2 EA
MORGAN M-1455A								
MANTEL @ LIVING								
ROOM FIREPLACE	1	EA			1 EA			1 EA
3/4"X 3 1/2" COLONIAL								
CASING @ DOORS	23	17'			391 LF			391 LF
3/4"X 3 1/2" COLONIAL	9	5			45 LF			
CASING @ SLIDERS	26	7			182 LF			
BIFOLDS & CASED OPNG.	1	19			19 LF			
					246 LF			246 LF
1"X5" @ CASED OPNG								
& BIFOLD JAMBS	1	126			126 LF			126 LF

Figure 9.19 (continued)

Means Forms

QUANTITY SHEET

PROJECT: SAMPLE PROJECT | SHEET NO. 6-19/25
ESTIMATE NO. 1
LOCATION: | ARCHITECT: HOME PLANNERS | DATE:
TAKE OFF BY: WJD | EXTENSIONS BY: WJD | CHECKED BY: KF

DESCRIPTION	NO.	DIMENSIONS				UNIT		UNIT		UNIT		UNIT
3/4" x 3 1/2" APRON												
@ WINDOWS & SHELVES	1	10			10	LF						
	1	12			12							
	4	3.5			14							
	3	6.5			20							
					56	LF					56	LF
3/4" x 4 3/4" STOOLS												
@ WINDOWS & SHELVES	1	10			10							
	1	12			12							
	4	3.5			14							
	3	6.5			20							
					56	LF					56	LF
3/4" x 3 1/2" CASING @												
DOUBLE HUNG WINDOW	3	18			54							
	12	7			84							
	2	12			24							
	3	7			21							
	6	5			30							
					213	LF					213	LF
7/16" x 1 1/4" STOPS												
@ WINDOW JAMBS	15	7			105							
	6	5			30							
	6	6			36							
					171	LF					171	LF
7/16" x 3 1/4" STOPS @												
WINDOW HEADS	3	3			9							
	2	12			24							
	3	7			21							
					54	LF					54	LF
3/4" AB PLYWOOD WITH												
3/4" x 3/4" EDGE												
BAND @ FOYER												
SHELVES	4	1	3		4	EA					4	EA
@ LINEN CLOSET	5	2	4		5	EA					5	EA
@ CLOSETS	2	1	4.5		2	EA					2	EA
	2	1	4.67		2	EA					2	EA
	1	1	3.25		1	EA					1	EA
1 1/2" DIAMETER CLOSET	2	4.5			9	LF						
POLE WITH BRACKETS	2	4.67			10	LF						
	1	3.25			4	LF					23	LF

Figure 9.19 (continued)

Means Forms

QUANTITY SHEET

PROJECT SAMPLE PROJECT

SHEET NO. 6-20/25

ESTIMATE NO. 1

LOCATION

ARCHITECT HOME PLANNERS

DATE

TAKE OFF BY WJD

EXTENSIONS BY: WJD

CHECKED BY: KF

DESCRIPTION	NO.	DIMENSIONS				UNIT		UNIT		UNIT		UNIT
1 1/2" DIAMETER HANDRAIL @ STAIR TO BASEMENT (ONE SIDE) WITH BRACKETS @ 32" O.C.	1	13			13	LF					13	LF
MORGAN #834 SUBRAIL @ STAIR	1	7.5			8							
	1	1.25			2						10	LF
					10	LF						
MORGAN #M-835 FILLET	1	7.5			8							
	1	1.25			2							
					10	LF					10	LF
MORGAN #M-821 RAIL	1	7.5			8							
	1	1.25			2							
					10	LF					10	LF
MORGAN #905-A ROSETTES	2	EA			2	EA						
MORGAN #M-770 NEWEL POSTS												
@ STARTING	1	EA			1	EA					1	EA
@ TOP LANDING	1	EA			1	EA					1	EA
MORGAN #M-836 BALUSTERS	16	3			16	EA					16	EA
MORGAN #M821 WALL RAIL W/ BRACKETS (2)	1	6			6	LF					6	LF
OAK STAIR TREADS @ MAIN STAIR 5/4" x 10 1/2"	15	3			15	EA					15	EA
OAK STAIR RETURN NOSING (5/4" x 10 1/2")	8				8	EA					8	EA
OAK TREAD COVE MOLDING (13/16" x 5/8")	8	1			8							
	15	3			45							
					53	LF					53	LF

Figure 9.19 (continued)

Means Forms

QUANTITY SHEET

PROJECT: SAMPLE PROJECT

SHEET NO. 6-21/25

ESTIMATE NO. 1

LOCATION

ARCHITECT: HOME PLANNERS

DATE

TAKE OFF BY: WJD

EXTENSIONS BY: WJD

CHECKED BY: KF

DESCRIPTION	NO.	DIMENSIONS				UNIT		UNIT		UNIT		UNIT
1"x10" OAK RISERS	16	3			16	EA					16	EA
1"x10" OAK CHEEK & SKIRT BOARDS	1	16			16	LF						
	1	8			8	LF						
					24	LF					24	LF
1x12 CHEEK & SKIRT BOARDS @ BASEMENT STAIR	2	14			28	LF					28	LF
5/4"x10 1/2" HARD PINE @ BASEMT STAIR TREADS	13	3			13	EA					13	EA
1"x10" PINE RISERS @ BASEMENT STAIR	14	3			14	EA					14	EA

Figure 9.19 (continued)

Means Forms

QUANTITY SHEET

SHEET NO. 6-22/25

PROJECT SAMPLE PROJECT

ESTIMATE NO. 1

LOCATION

ARCHITECT HOME PLANNERS

DATE

TAKE OFF BY WJD

EXTENSIONS BY: WJD

CHECKED BY: KF

DESCRIPTION	NO.	DIMENSIONS			UNIT	UNIT	UNIT	UNIT
KITCHEN CABINETRY & COUNTERS								
CABINETRY TO BE "ACME" DELUXE SERIES IN OAK								
WALL UNITS (12" TYPICAL)								
36"x 15" OVER REFR.	1				1 EA			1 EA
24"x 30" LEFT & RIGHT OF RANGE (SINGLE DR)	2				2 EA			2 EA
27"x 18"x 24" DEEP OVER PANTRY & BROOM CLOSET	2				2 EA			2 EA
15"x 30" BETWEEN BROOM CLOSET & PANTRY	3				3 EA			3 EA
1½"x 30" FILLERS @ EA. SIDE OF 15"x30"	2				2 EA			2 EA
BASE UNITS (24" TYP)								
27" BASE W/ 2 DOORS & 2 DRAWER L&R OF RANGE	2				2 EA			2 EA
15" BASE W/ DRAWER BETWEEN PANTRY & BROOM CLOSET	3				3 EA			3 EA
36" SINK BASE	1				1 EA			1 EA
24" BASE W/ DRAWER @ ISLAND	1				1 EA			1 EA

Figure 9.19 (continued)

Means Forms

QUANTITY SHEET

SHEET NO. 6-23/25

PROJECT SAMPLE PROJECT

ESTIMATE NO. 1

LOCATION

ARCHITECT HOME PLANNERS

DATE

TAKE OFF BY WJD

EXTENSIONS BY: WJD

CHECKED BY: KF

DESCRIPTION	NO.	DIMENSIONS				UNIT		UNIT		UNIT		UNIT
42" x 12" DEEP BASE UNIT	2				2	EA					2	EA
36" x 3/4" FINISHED END PANELS @ ISLAND	2				2	EA					2	EA
27" x 24" DEEP x 66" HIGH PANTRY UNIT	1				1	EA					1	EA
27" x 24" DEEP x 66" HIGH BROOM CLOSET	1				1	EA					1	EA
1 1/2" x 34 1/2" FILLER STRIPS @ BROOM CLOSET & PANTRY BASE	2				2	EA					2	EA
COUNTERTOP												
SQUARE EDGED PLASTIC LAMINATE COUNTER TOP (1 1/2" THICK)												
@ ISLAND 87 1/2" x 38" ALL FINISH EDGES	1	7.29	3.16		23	SF					23	SF
@ L & R OF RANGE 27" x 25" w/ 3 EDGES FINISHED	2	2.25	2.1		10	SF					10	SF
BETWEEN BROOM CLOSET & PANTRY 48" x 25" W/ 1 EDGE FINISHED	1	4	2.1		9	SF					9	SF
4" MATCHING BACKSPLASH												
@ RANGE BOTH END FINISH	2	2.25			2	EA					2	EA
@ B.C. & PANTRY (PLAIN END)	1	4			1	EA					1	EA
@ " " RETURNS (FINISH ONE END)	2	2.1			2	EA					2	EA

Figure 9.19 (continued)

Means Forms

QUANTITY SHEET

SHEET NO. 6-24/25

PROJECT SAMPLE PROJECT

ESTIMATE NO. 1

LOCATION

ARCHITECT HOME PLANNERS

DATE

TAKE OFF BY WJD

EXTENSIONS BY: WJD

CHECKED BY: KF

DESCRIPTION	NO.	DIMENSIONS				UNIT		UNIT		UNIT		UNIT
CUT OUT FOR SINK												
@ ISLAND	1				1	EA					1	EA
1" ⌀ BRASS KNOBS @												
CABINETRY	37				37	EA					37	EA
BATH VANITIES												
36" x 22" VANITY @												
POWDER ROOM												
"ACME" DELUXE IN OAK	1				1	EA					1	EA
36" x 22" VANITY @												
COMMON BATH "ACME"												
DELUXE W/ WHITE FINISH	1				1	EA					1	EA
24" x 22" VANITY @												
MASTER BATH "ACME"												
DELUXE W/ WHITE												
FINISH	1				1	EA					1	EA
VANITY TOPS												
3'-3" x 23" PLASTIC												
LAMINATE TOP (1 1/2")	1	3.25	1.92		7	SF					7	SF
@ POWDER ROOM												
3'-1" x 23" P.L. TOP @												
COMMON BATH	1	3.1	1.92		6	SF					6	SF
2'-1" x 23" PL TOP @												
MASTER BATH	1	2.1	1.92		4	SF					4	SF
4" BACKSPLASH TO MATCH												
@ POWDER RM (BACK)												
@ " " RETURN	1	3.25			1	EA					1	EA
(ONE FINISH END EACH)	2	1.92			2	EA					2	EA

Figure 9.19 (continued)

Means Forms

QUANTITY SHEET

PROJECT: SAMPLE PROJECT

SHEET NO. 6-25/25

ESTIMATE NO. 1

LOCATION

ARCHITECT: HOME PLANNERS

DATE

TAKE OFF BY: WJD

EXTENSIONS BY: WJD

CHECKED BY: KF

DESCRIPTION	NO.	DIMENSIONS				UNIT		UNIT		UNIT		UNIT
4" BACKSPLASH @												
COMMON BATH												
@ BACK w/ 1 FIN. END	1	3.1			1	EA					1	EA
@ RETURN w/ 1 FIN. END	1	1.92			1	EA					1	EA
@ MASTER BATH												
@ BACK w/ 1 FIN. END	1	2.1			1	EA					1	EA
@ RETURN w/ 1 FIN. END	1	1.92			1	EA					1	EA
CUT OUT FOR VANITY												
SINKS	3				3	EA					3	EA

Figure 9.19 (continued)

Chapter Ten

THERMAL AND MOISTURE PROTECTION

(Division 7)

Chapter Ten

THERMAL AND MOISTURE PROTECTION

(Division 7)

Division 7 includes the classifications of work that shield the structure from the elements. *Waterproofing* and *dampproofing* include the application of coatings, both below and above grade, to prevent the migration of moisture through the structure. *Insulation* includes the installation of materials to reduce the transmission of heat through the exterior envelope. *Roofing* concerns materials and methods that provide a watertight surface for the uppermost surface of the structure. In addition to the actual roofing work, metal flashings are used to tie the roofing surfaces to adjacent or abutting surfaces. *Siding* provides protection from the elements on the vertical surfaces of the structure. To provide a water- and moisture-tight envelope, caulking and sealants are also used to fill in spaces and seal surface areas. Miscellaneous items such as skylights, roof accessories, gutters, and downspouts are also part of Division 7.

Most of the work of Division 7 can be found on the architectural drawings. The contractor should pay close attention to the roofing plan for the location of roofing materials. Details and sections of the roof system should also be studied for flashing and sheet metal details, insulation at the roof envelope, and roof accessories. Wall sections and details provide information concerning the thermal insulation and vapor barrier at the exterior walls. Cross-sections indicate the insulation materials at the floor level. Wall and foundation sections and details indicate surfaces to be waterproofed, dampproofed, or insulated, often below grade. The specifications should be studied thoroughly to determine the appropriate products and methods of installation.

Waterproofing

The purpose of waterproofing is to prevent the penetration of water through the exterior surfaces of the structure. Waterproofing is most often done below grade at the foundation level to prevent the transmission of water through the foundation walls or slab. When reading a set of drawings care should be taken to determine the limits of waterproofing materials.

There are three basic classifications for waterproofing. The *integral* method involves the use of special additives mixed with concrete for use in poured foundations. *Membrane waterproofing* is the application of a waterproof membrane to the surface of the protected area. Finally, *metallic waterproofing* involves the use of a fine iron powder mixed with oxidizing agents and applied to the surface area to be protected. The compound fills the pores of the concrete or masonry and, as the iron particles rust, they expand to form an impervious barrier to the migration of water.

Again, the specifications should be thoroughly reviewed for methods and products to be used.

Units for Takeoff

The standard units for takeoff for both the membrane and metallic methods are SF, which can be extended to the square (100 SF). The SF area to be protected is determined by the amount of surface area that will be below grade. For foundation applications, this is the perimeter of the foundation multiplied by the distance from the finish grade at the exterior to the bottom of the footing.

For the integral method, the quantity of admixture to be added to the concrete mix depends on the manufacturer's recommendations for the product used. For takeoff purposes, this is often reflected as an additional cost for the concrete.

Other Takeoff Considerations

Quantities for each method should be listed separately in the takeoff. The thickness or layers specified for both the metallic and membrane methods should be noted, as this has a major effect on the cost of the process. The contractor should also note any required surface preparation such as cleaning or roughing the surface.

Dampproofing

Dampproofing is typically applied to the foundation area below grade. It is not intended to resist water pressure and should not be confused with waterproofing. Dampproofing is used to prevent the penetration of moisture through the foundation wall in below-grade applications.

The most common methods of dampproofing include the application of a bituminous-based tar-like coating on the protected areas. The bituminous materials can be sprayed, painted, or troweled on, to provide a uniform coverage of the areas below grade. Another method involves the plastering of a cement-based mixture, called *parging*, over the surface area to be protected.

Units for Takeoff

The standard units for takeoff for all types of dampproofing are SF, which can be extended to squares (100 SF). The procedures for determining the areas are the same as those discussed in the previous section for waterproofing. Quantities for bituminous applications should be separated according to method of application, as this will affect the labor cost. Parging should be listed in the takeoff by the thickness and number of coats. Standard thicknesses range from 1/4" to 1".

Other Takeoff Considerations

The contractor should include in the takeoff any special preparations such as cleaning or roughing the surface area, as noted in the specifications or required by the manufacturer. Quantities for surface preparation are listed by the SF and should be noted as part of the takeoff. Certain dampproofing products may require the application of a primer coat prior to the actual dampproofing application.

Insulation and Vapor Barriers

Insulation can be defined as the use of materials to reduce the transmission of heat, cold, or sound. In residential light commercial construction, insulation is used to resist heat loss to the exterior of the structure or to unheated areas within the structure. Insulation materials are rated for their thermal resistance, and are expressed as the material's *R-value*. The higher the R-value, the more effective the insulation.

Insulation materials are available in a variety of forms, sizes, thicknesses, and R-values. In wood and light-gauge metal frame construction, blankets or rolls of insulation, referred to as *batts*, are installed between the studs or joists at the exterior walls, roof, or floor of the structure. Batt insulation is available with an attached paper or foil facing that acts as a *vapor barrier*. Vapor barriers prevent the condensation of moisture, caused by high relative humidities within the building and low exterior temperatures. Other vapor barriers, such as *polyethylene sheeting*, can be used. Polyethylene is a plastic film material available in a variety of thicknesses designated by mils (see Chapter 6), the most common of which are 4- and 6-mil thicknesses. Vapor barriers are typically applied at the warm side of the wall, floor, roof, or ceiling construction.

Rigid forms of insulation are used in applications that require a rigid shape to be maintained, such as under concrete slabs, at the exterior of foundation walls, for roof decks, and over the exterior sheathings under the siding materials. Rigid insulation is available in a variety of sizes, shapes, R-values, and compositions. Some of the more common types include expanded polystyrene, often called "beadboard," extruded polystyrenes such as Styrofoam™, and urethane insulation.

Loose fill insulation in the form of mineral wool (a form of molten rock) and expanded volcanic rock materials, such as perlite and vermiculite insulation, are often used to insulate cavity spaces, cells of concrete blocks, and attic spaces. These can be installed by spraying, pouring by hand, or machine blowing.

Insulation requirements are found on building cross-sections, wall sections, and details, particularly on the sections through the exterior envelope of the structure. Figure 10.1 shows a section and various details illustrating the location of insulation.

Units for Takeoff

Batt or roll insulation is taken off and listed by the SF, and should be listed separately according to the width of the space it will occupy. For wood or metal framing, the width is the space between each stud or joist. For example, wood studs spaced at 16″ on center require 15″-wide material. Materials should also be separated by varying R-values and vapor barrier facings, such as kraft paper and foil facing.

Rigid insulation is taken off according to the type of material used, application, method of installation, R-value, and size, and is expressed as SF. The contractor may elect to convert the SF to sheets of the individual product based on the size needed.

Loose fill insulation is taken off by the SF per inch of thickness, and can be extended to cubic feet. This is done by multiplying the area to be covered (in SF) by the depth of the material (in feet).

Polyethylene vapor barriers are taken off by the SF and separated according to thickness in mils.

Calculating the areas for various types of insulation involves becoming familiar with the details showing insulation and cross-referencing the plan

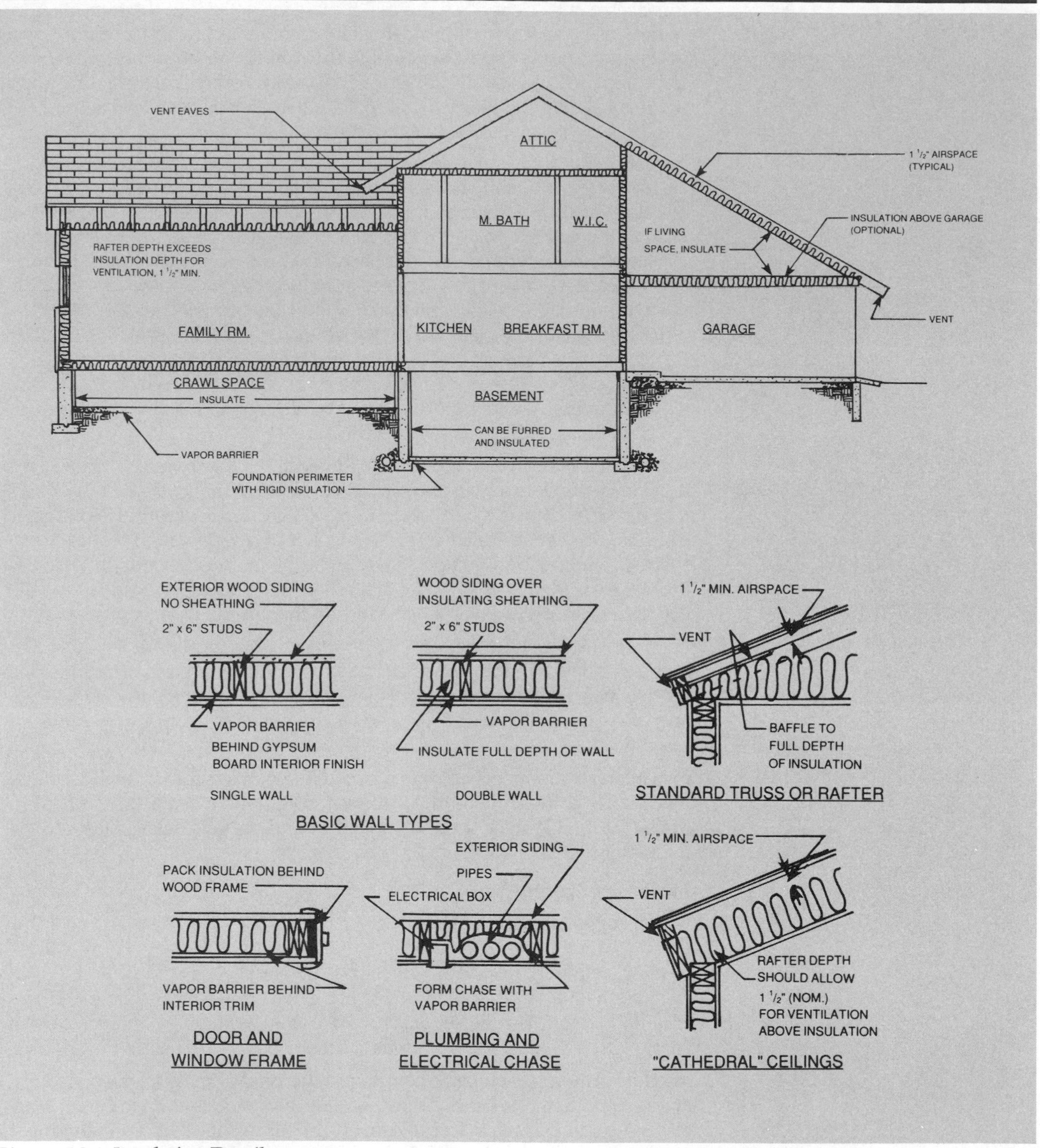

Figure 10.1 Insulation Details

views and elevations for dimensions of the space to be insulated. In addition, the contractor must study the specifications for the type of insulation required.

Other Takeoff Considerations

The contractor should study the architectural drawings carefully to determine the thickness of the insulating materials. Most insulation is specified by the material's R-value, which may translate to a thickness greater than the depth of the available space. This commonly occurs in the framing members (rafters) of cathedral ceilings. Special high-density batt insulation is available that provides a higher R-value per inch of thickness. Where needed, this should be noted on the takeoff, as it is considerably more expensive.

Many roofing systems include insulation as part of the roofing specifications. The contractor should be careful not to duplicate quantities of roof insulation that are considered part of the roof system.

Air Infiltration Barriers

Air infiltration barriers provide resistance to drafts caused by wind and water penetration. Air barriers are installed at the exterior of the sidewall sheathing under the exterior siding. There are a variety of products available, such as TYVEK® Housewrap. The efficiency of the building insulation is influenced by humidity (moisture in the air) and the exterior temperature. The best possible scenario is to have a drywall cavity free from moisture; however, the building needs to "breathe". Today, many designers opt for synthetic wraps such as TYVEK® Housewrap, manufactured by Dupont. TYVEK® is composed of high-density polyethylene fibers bonded together under heat and pressure. Less sophisticated products, such as asphalt felt paper, are also used.

The contractor should refer to the architectural wall sections for the location, and the exterior elevations for the limits, of air infiltration barriers. Specifications should be reviewed for the particular product(s) required.

Units for Takeoff

Air infiltration barriers are taken off by the SF. This is done by measuring the SF area of the exterior wall sheathing to be covered. The contractor may convert the SF area to the roll size, according to the specific manufacturer, or as dictated in the specifications.

Other Takeoff Considerations

Deductions for window or door openings are usually not taken into account when quantifying air infiltration barriers. The contractor should allow additional material for overlaps as required by the specific manufacturer.

Siding

Exterior siding is available in a variety of materials, most commonly wood, metal, and vinyl. Wood siding is manufactured in bevel, shiplap, and tongue-and-groove patterns to be installed horizontally as exterior cladding. Bevel siding, commonly referred to as *clapboard*, is installed with exposures ranging from 3″ to 6″ (see Figure 10.2). *Exposure* refers to the portion of the clapboard that is exposed to the weather. Clapboard is available in lengths ranging from 3′ to 20′ in 1′-0″ increments. Cedar is the most common wood for clapboard siding, but redwood, pine, fir, and spruce siding are also manufactured.

Other types of wood siding include shakes and shingles installed at the exterior sidewall with typical exposures from 4″ to 9″. Shakes and shingles range in length from 12″ to 24″ long, and can be from 2-1/2″ to 14″ in random widths or sawn to widths of 4″, 5″, or 6″. Shakes and shingles are most often red or white cedar, and vary in grade and price. Figure 10.3 illustrates cedar shingle siding.

Alternate forms of siding meant to resemble wood clapboard are also available, manufactured from metals such as thin-gauge steel and aluminum, and vinyl siding composed of extruded polyvinyl chloride (PVC). Most of this siding is installed horizontally with preset exposures from 4″ to 8″. Most panels are approximately 8″ to 9″ in height and 12′ to 12′-6″ in length. Finishes range from smooth to wood-grain simulated textures, in a variety of colors. Preformed siding requires accessories for proper installation that must be included as part of the takeoff. Accessories include inside and outside corners, "J" channels for termination of siding, horizontal starter strips, and finish trims. In addition, soffit and fascia pieces and trims are also manufactured for a complete installation. Figure 10.4 illustrates preformed siding and its various components.

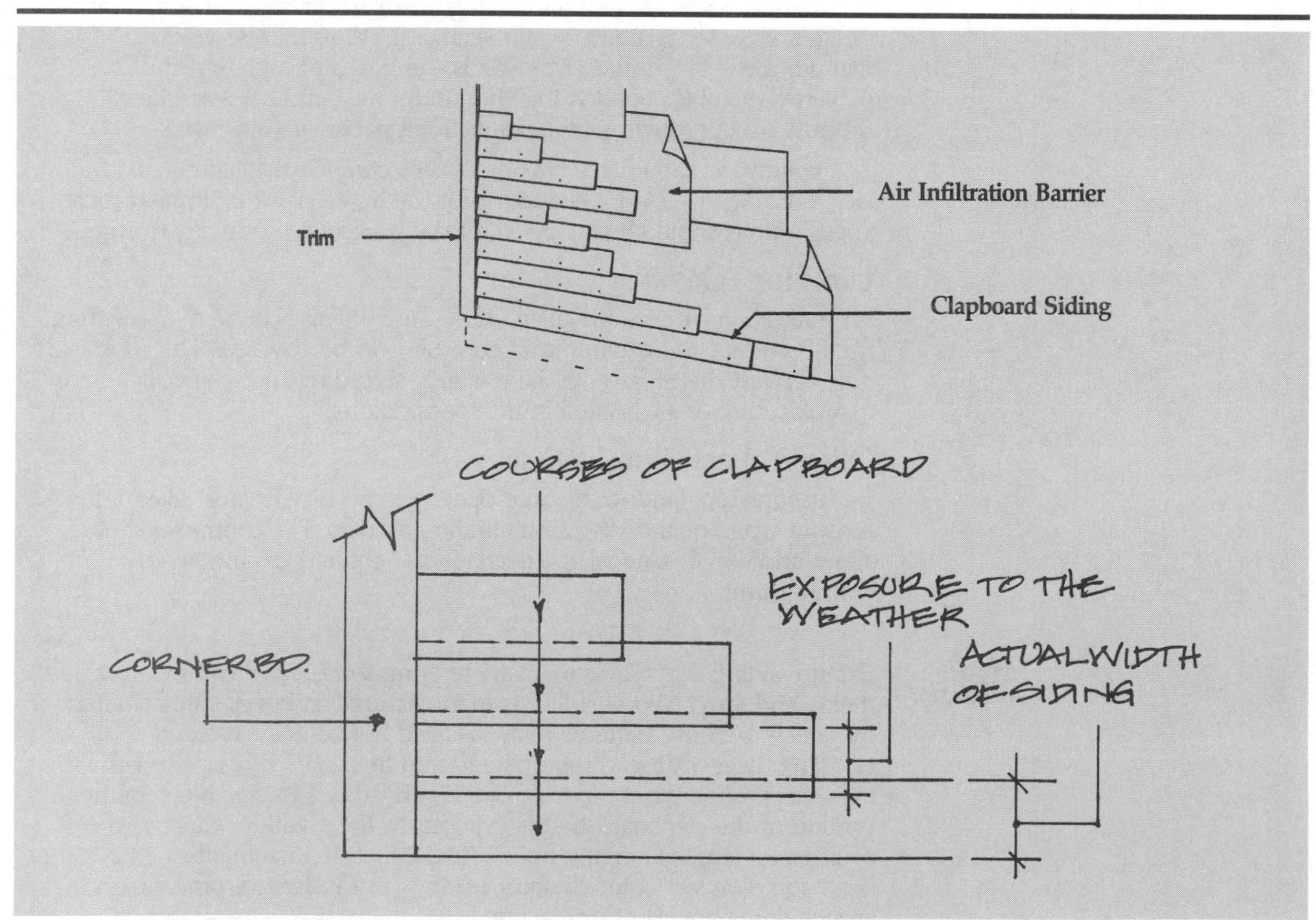

Figure 10.2 Clapboard Siding

Custom shapes and architectural details like cornice trim, dentil moldings, and round and oval shapes are also generally available for complementing any of the manufactured sidings.

Architectural wall sections showing details of the siding should be reviewed to determine its vertical dimensions. Exterior elevations should be used for taking off SF quantities. The specifications must be reviewed for the type, species, and exposure in the case of wood siding, and the manufacturer, model, color, and texture of preformed siding. This information must be noted in the takeoff for accurate pricing of both the product and the installation.

Units for Takeoff

Wood siding, including clapboard, shakes, and shingles, is taken off by the SF and is extended to the square (100 SF). The takeoff should note the

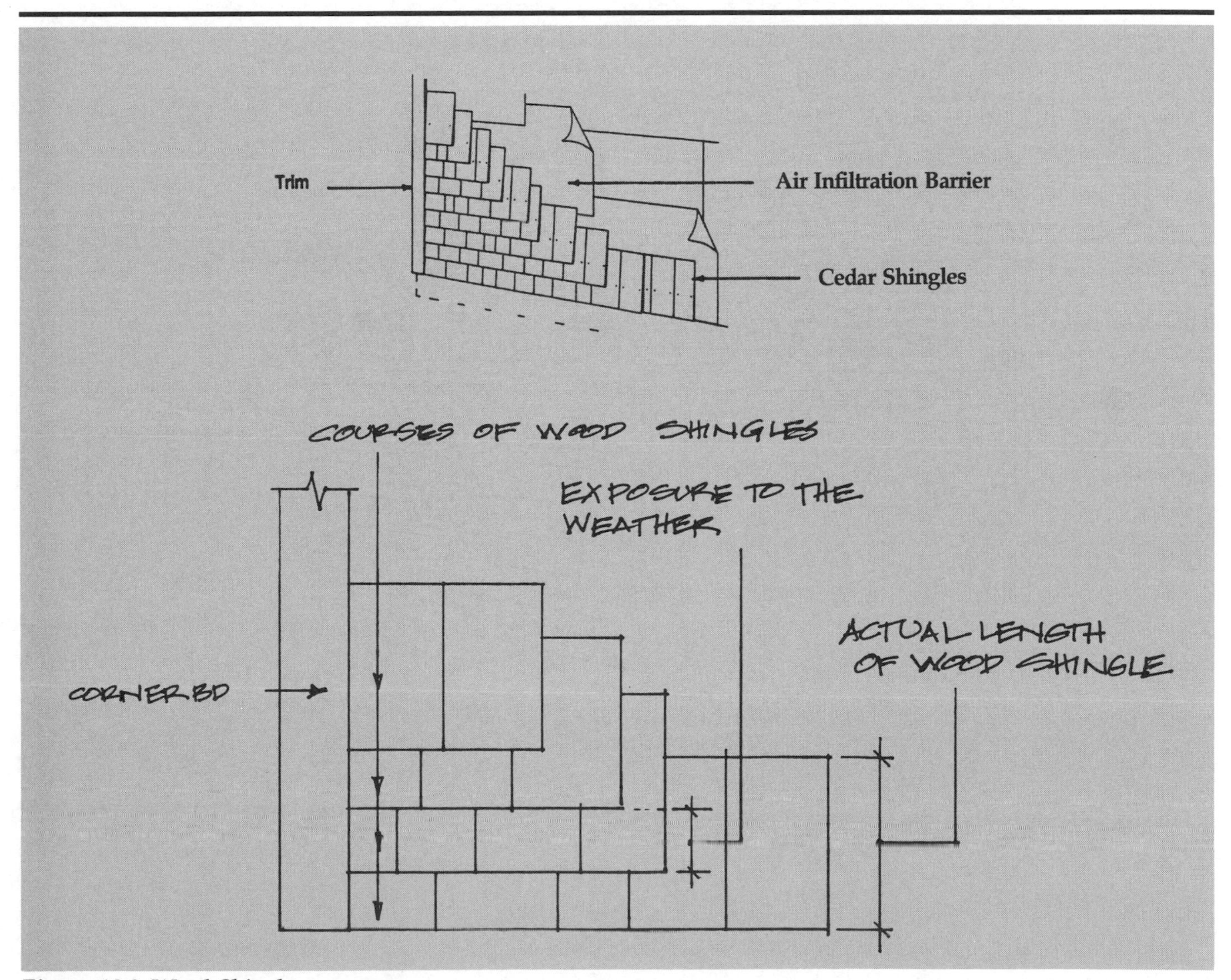

Figure 10.3 Wood Shingle

grade and species of the wood siding, as well as the exposure. Exposure is the surface area of the material that is open to the weather. The actual quantity of siding per square is based on this principle. Changing the exposure from 5" to 4" for white cedar shingles, for example, would mean a 20% increase in material.

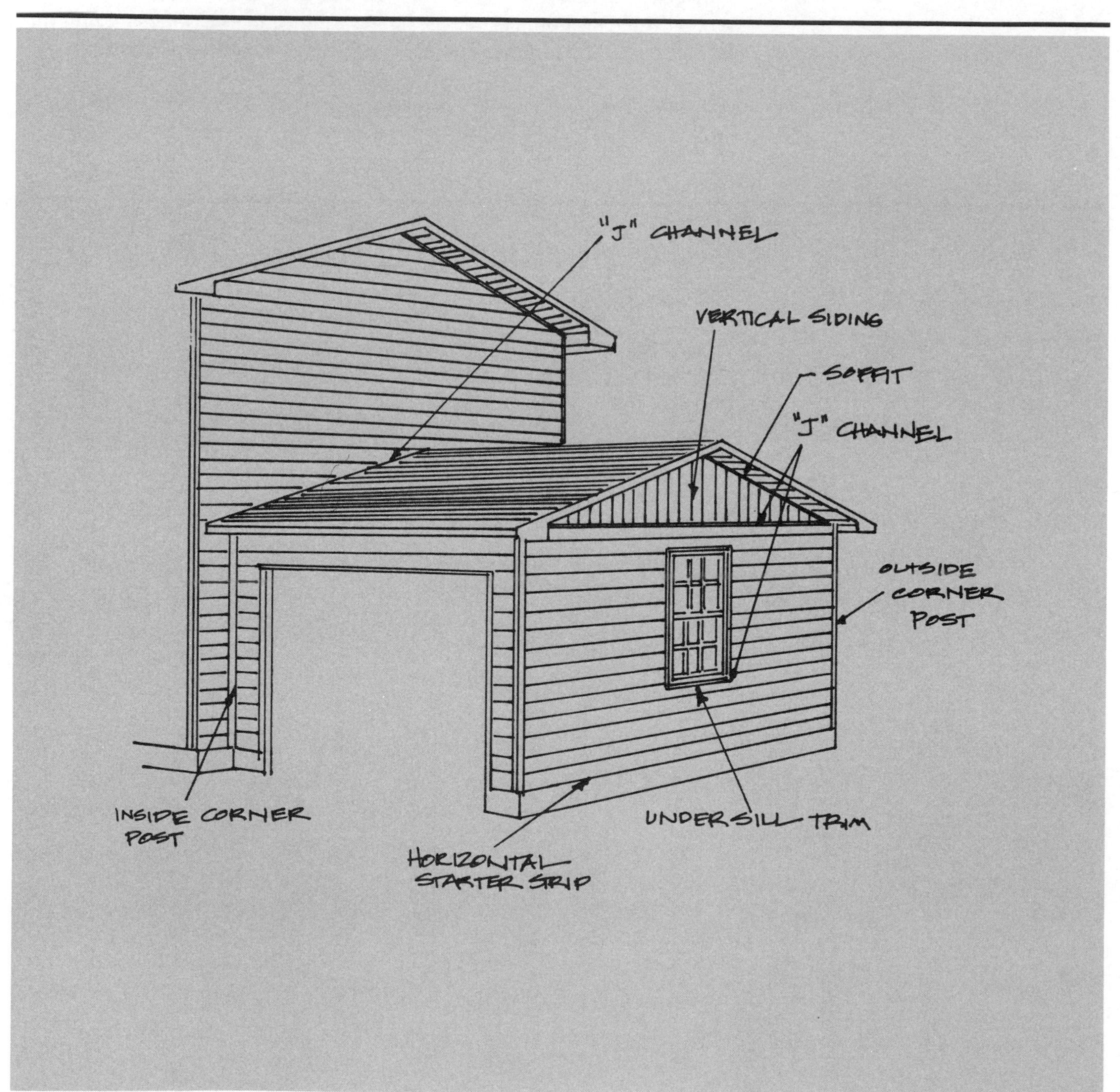

Figure 10.4 Vinyl Siding Trims

Preformed metal and vinyl siding is taken off by the SF and converted to the square (100 SF). In addition, trim pieces must be included in the takeoff. Trims such as starter strips, inside and outside corners, and "J" channels, are taken off by the LF and may be converted to the manufacturer's individual sales length.

Accessories such as soffit and fascia materials are taken off by the SF, with trim pieces such as "F" channels taken off by the LF. Decorative vinyl shutters for windows are taken off by the individual piece according to size, style, and color.

Other Takeoff Considerations

Allowances for waste must be included for wood siding. The contractor must consider the species and grade of the wood and the type of application. Generally speaking, as the grade of product decreases, the amount of waste increases. Applying waste factors to wood siding should be left to the experienced contractor. The final or net quantity of siding materials should include all deductions for openings such as doors or windows. In general, openings smaller than 4 SF are not deducted.

All siding is installed over the exterior sheathing and air infiltration barrier. Wood clapboards or shingles installed without cornerboards require mitering. The length of exterior corners to be mitered should be taken off separately and listed by the LF. This type of application is considerably more labor-intensive, and must be noted for accurate pricing. When an exterior wall intersects a roof there is generally flashing, and custom cutting of the siding material is required. This fitting or scribing should also be calculated in LF of the abutting surfaces.

Roofing

A roofing system consists of many different parts that work together to form a watertight envelope at the exterior of the building. The major components of the roof system are: the insulation (applied to the roof deck), the waterproofing membrane or layer, the protective surfacing, flashings and counterflashings, and metal perimeter termination devices. Figure 10.5 illustrates the standard terminology associated with roofing systems.

The contractor should refer to the architectural drawings, specifically the cross-section, roof plan, and larger scale plans showing the details at the roof line. Specifications should also be studied for the type of roofing system, as well as information on the individual components such as the membrane, insulation, flashings, etc. Figure 10.6 shows a typical roofing plan and details for a shingle roof that may be encountered in a set of residential drawings.

There are four basic types of roof systems for the purpose of takeoff: built-up, single-ply, metal, and shingles and tiles. Each of these types will be discussed in detail in the following sections. A combination of one or more roof systems in one project is not uncommon.

Built-up Roof Systems

A built-up roof (as shown in Figure 10.7) is comprised of three elements: felt, bitumen, and surfacing. The *felts*, which are made of glass, organic, or polyester fibers, serve much the same purpose as reinforcing steel in concrete. The felts are necessary as tensile reinforcement to resist the pulling forces in the roofing material. Felts installed in layer fashion allow more bitumen to be applied to the whole system. *Bitumen*, either coal-tar pitch or asphalt, is the "glue" that holds the felts together. It also serves as waterproofing material. The *surfacings* are smooth gravel or slag, mineral granules, or a mineral-coated cap sheet. Gravel, slag, and mineral granules may be embedded into the still-fluid flood coat. The surfacing materials protect the membrane from mechanical damage. Built-up roof systems can contain two, three, or four plies, or layers, of felt with asphalt bitumen or coal-tar pitch. All systems may be applied to rigid deck insulation or directly to the structural roof deck.

In addition, flashings, counterflashings, metal gravels, stops, and treated wood cants are also needed to complete the system.

Units for Takeoff

The individual components in the built-up roof system are taken off separately. Roof insulation and felts are taken off by the SF and can be extended to the square (100 SF). Bitumen is taken off based on the coverage required in the specifications. This can be listed by the weight (as based on the weight per SF specified by the manufacturer). The mineral surfacing is taken off by the weight specified per SF, and converted to total weight in pounds or tons. Flashings, counterflashings, gravel stops, and treated wood or fiber cants are taken off and listed by the LF. All items should be separated according to type of metal, gauge of metal, and overall size of the flashings.

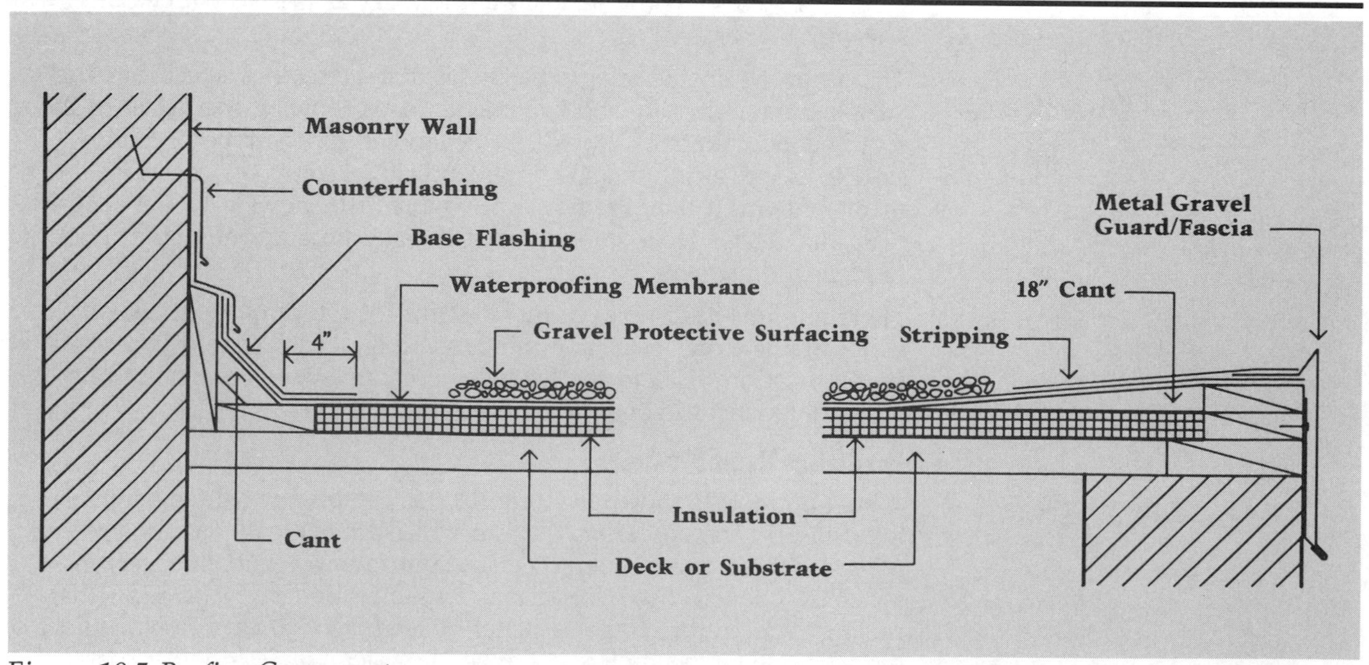

Figure 10.5 Roofing Components

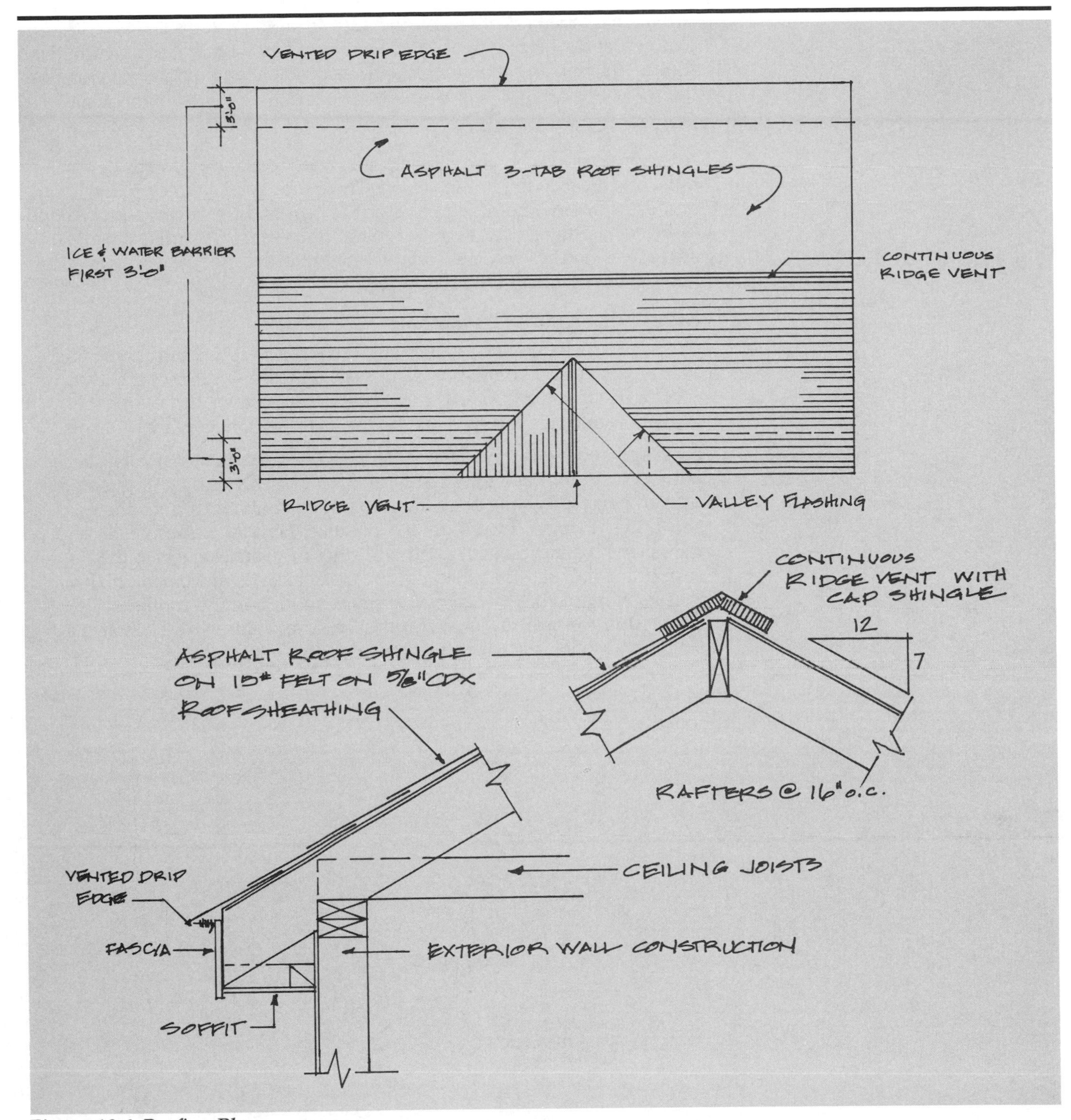

Figure 10.6 Roofing Plan

An alternate method of takeoff includes grouping the insulation, felts, bitumen, and surfacing materials as one unit, based on SF, extended to the square (100 SF). The contractor should select the method that best expresses the roof system.

Other Takeoff Considerations

Additional items such as collars, sleeves, or penetrations through the roof must also be included in the takeoff. These should be counted individually and listed as EA according to size of the item and use. Items supplied and installed by other trades, such as rooftop curbs for HVAC units or smoke and access hatches that are to be flashed into the roof system, may also need to be included in the roof takeoff.

Single-ply Roofing Systems

Single-ply or elastomeric roofing falls into three categories: thermosetting, thermoplastic, and composites. The system includes the use of a PVC or rubber sheet material, called a *membrane*, applied over the top of the insulation, or approved substrate, as the waterproofing surface.

Single-ply roofing can be applied loose-laid and ballasted, partially adhered, and fully adhered. *Loose-laid* systems involve fusing or gluing the side and end lap of the membrane to form a continuous nonadhered sheet, held in place by 1/4" to 1/2" of stone. *Partially adhered* single-ply membrane is attached with a series of strips or plate fasteners to the supporting roof substrate. Because the system allows movement, no ballast is usually required. *Fully adhered* systems are uniformly, continuously adhered to the manufacturer's approved base. Figure 10.8 illustrates a single-ply roofing system and its components.

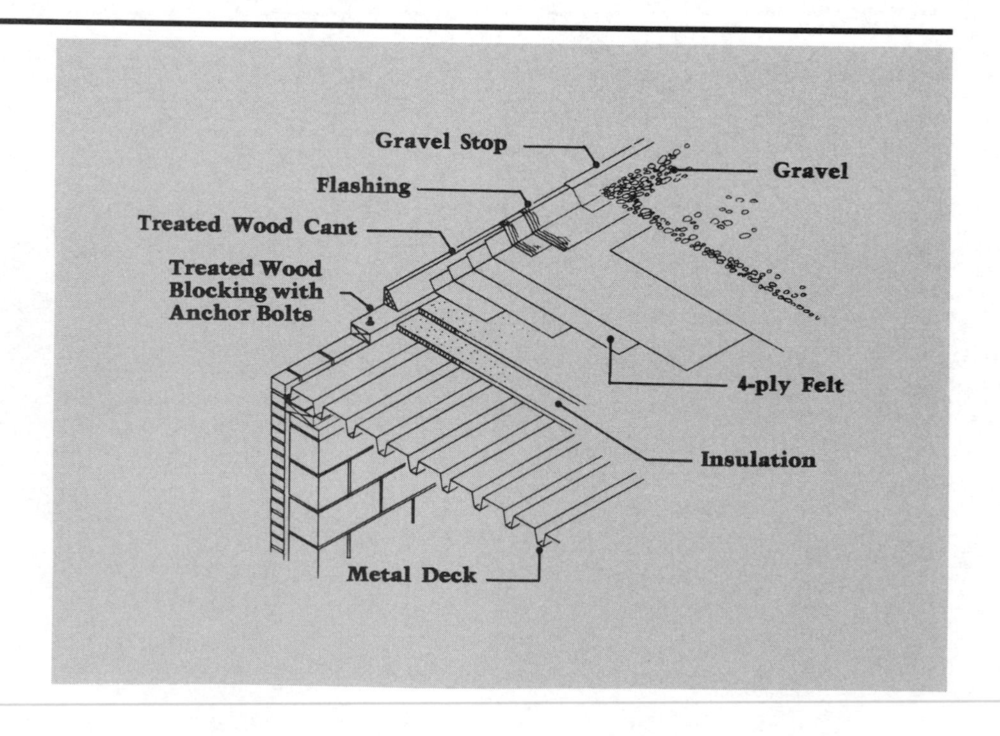

Figure 10.7 Components of a Built-up Roof

Units for Takeoff

The takeoff procedure and units are the same for single-ply roofing as for built-up roof systems. SF can be extended to the square (100 SF). The components of the single-ply system are the insulation, the membrane, a hardboard base (if applicable), and the metal fascia. Metal fascias, flashings, counterflashings, and termination strips are taken off by the LF.

Other Takeoff Considerations

The treated wood nailers shown on both the built-up and single-ply systems are provided for support at the edge of the rigid insulation and for nailing the fascia or gravel stop. This work is typically done by other trades for the roofer. The contractor may refer to the practices of local roofing contractors to determine if this item is typically part of the roofer's work in the locality.

Refer back to "Other Takeoff Considerations" for built-up roof systems for more information.

Metal Roof Systems

For estimating purposes, metal roofing systems may be divided into two groups: *preformed* and *formed* metal. Preformed metal roofs, available in long lengths of varying widths and shapes, are constructed from aluminum, steel, or composition materials such as fiberglass. Aluminum roofs are generally prepainted or can be left natural. Steel roofs are usually galvanized or painted; most manufacturers require the product to be factory-painted in order to warranty the finish of the product.

Preformed metal roofing is installed on sloped roofs according to the manufacturer's recommendation as to the minimum pitch required. Lapped ends may be sealed with a preformed sealant to match the deck

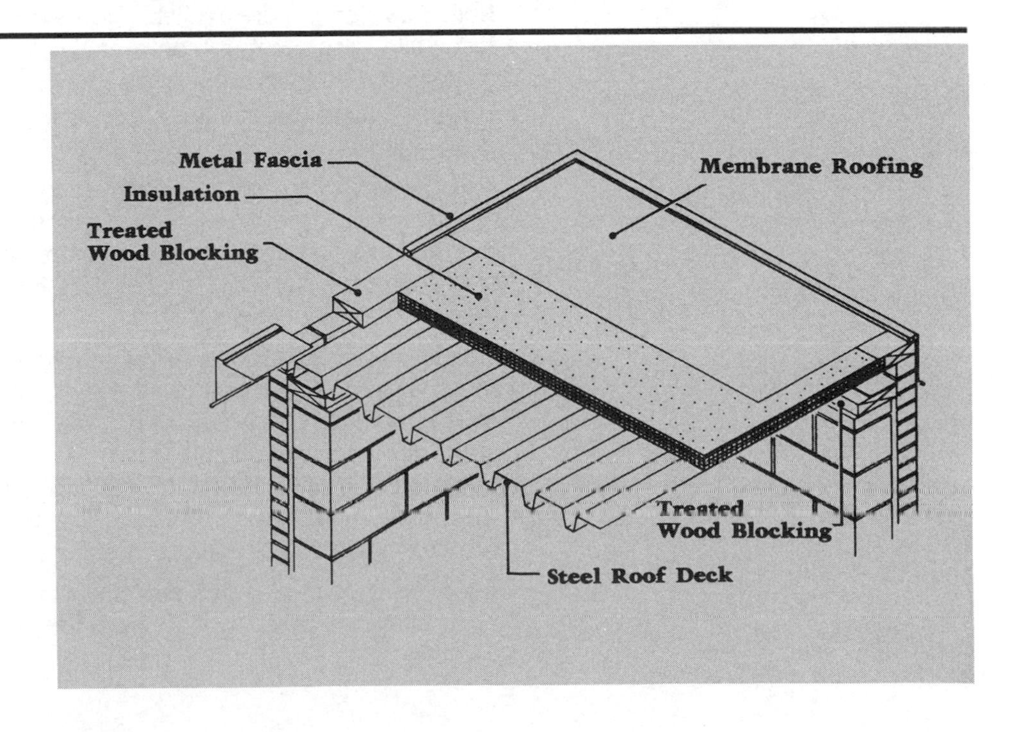

Figure 10.8 Single-ply Roofing

configuration. Preformed roofing is fastened to the supporting members with self-tapping screws with attached neoprene washers to prevent leakage.

Formed metal roofing, typically selected for aesthetic reasons, is installed on sloped roofs that have been covered with a base material such as plywood or concrete. Typical materials include copper, lead, and zinc alloy. Flat sheets are joined by tool-formed batten-seam, flat-seam, and standing-seam joints.

Units for Takeoff

Preformed metal roofing is taken off by the SF, and can be extended to the individual pieces required. Trim pieces and preformed sealant material are taken off by the LF. Quantities can be determined by calculating the SF of roof surface to be covered with the manufacturer's recommended allowances for side and end laps.

Formed metal roofing is taken off by the SF, and can be extended to the square (100 SF). Trim pieces such as battens for the seams and finish end pieces can be taken off by the LF or by the individual pieces required.

Other Takeoff Considerations

Formed metal roofing often requires shop drawings for fabrication, which should be listed in the takeoff.

Takeoff for both types of metal roofing should be separated according to type of metal used, finish of the metal, and particular application.

Roof Shingles and Tiles

Shingles and tiles are popular materials for covering sloped roofs. Both are *watershed* materials, which means they are designed to direct water away from the building by means of the slope or pitch of the roof. Shingles and tiles are used on roofs with a pitch of 3″ or more per foot. Shingles are installed in layers with staggered joints over roofing felt underlayment. Nails or fasteners are concealed by the course above. Shingle materials include wood, asphalt, fiberglass, metal, and masonry tiles. Asphalt and fiberglass shingles are available in a variety of weights and styles, with three-tab being the most common. Figure 10.9 illustrates an asphalt shingle roof system and its components.

Wood shingles may be either shingle or shake grade; cedar is the most common species of wood used. Metal shingles are either aluminum or steel and are generally available prefinished. Slate and clay tiles are available in a variety of shapes, sizes, colors, weights, and textures. These are typically heavier materials and therefore require specialized installation techniques, and a stronger structural roof system to support the added loads imposed by the weight of the tile.

In addition to the shingle or tile materials, special metal trim pieces are required to protect the edge of the roof deck and allow water to drip free of the roof edge. *Drip edge* is a corrosion-resistant metal, typically aluminum; it may be omitted with wood shingles or slates when the edge of the shingle or slate projects beyond the roof edge. Metal flashings for the valleys of shingled and tiled roofs may also be specified. Valley flashing can be lead-coated copper, copper, or zinc alloy.

To provide venting of the attic space, ridge vents are sometimes used. They allow the transfer of air from the attic or rafter space to the outside, to prevent the build-up of moisture along the underside of the roof sheathing. Ridge-venting materials are available in a variety of styles and compositions.

Shingle and tile roof systems require special shingles or tiles, called *cap shingles* at the ridge or hip of the roof. A cap shingle may be a regular three-tab shingle modified for use as a cap, as in the case of asphalt or fiberglass shingles, or a special prefabricated cap (as in the case of some clay tile or metal tile designs).

Special membrane material installed under the first couple of courses, and at the hips and valleys of the shingle or tile roof, is called an *ice/water barrier*. Most of these products are a bitumen-based self-adhering membrane for use in cold climates where ice and water may dam along the eaves and valleys and cause water to back up under the roof shingles.

Units for Takeoff

Shingle and tile roofing is taken off by the SF and extended to the square (100 SF). Underlayment felts are taken off by the SF and can also be converted to squares. Wood shingles and slate or similar tiles are taken off by the SF and can be extended to the square. Special hip and ridge tiles or cap shingles are taken off by the LF or can be converted to the individual piece, as in the case of metal or clay tiles. Drip edge is taken off by the LF, and can be converted to the individual piece and listed as EA. Ridge vent is taken off by the LF and can be converted to the individual piece, listed as EA or roll. Ice/water barriers are taken off by the SF and can be extended to the manufacturer's size roll.

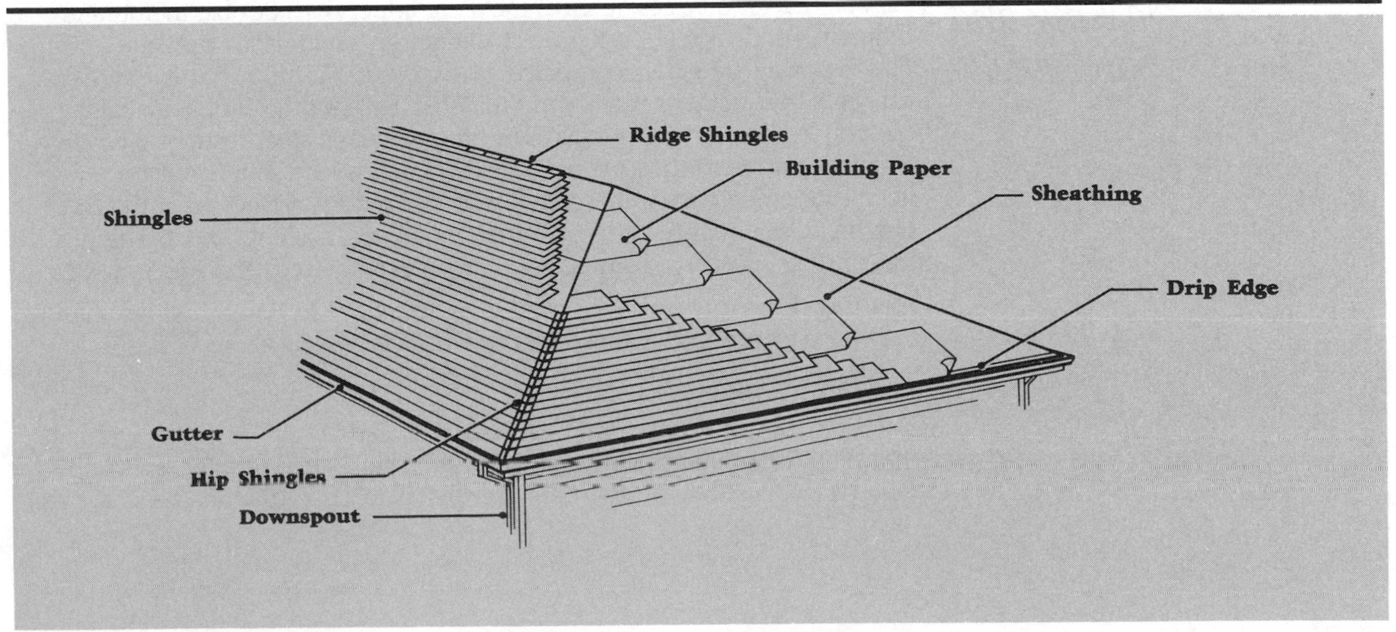

Figure 10.9 Shingled Roof

Other Takeoff Considerations

The contractor should study the specifications for the particular product and the required accessories.

Since slate is a natural stone product, it has a substantial weight per square. Therefore, the takeoff should include the requirement for freight or shipping to the job site. Slate requires the use of nonferrous nails, usually copper in composition. This should be noted as part of the takeoff because of the higher-than-usual cost.

Most shingle or tile roofing requires some form of starter course. For most types of shingle roofs, the starter course is the same as the roofing materials and should be allowed for in the materials takeoff. Some of the more elaborate "architectural" asphalt or fiberglass roof shingles require the use of special shingles for the ridge and hip caps. The LF measure of these shingles can be converted to the number of bundles, based on the individual product.

Ridge vent material is sold in rolls and pieces from 4′ to 10′ in length. Some are sold complete, and others require the use of a cap shingle to provide a watertight installation. The contractor should be familiar with the individual product and its accessories.

The quantity of shingles (in squares) may be converted to bundles, based on the number of bundles per square. Asphalt or fiberglass shingles are typically 3, 4, or 5 bundles per square, depending on the manufacturer and model.

Drip edge is available both vented and non-vented. The contractor should review the drawings and specifications for the location and quantity of each type.

Since shingle and tile roofing is installed on a sloped surface, staging and scaffolding may be required when the pitch of the roof precludes workers from walking on the surface. These items should be taken off and included as part of the work.

Flashing and Sheet Metal

Flashing refers to pieces of sheet metal, or impervious flexible membrane material, used to seal and protect joints in a building to prevent leaks. Flashing may be either concealed or exposed. Flashing that is concealed can be sheet metal or membrane material. Exposed flashing is usually copper, lead-coated copper, aluminum, galvanized steel, lead, or zinc alloy. Flashing can be formed on site, or preformed in the shop prior to installation. Flashing is not solely associated with construction of the roof, but includes such items as window/door head flashing and flashing for masonry applications. Figure 10.10 illustrates some typical flashing details.

Flashing details are often depicted in architectural section drawings. Plan view and elevation drawings may also indicate the locations of flashing.

Units for Takeoff

Flashing is taken off by the LF for materials that are 12″ or less in width, or that are preformed or fabricated to specific shapes and lengths. Flashing materials or stock wider than 12″ may be taken off by the SF. In either case,

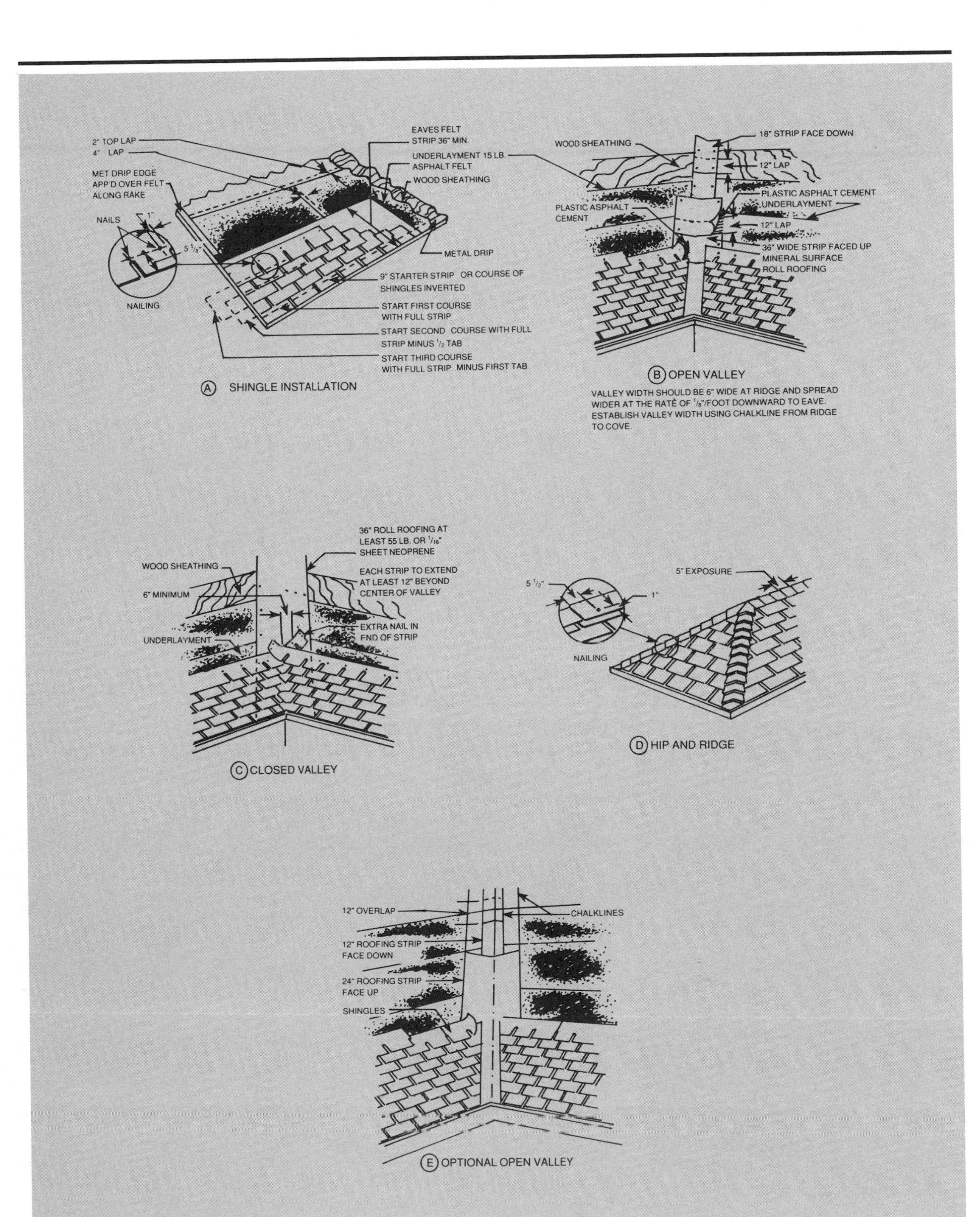

Figure 10.10 Flashing Details

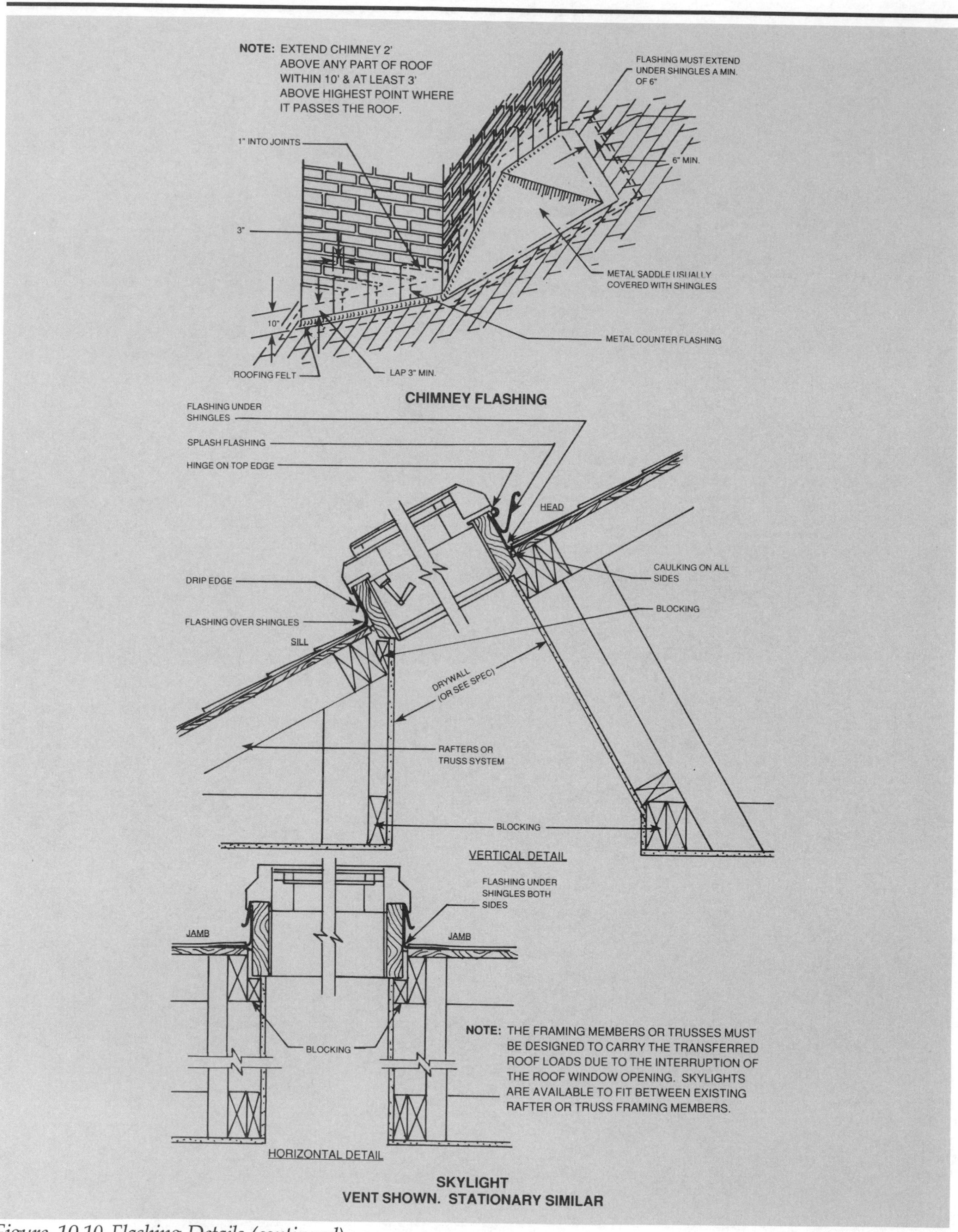

Figure 10.10 Flashing Details (continued)

flashings should be listed in the takeoff according to the metal's type, size, application, gauge or thickness, and any coating or specified prefinishing.

Other Takeoff Considerations

The contractor should be careful not to duplicate flashing that may be considered part of the roof or other building system. Careful review of the specifications and drawings should prevent this problem.

Allowances for the overlap of continuous preformed flashings (as in the case of valleys) should be included in the takeoff. The contractor should review the specifications for the required overlap of individual pieces.

Roof Accessories

Many items provided and installed by other trades are required to be flashed into the roof. Some examples are skylights, roof hatches, pipes and conduits through the roof, and roof curbs for HVAC units. In addition, such items as gutters, downspouts, and expansion joints are considered roof accessories. Figure 10.11 illustrates some typical roof accessories.

Roof accessories can be found on many different drawings. Roof plans and sections should be studied as well as elevations for the locations of such items as skylights, roof hatches, gutters, and downspouts. Mechanical and

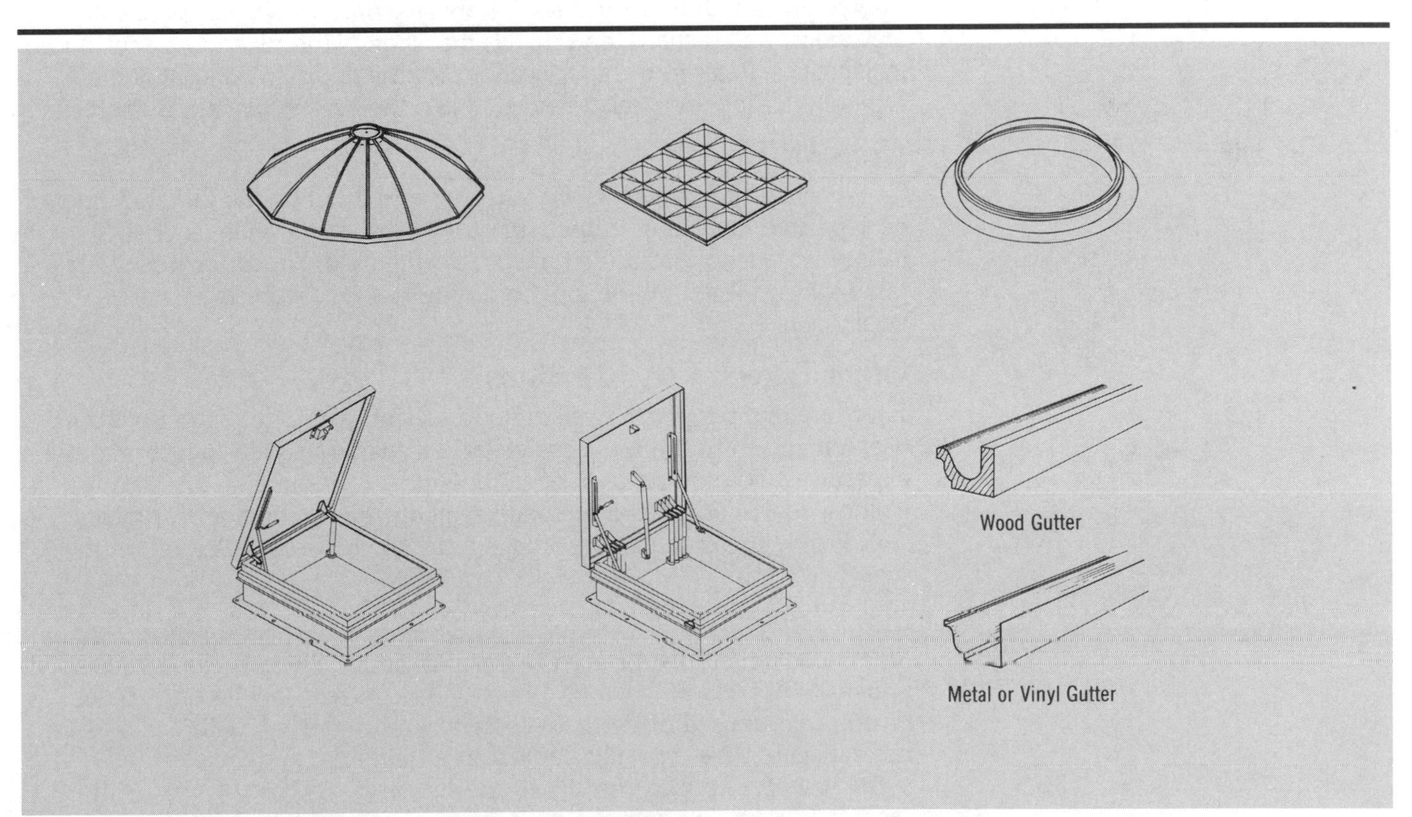

Figure 10.11 Roof Accessories

electrical drawings should be cross-referenced for the quantity and locations of curbs, piping, and conduits.

Units for Takeoff

Gutters and downspouts are taken off by the LF, and should be separated according to composition, size, and application (gutter or downspout), as well as special colors or finishes. The method of gutter installation should be noted for accurate pricing. Simple installations may include gutters and fasteners, while a more complicated system featuring trim accessories and fittings may require an itemized list of components for pricing.

Skylights, roof hatches, flashings for pipes, flashing of roof curbs, and most roof accessories in general are taken off by the individual piece (EA) according to size, function, and any special characteristics that may affect the price of the item.

Caulking and Sealants

Sealants and caulking compounds are used to provide a water, vapor, and air barrier between joints or gaps of adjacent but dissimilar materials. A classic example is the joint between a steel door frame and a masonry wall opening. Caulking and sealants are manufactured for a full range of applications, such as interior or exterior use, ability to expand and contract, service temperature range, paintability, and compatibility with the material to be sealed.

Joint sealants are normally applied over a backup material that controls the depth of the joint. They serve as bond breaks to allow free movement of the joint and prevent water penetration. The backup material, called *backer rod,* is available in a variety of compositions, such as butyl, neoprene, polyethylene, or rubber.

The contractor should refer to the architectural drawings, specifically elevations, wall sections, and details that show window and exterior door installations. Refer to the specifications for the appropriate location and type of caulking and sealants needed and their respective application.

Units for Takeoff

Caulking and joint sealants are taken off and listed by the LF. They should be separated according to the size of the bead, application (such as interior or exterior), and the type of material used. Backup materials are taken off and listed by the foot according to size, composition, and application.

Other Takeoff Considerations

There are many types of caulking and sealants available for construction use, ranging from the inexpensive latex caulking used by painters to the expensive two-part elastomeric compounds. Determining the quantity in gallons, quarts, or tubes quite often requires the use of charts or tables supplied by the manufacturer that specify the LF quantity based on the size of the bead.

Stucco

Stucco is a cement plaster used to cover exterior wall and ceiling surfaces; it is usually applied to a wood, metal lath, concrete, or masonry base. Stucco is comprised of Portland cement, lime, and sand, with water as the mixing agent. It is typically applied in a three-coat system. The *scratch coat* is the first coat applied directly to the substrate. The brown coat is the second coat applied over the scratch coat. The final coat is the *finish coat.*

Stucco can be applied directly to the substrate, as in the case of masonry or concrete walls, or can be applied over a metal lath installed on a wood surface. Stucco provides a durable, weather-resistant surface that is virtually maintenance-free and impervious to moisture. The finish coat can be trowelled in a variety of textures, then painted.

The completed stucco system requires special trim pieces to achieve square and plumb corners at the intersection of wall and ceiling surfaces. Other trims are also needed at the intersection of stucco with other materials, similar to J-bead. Expansion and control joints, used to control cracking, are also required.

Architectural drawings should be reviewed for the location of stucco surfaces, with special attention to exterior elevations, exterior reflected ceiling plans, and sections and details. Specifications should provide the particular mixing proportions of water, Portland cement, sand, and lime to achieve the required strength, as well as the texture and thickness of the finish coat. Specifications also note the type and location of the various trim pieces and metal lath (if applicable).

Units for Takeoff

Stucco is taken off by the SF by measuring the length and width of the various surfaces to be covered. SF can then be converted to SY for pricing. Quantities of metal lath can be determined from the SF area to be covered and converted to the standard-sized sheets of the particular product. Allowances for overlap at the sides and ends of the individual sheets should also be included in the takeoff, based on the manufacturer's recommendations or the specifications. Trims such as inside and outside corner beads, control joints, and J-beads are taken off by the LF and can be converted to the manufacturer's standard unit (typically 8′ or 10′ lengths). The contractor may elect to convert SY of stucco to the individual components of Portland cement, sand, and lime by weight, as dictated by the proportions in the specifications.

Other Takeoff Considerations

Because it uses water as a mixing agent, stucco is subject to freezing, and may require temporary enclosures and heating during its curing period (see "Temporary Protection and Heat" in Chapter 6.) This should be noted in the takeoff.

In addition, temporary staging or scaffolding may also be necessary to gain access to the work (see "Staging and Scaffolding for Masonry Work" in Chapter 7).

Exterior Insulation and Finish System

Exterior insulation and finish system (EIFS), sometimes referred to as *synthetic stucco*, is an exterior siding material that has the durability of stucco and thermal insulation value. It is composed of expanded polystyrene insulation board (see the "Insulation and Vapor Barriers" section earlier in this chapter) and a cementitious base coat applied in varying thicknesses, with a synthetic woven mesh that acts as a reinforcement. The top or finish coat is an acrylic stucco, available in a variety of textures and colors. The insulation is adhered or mechanically fastened to the sidewall substrate of plywood, masonry, or gypsum sheathing. The base coat is troweled on, similar to a conventional stucco system, and the mesh reinforcing is embedded in the base coat. The finish coat is applied after the base coat has had sufficient time to dry.

The contractor should refer to the architectural drawings for sections and details of the system, with attention to the details at the abutting surfaces of dissimilar materials, door frames, windows, and wood surfaces. Control and expansion joints are required as per the manufacturer's recommendations to control expansion and contraction associated with the product. Exterior elevations are used to determine the limits of the EIFS. The specifications should indicate the individual product specified as well as the manufacturer's standard application and installation procedures.

Units for Takeoff

Exterior insulation and finish systems are taken off by the SF. Separation into individual components such as polystyrene insulation, base coat, reinforcing mesh, and finish coat may be required for more accurate pricing, depending on the individual product. All components are taken off by the SF. Base and finish coats may require conversion of the SF quantities to gallons of material needed for the particular thickness specified. Control and expansion joints are taken off by the LF and should be listed separately. Expanded polystyrene insulation of varying thicknesses should be listed according to the thickness (in inches) and the application, such as soffits, fascias, walls, and ceilings. Finish coats with different textures, colors, or thicknesses should also be listed separately.

Other Takeoff Considerations

EIFS includes a wide variety of products that are similar in composition, design, and performance. To avoid mistakes in pricing and takeoff, the contractor should be familiar with the individual product specified.

Since EIFS is subject to freezing during its curing period, the contractor may need to consider provisions for temporary heating or enclosure (see the "Temporary Protection and Heat—Concrete" section in Chapter 6). In addition, staging or scaffolding may be necessary for proper access to the work (see the "Staging and Scaffolding" section in Chapter 7).

Summary

The takeoff sheets in Figure 10.12 are the quantities for the work of Division 7—Thermal and Moisture Protection for the sample project. The takeoff for items such as insulation and beveled wood siding include deductions for openings in the exterior wall. This provides a more accurate quantity of each item for the pricing of labor.

Division 7 Checklist

The following list can be used to ensure that all items for this division have been accounted for.

Waterproofing

- ☐ Integral waterproofing
- ☐ Membrane waterproofing
- ☐ Metallic waterproofing
- ☐ Preparation of substrate

Dampproofing

- ☐ Bituminous dampproofing
- ☐ Parging
- ☐ Priming or preparation of substrate

Insulation and Vapor Barriers

- ☐ Unfaced batt insulation
- ☐ Faced batt insulation
- ☐ Rigid insulation
- ☐ Loose fill insulation
- ☐ Vapor barriers

Air Infiltration Barriers

Siding

- ☐ Wood shingles
- ☐ Wood siding (vertical and horizontal)
- ☐ Vinyl siding (vertical and horizontal)
- ☐ Metal siding (vertical and horizontal)

Roofing

- ☐ Built-up roof systems
- ☐ Single-ply roof systems
- ☐ Metal roof systems
- ☐ Shingle and tile roofs
- ☐ Specialized staging for roof work (roof brackets, planks, etc.)

Roof Accessories

- ☐ Vented and non-vented drip edge
- ☐ Ridge vents
- ☐ Skylights and roof hatches
- ☐ Ice and water barriers
- ☐ Gutters and downspouts

Flashing and Sheet Metal

- ☐ Flashing and counterflashing
- ☐ Valley flashing
- ☐ Window and door flashings
- ☐ Gravel stops
- ☐ Metal fascias
- ☐ Metal panning at wood trims
- ☐ Flashing of work installed by other trades
- ☐ Custom roofs for bay windows, etc.

Caulking and Sealants

- ☐ Latex caulking
- ☐ Sealants
- ☐ Backer rod

Exterior Insulation and Finish System

- ☐ Polystyrene insulation
- ☐ Base coat
- ☐ Reinforcing mesh
- ☐ Finish coat
- ☐ Expansion and control joints

Means Forms

QUANTITY SHEET

PROJECT: SAMPLE PROJECT	SHEET NO. 7-1/5
	ESTIMATE NO. 1
LOCATION:	ARCHITECT: HOME PLANNERS	DATE:
TAKE OFF BY: WJD	EXTENSIONS BY: WJD	CHECKED BY: KF

DESCRIPTION	NO.	DIMENSIONS				UNIT		UNIT		UNIT		UNIT
DIVISION 7: THERMAL & MOISTURE PROTECTION												
WATERPROOFING												
CLEAN & PREP CMU												
WALLS BELOW GRADE												
@ CRAWL SPACE	1	45.33	3		136	SF						
@ MAIN HOUSE FOUND.	1	124	7.33		910	SF						
					1046	SF					1046	SF
ASPHALT WATERPROOFING												
@ CMU (ABOVE)					1046	SF					1046	SF
INSULATION & VAPOR BARRIERS												
4MIL POLYETHELENE												
VAPOR BARRIER @												
CRAWLSPACE	1	13	15.33		200	SF					200	SF
6" R19 BLANKET												
UNFACED INSULATION												
@ STUDY FLOOR	1	14	17		238	SF					238	SF
@ MAIN HOUSE FLOOR	1	32	30		960	SF						
DEDUCT FOR STAIR	1	3	11		(33	SF)						
" " FIREPLACE	1	3.67	11.67		(43	SF)						
					1122	SF					1122	SF
6" R19 KRAFT												
FACED INSULATION @												
EXT. WALLS 1ST FLOOR	1	156	10.1		1576	SF						
DEDUCT FOR OPNGS	3	6	6		(108	SF)						
	2	3	7		(42	SF)						
	3	3	6		(54	SF)						
	4	2	6		(48	SF)						
	2	5	6		(60	SF)						
					1264	SF					1264	SF

Figure 10.12

Means Forms

QUANTITY SHEET

SHEET NO. 7-2/5

PROJECT SAMPLE PROJECT

ESTIMATE NO. 1

LOCATION

ARCHITECT HOME PLANNERS

DATE

TAKE OFF BY WJD

EXTENSIONS BY: WJD

CHECKED BY: KF

DESCRIPTION	NO.	DIMENSIONS				UNIT		UNIT		UNIT		UNIT
6" R19 KRAFT FACED												
INSULATION @ 2ND												
FLR. EXT. WALLS	1	83	8.12		675	SF						
	1	32	9		288	SF						
DEDUCT FOR OPNGS	3	6	5		(90	SF)						
	1	3	7		(21	SF)						
					852	SF					852	SF
9" R-30 UNFACED												
INSULATION @ STUDY												
CEILING	1	21	14		294	SF					294	SF
9" R-30 UNFACED												
INSULATION @ MAIN												
HOUSE CEILING	1	25	32		800	SF						
DEDUCT FOR CHIMNEY	1	2	5.33		(11	SF)						
					789	SF					789	SF
AIR-INFILTRATION BARRIER												
@ EXT. WALLS UNDER												
SIDING	1	14	11		154	SF						
	1	32	13		416	SF						
	1	30	13		390	SF						
	2	16	16		512	SF						
	2	14	4		112	SF						
	1	41	9		369	SF						
	1	27	6		162	SF						
					2115	SF					2115	SF
WOOD SIDING												
BEVEL WOOD SIDING												
W/ 6" EXPOSURE												
@ FRONT	2	2	10		40							
	1	31	11.5		357							
	1	4.5	7		32							
DEDUCT FOR WINDOWS	(1	8.5	3.25)		(28	SF)						
" "	(2	3.25	7.5)		(49	SF)						
					352	SF					352	SF

Figure 10.12 (continued)

Means Forms

QUANTITY SHEET

PROJECT SAMPLE PROJECT			SHEET NO. 7-3/5	
			ESTIMATE NO. 1	
LOCATION	ARCHITECT HOME PLANNERS		DATE	
TAKE OFF BY WJD	EXTENSIONS BY: WJD		CHECKED BY: KF	

DESCRIPTION	NO.	DIMENSIONS				UNIT		UNIT		UNIT		UNIT
BEVEL SIDING (CONT)												
@ REAR	1	13.5	7		95	SF						
	1	7	9		63	SF						
	2	13.5	3		81	SF						
	4	8.5	1		34	SF						
	1	11.5	4.5		52	SF						
					352	SF					352	SF
@ LEFT SIDE	1	16.5	13		215	SF						
	1	10	4.5		45	SF						
	1	1	8		8	SF						
	1	13	13		169	SF						
DEDUCT FOR OPNG	1	7.5	6	SF	(45	SF)						
	1	2	2.5	SF	(5	SF)						
	1	9.5	11		105	SF						
	1	2	22		44	SF						
	1	14	4		56	SF						
	1	2.5	13		33	SF						
	1	4.5	8		36							
	1	4.5	4.5	.5	11	SF						
					672	SF					672	SF
@ RIGHT SIDE	1	20.33	8		163	SF						
	1	10.25	10.25		105	SF						
	1	13.5	13		176	SF						
	1	11	11	.5	61	SF						
	1	12.5	12.5	.5	79	SF						
	2	5.5	5.5	.5	31	SF						
	1	15	4		60	SF						
					675	SF					675	SF
											2051	SF
ALLOWANCE FOR WASTE												
ON BEVEL SIDING (15%)	2051	.15			308	SF					308	SF

Figure 10.12 (continued)

Means Forms

QUANTITY SHEET

PROJECT: SAMPLE PROJECT	SHEET NO. 7-4/5
	ESTIMATE NO. 1
LOCATION:	ARCHITECT: HOME PLANNERS
DATE:	
TAKE OFF BY: WJD	EXTENSIONS BY: WJD
CHECKED BY: KF	

DESCRIPTION	NO.	DIMENSIONS				UNIT		UNIT		UNIT		UNIT
ROOFING + FLASHINGS												
3-TAB 25 YR ASPHALT ROOF SHINGLE (3 BUNDLES/SQ)												
@ ATTIC STORAGE	1	14	22		308	SF						
	1	14	23		322	SF						
@ GARAGE	2	14	16		448	SF						
@ MAIN ROOF	1	32.5	23		748	SF						
	2	2.5	25		125	SF						
	1	27	19		513	SF						
	1	27	4		108	SF						
					2572	SF					26	SQ
ADD FOR STARTER COURSE	1	149									149	LF
RIDGE CAP SHINGLES	1	55									55	LF
15# ASPHALT FELT UNDER SHINGLES W/10% FOR OVERLAP	2572	1.10			2892						2892	SF
METAL DRIP EDGE @ FASCIA & RAKES	1	316									316	LF
5"x7" STEP FLASHING @ CHEEKS												
@ GARAGE	1	16			16							
@ ATTIC STORAGE	1	45			45							
@ DORMER	1	17			17							
					78	LF					78	LF
8" ALUMINUM ROLL FLASHING @ DORMER	1	27									27	LF
METAL ROOF @ WINDOW SILL @ DORMER 7'-6" x 18" W/ 18" RETURN	2	7.5	1.5		2	EA					2	EA

Figure 10.12 (continued)

Means Forms
QUANTITY SHEET

PROJECT	SAMPLE PROJECT		SHEET NO.	7-5/5
			ESTIMATE NO.	1
LOCATION		ARCHITECT: HOME PLANNERS	DATE	
TAKE OFF BY	WJD	EXTENSIONS BY: WJD	CHECKED BY:	KF

DESCRIPTION	NO.	DIMENSIONS				UNIT		UNIT		UNIT		UNIT
8" COUNTER FLASHING @ CHIMNEY	1	18			18	LF					18	LF
WINDOW HEAD FLASHINGS (FRONT) 3'-6" x 5 1/2" w/ 1" RET.	5	3.5			18	LF					18	LF
1 1/4" HEAD FLASHING @ REAR & SIDE WIND. & LOUVER VENTS	1	99			99	LF					99	LF
3/4" ZEE FLASHING @ 1X STOCK	1	17			17	LF					17	LF
8" ALUMINUM FLASHING @ DOOR THRESHOLDS	1	17			17	LF					17	LF
4" x 5" ALUMINUM GUTTERS w/ HANGER @ 12" o/c	2	32			64	LF						
	3	14			42	LF						
	1	16			16	LF						
	1	27			27	LF						
					149	LF					149	LF
2" x 3" ALUMINUM DOWN SPOUTS (CONDUCTOR)	2	5			10	LF						
	2	11			22	LF						
	3	12.5			38	LF						
	1	8			8	LF						
	1	10.5			11	LF						
					89	LF					89	LF
DOWN SPOUT ELBOWS	6				6	EA					6	EA
CONCRETE SPLASH BLOCKS	6				6	EA					6	EA
16" x 8" ALUMINUM SCREENED VENTS @ CRAWL SPACE	2				2	EA					2	EA

Figure 10.12 (continued)

Chapter Eleven

DOORS AND WINDOWS

(Division 8)

Chapter Eleven

DOORS AND WINDOWS
(Division 8)

Division 8 includes windows, interior and exterior doors and frames, finish hardware, special doors (such as garage doors), and glass and glazing. Hardware for this division includes items such as hinges, locksets, passage sets, thresholds, weatherstripping, door closers, and panic devices.

Both doors and windows are available in a multitude of sizes with various functions, insulative values, finishes, and glass types.

The size, location, quantity, and specific information concerning each individual door, door frame, window, or hardware item are obtained from the architectural drawings. The plan view and elevation view drawings provide information on the location and operation of the doors and windows, as well as the quantity of each. Figure 11.1 illustrates some of the more common operations of doors as they would be seen in plan view.

The type and specific operations of windows may need to be clarified by the use of details. Figure 11.2 illustrates some common window types.

The architectural drawings include both a Door Schedule and a Window Schedule (refer to the "Schedules" section in Chapter 1). The schedules are laid out in block column form and list all the information concerning each item. In addition, many projects employ detailed drawings of the head, sill, and jambs of the doors, frames, or windows for the purpose of clarification.

Door Schedules normally list each door opening by a specific number, or designation, called the door *mark*. The mark can also be used to identify the location of the door in the structure. Beside the mark is the door size and material; a designation for the type and material; composition; size of the frame; fire rating requirements (if any); any louver or vision panels required; the hardware set; and a special "remarks" column for specific instructions. Figure 11.3 is a typical Door Schedule that shows some of the corresponding details of the head, jamb, and sill.

Windows are laid out in much the same manner, using designation (usually by letter), the size and material of the window, occasionally the manufacturer and model number, the type or function of the window (e.g., fixed, double-hung, casement, awning), and the glazing requirements. The Window Schedule sometimes lists the size of the *rough opening* for the installation of the window rather than the size of the window itself. Figure 11.4 is a typical Window Schedule with related details.

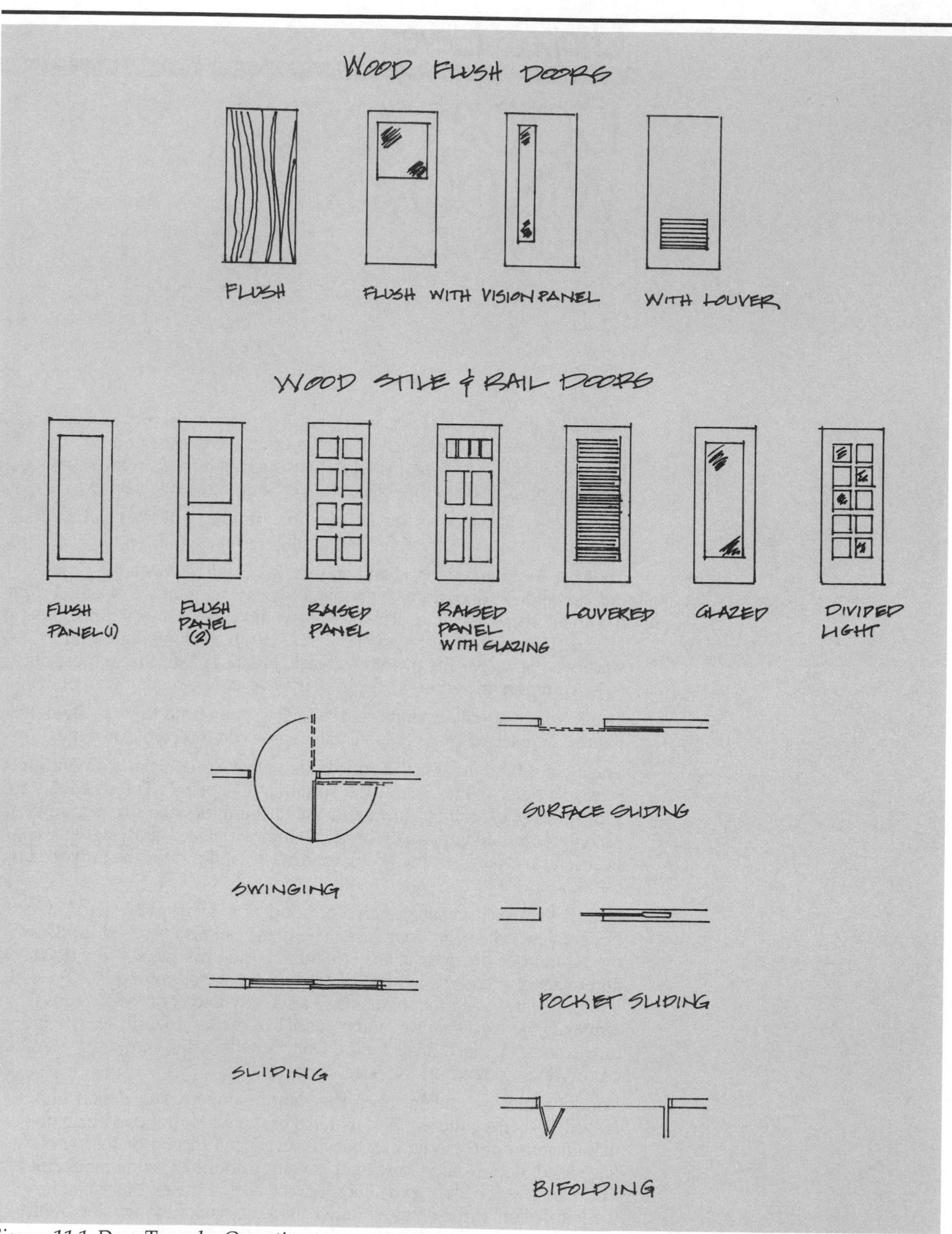

Figure 11.1 Door Types by Operation

For the proper operation of the door, special hardware, called *finish hardware,* is required. The Finish Hardware Schedule is different from window and door schedules in that it does not always appear in column form on the architectural drawings. This information is sometimes included in the finish hardware section of the specifications. The Hardware Schedule is organized by listing the items needed to outfit a particular door. The listing is called the *hardware set*, and is typically noted by a number. Each piece of hardware in the particular set is stated by manufacturer, model number, size, and color or appearance, called the *finish.* The finish should be noted as part of the hardware takeoff, as it has a major impact on the materials pricing. The following is a sample description of a simple hardware set as it might be encountered in a set of specifications.

Hardware Set #4

- 1-1/2 pair 4-1/2″ Stanley CB Series
- Corbin 977L-9500 Series Mortise Lockset
- LCN 4010 CUSH Series Closer
- Ives 436 B Floor Stop
- Ives #20 Silencers

"Hardware Set #4" does not necessarily refer to only one door, but may actually refer to several different doors on one job.

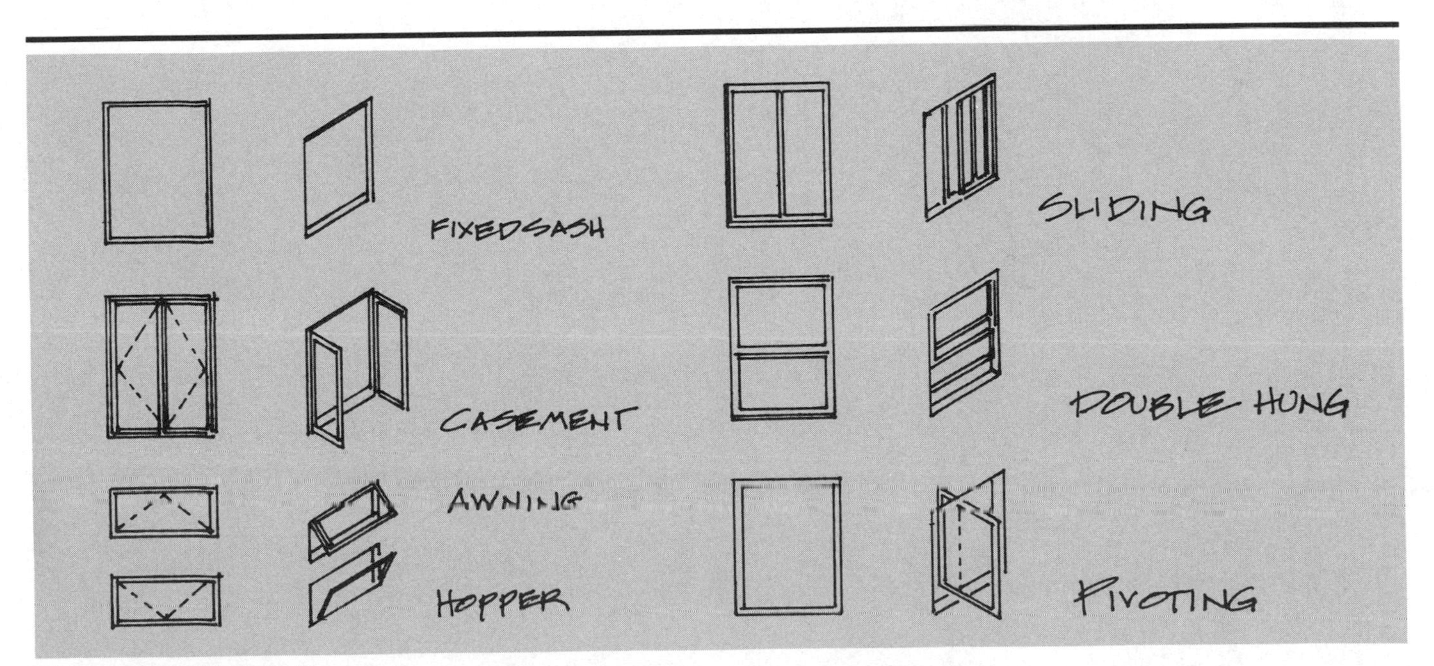

Figure 11.2 Window Types

DOOR SCHEDULE								
NO.	SIZE	TYPE	MATL.	FRAME	THSH.	CLOSER	HARDWARE	REMARKS
101	$3^{0} \times 7^{0} \times 1\frac{3}{4}''$	A	WD/GL	WD.	ALUM.	✓	BRASS PUSH BAR/PULL LOCKSET	MORGAN M-5911
102	$3^{0} \times 7^{0} \times 1\frac{3}{4}''$	A	"	WD	"	✓	"	"
103	$3^{0} \times 7^{0} \times 1\frac{3}{4}''$	A	"	WD.		✓	"	"
104	$3^{0} \times 6^{8} \times 1\frac{3}{8}''$	B	WD	WD	MARBLE		PRIVACY SET	MORGAN 5 CROSS PANEL
105	$3^{0} \times 6^{8} \times 1\frac{3}{8}''$	B	WD.	WD	"		"	"
106	BY WALK-IN MANUFACTURER							2'6" WIDE MAX.
107	$3^{0} \times 6^{8} \times 1\frac{3}{8}''$	D	WD	WD			SPRING HINGE	BY OWNER
108	$3^{0} \times 7^{0} \times 1\frac{3}{4}''$	C	H.M.	P.M.		✓	EXISTING	EXIST H.M. DOOR
109	$3^{0} \times 7^{0} \times 1\frac{3}{4}''$	C	H.M.	P.M.		✓	"	"

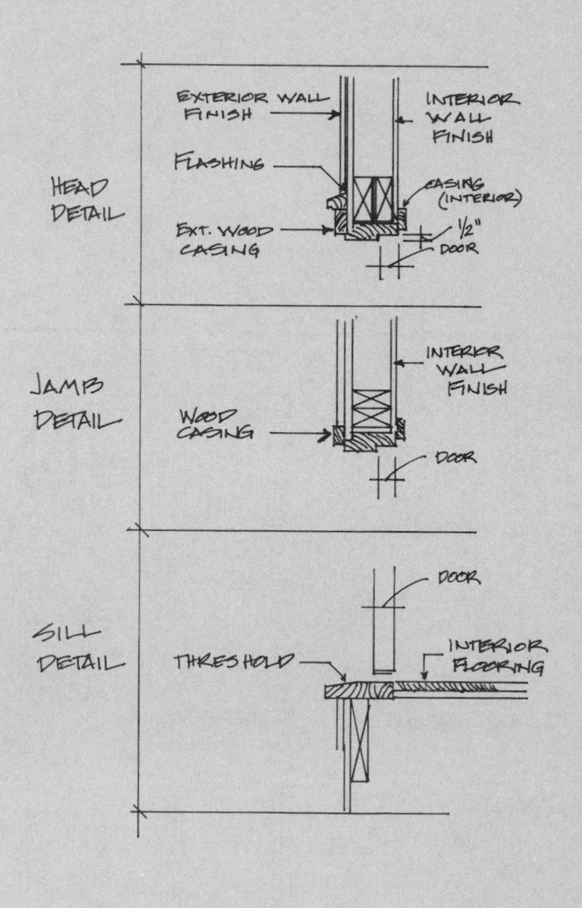

Figure 11.3 Wood Door Frame Detail

WINDOW SCHEDULE						
MARK	MANUF/MODEL	TYPE	ROUGH OPEN.	GLASS	JAMB	REMARKS
A	ANDER/C24	CASEMENT	4'-0½" x 4'-0½"	HIGH PERFORM.	4 9/16"	GRILLES, SCREENS
B	ANDER/C34	CASEMENT	6'-0½" x 4'-0½"	"	"	" "
C	ANDER/CW14	CASEMENT	2'-4⅞" x 4'-0½"	"	"	" "
D	ANDER/CW25	CASEMENT	4'-9" x 5'-0⅜"	"	"	" "
E	ANDER/24210	DOUBLE HUNG	2'-6⅛" x 3'-1¼"	"	"	" "
F	ANDER/2842	DOUBLE HUNG	2'-10⅛" x 4'-5¼"	"	"	" "
G	ANDER/30-364618	30° BAY	7'-0" x 4'-10¾"	"	"	" "
H	ANDER/A330	AWNING	3'-0½" x 3'-0½"	"	"	" "
I	ANDER/AW31	AWNING	3'-0½" x 2'-4⅞"	"	"	" "
J	ANDER/CW13	CASEMENT	2'-4⅞" x 3'-0½"	"	"	" "
K	ANDER/C12	CASEMENT	2'-0⅝" x 2'-0⅝"	"	"	" "

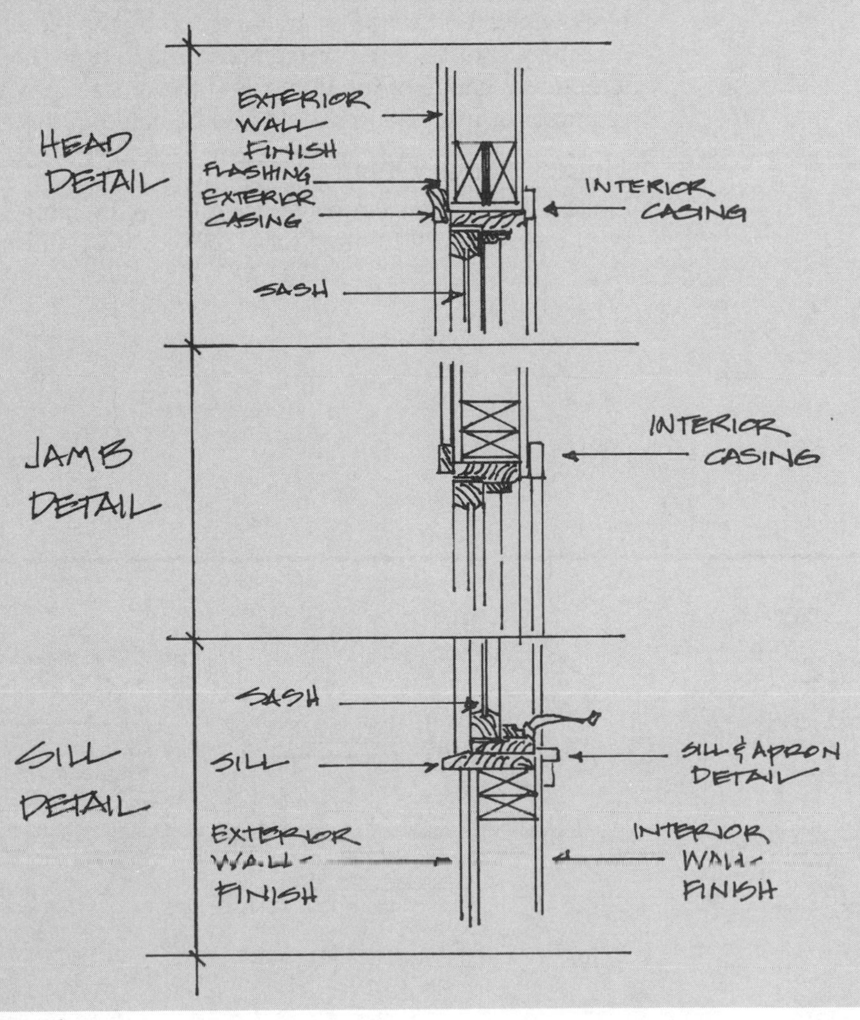

Figure 11.4 Window Details

Hollow Metal Doors and Frames

Hollow metal doors and frames are typically used for commercial projects that require a more durable, heavy-duty door and frame system, and where aesthetics are not a real concern.

Hollow Metal Frames

Hollow metal frames are formed of 18-, 16-, and 14-gauge steel, and are made to accommodate 1-3/8" and 1-3/4" wood or metal doors. Hollow metal frames are available in a variety of standard wall thicknesses, sometimes called *throat,* typically 4-3/4", 5-3/4", 6-3/4", and 8-3/4". They are available prefinished, galvanized, primed, or unfinished. Hollow metal frames can be installed in wood frame walls, masonry walls, metal stud and drywall walls, and walls that combine wood, steel, and masonry. They are available in two standard levels of fabrication: *knockdown,* where the frame is disassembled into the two jambs and the head piece and assembled on site, and *welded assembly,* where the frame is welded (at the factory) at the corners to produce a rigid, square, and true frame for site installation. Frames can be installed with the frame wrapped around the wall thickness, as in the case of interior partitions, or with the frame butted up to the jamb and head, as in the case of metal frames installed with a masonry or concrete rough opening. Figure 11.5 illustrates some typical hollow metal door frames.

Units for Takeoff

Hollow metal door frames are taken off by the piece (EA). Takeoff quantities should be separated according to type (knockdown or welded), size, finish, gauge of the frame, and throat size. Any special fabrications required for installation should also be noted in the takeoff.

Hollow Metal Doors

Hollow metal doors are constructed of 16-, 18-, and 20-gauge face sheets, with interior metal framing for a 1-3/8" or 1-3/4" finished thickness. Hollow

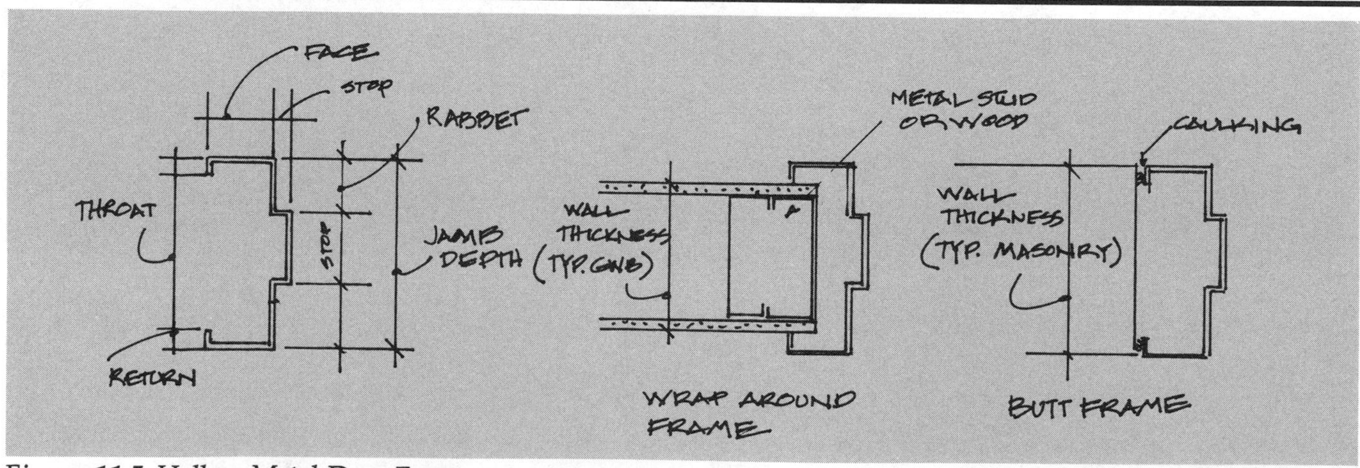

Figure 11.5 Hollow Metal Door Frame

metal doors are available in a variety of styles, including flush, small vision panels, full or half glass, and louvered.

Units for Takeoff

Doors are taken off by the individual piece (EA), according to size, thickness, type, gauge of metal, and finish. Any special preparations to the door, such as for a deadbolt, should also be noted. Double door sets are noted as a pair (PR) in the takeoff, with the aforementioned qualifications.

Other Takeoff Considerations

In accordance with most building codes, certain locations throughout the building will be required to be fire-rated. Fire rating refers to the door and frame's capacity (label) to slow the transmission of fire. Typical labels are *C Label* for a 3/4-hour rating, *B Label* for a 1-1/2- to 2-hour rating, and *A Label* for a 3-hour rating. Other restrictions and qualifications also govern label doors and frames; the contractor should review the plans and specifications carefully. Label doors and frames are considerably more expensive than other doors.

Wood Doors and Frames

Wood doors are available in a variety of types, materials, sizes, and thicknesses. They can be supplied separately for installation in hollow metal frames, or can be specified *prehung*. Prehung refers to a packaged unit comprised of a finished door on a frame, complete with trim and hardware. Wood doors can be generally classified in one of two categories. The first, *flush doors*, are flat slab doors that are either *solid-core* or *hollow-core*. Solid-core wood doors are made of either particle board or mineral core composition with a wood veneer facing. They are used where increased fire resistance, sound insulation, and dimensional stability are specified. Hollow-core doors are used for interior applications, and have a honeycombed cardboard core with a wood veneer facing. They are subject to warping and have no real thermal or sound-insulating value.

The second major classification of wood doors is the *stile and rail door*. Stile and rail doors are constructed of vertical (stiles) and horizontal (rails) members that provide framework for wood, glass, or louver center panels. Wood doors are manufactured in widths ranging from 2′ to 4′, with smaller widths available in some styles. Heights range from 6′-6″ to 9′-0″ and thicknesses are 1-3/4″, 1-3/8″, and 1-1/8″ for some residential-grade bifold and sliding closet doors.

Wood frames are available for varying door and wall thicknesses and can be purchased in sizes and species of wood to match most wood doors. They are made with either stain-grade woods or materials suitable for painting only. Both doors and frames can be purchased already prepared (cut for hinges, lockset, deadbolts, etc.) or blank for field preparation.

Units for Takeoff

Wood doors are taken off by the piece (EA) and listed according to size, thickness, composition or species of wood, type of door (flush or stile and rail), and any special features such as fire-rating label, vision panels, and preparation for special hardware. Wood frames are taken off by the piece and specified according to size of the door they fit, wall thickness, species and quality of wood (paint- or stain-grade), and any preparation to the frame, such as for hinges, lockset, and deadbolt. Prehung units are taken off by the piece, according to the size and type of the door itself, wall thickness, species and quality of both door and frame wood, labels, vision panels, and so forth. Double doors are taken off by the pair (PR) with the above-mentioned qualifications.

Other Takeoff Considerations

Occasionally the *handing*, or direction of swing, of the door may affect the price, especially for exterior residential door units that swing out instead of operating with the normal in-swing function. Similar requirements should be noted in the takeoff.

Because prehung units come with the casing attached or included, it is necessary to specify the type of casing for accurate pricing of the unit.

Sliding Glass Doors

Sliding glass doors, commonly referred to as *sliders*, are a combination of fixed and operable panels incorporated into a frame that acts as a track for the sliding of the operable panel. They typically consist of a stile and rail door with a full glass panel, sold as a single unit. Slider units range from 5′ to 12′ in width and from 6′-8″ to 8′-0″ in height. The stiles and rails can be wood or aluminum, or can be covered with a metal or vinyl coating, referred to as *cladding*, at the exterior. The panels are insulated, tempered safety glass. Slider units are noted on the plan view and exterior elevation drawings. The contractor should refer to the specifications and the door schedule for the information necessary to price the door. Similar units that swing, as opposed to slide, are also available; both panels may operate as a single operable unit.

Units for Takeoff

Sliding glass door units are taken off by the piece (EA). "Piece" refers to the entire unit or assembly of doors, hardware, and track. The slider should be specified in the takeoff by manufacturer and model number when this is sufficient for accurate pricing. If the manufacturer and model number is not specified, the unit should be listed by the size (width x height), composition, wall thickness the unit will occupy, type of glazing, screens or grilles, hardware, and, finally, the color of the finish (if applicable). Swinging glass door units are qualified the same way, with special notation as to the number and location of operable panels.

Other Takeoff Considerations

There is a wide selection of sliders on the market, with dramatically different prices and quality. Some are shipped assembled ready for installation, while others require some assembly prior to installation. Where required, assembly should be noted in the takeoff, as this will affect the installation cost. The same precautions should be taken for swinging glass door units.

Special Doors

Special doors include folding doors, pocket doors, overhead garage doors, and surface sliding doors. *Folding doors* are accordion or bifold doors that fold or stack against a wall or jamb. Folding doors over 150 SF in area are referred to as *folding partitions*. Both folding doors and partitions are typically provided with tracks and all related hardware for installation. They are available in a variety of styles, compositions, sizes, and finishes. Available finishes include wood, fabric, and vinyl. Folding doors and partitions are often used to separate space within a room and therefore may require that they resist the transmission of sound, expressed as the door's STC (sound transmission class). This is analogous to the R-value of insulation. In general, the higher the sound-insulating value, the more expensive the door.

Pocket doors are installed within the framework of the wall so that, when opened, the door can be stored within the wall cavity. They are used in residential applications where special constraints preclude the use of swing or bifold doors. Pocket doors are provided as a package unit with all necessary hardware and track, and are available in a variety of compositions, finishes, styles, and sizes.

Steel insulating door units for residential entrances are comprised of thin steel sheets over a wood-and-foam insulating core. They are typically provided prehung in a wood frame with an integral aluminum threshold, bored for locksets and/or deadbolts. Steel door units are available with designs either embossed in the face sheets or surface-applied, in many styles and sizes. Entry units with fiberglass face sheets and similar core construction are also available. Sizes range from 2′-8″ to 3′-0″ in width and 6′-6″ or 6′-8″ in height. Steel doors are provided primed for field-applied paint, and fiberglass units are unfinished, ready for field-applied stains or paints.

Overhead doors such as garage doors are sectional panels attached at the top and bottom rails by special hinges. The doors are constructed of wood stiles and rails with hardboard flush inserts, or thin steel face sheets over a steel frame and foam insulating core. Face sheets can be flush or may have an embossed design. Special designs are also manufactured that include glass lights within the individual panels. Ball-bearing rollers permit the individual sections of the door to travel along a flanged steel track attached at the door jambs. The doors are spring balanced for ease of manual operation or can be mechanically operated by means of an electric garage door opener. Wood door finishes are either unfinished or primed for field painting. Metal doors are usually provided prefinished in the manufacturer's standard colors. Doors are manufactured for standard openings ranging from 8′ to 18′ in width and 7′ to 12′ in height. Custom sizes are also available, but may constitute an additional cost. Overhead doors are typically provided with the necessary hardware for installation, as well as locking mechanisms for security.

Aluminum storm doors are lightweight, prefinished aluminum doors with interchangeable glass sashes and screens, set within a matching aluminum frame. The storm unit is surface-applied by screwing through flanges, attached to the frame jambs and head, to the surface of the exterior door trim. Storm doors provide protection of the entry unit from the weather and allow air passage through the screen in warm weather. Handles, latches or locksets, and hydraulic closers are included as part of the unit. Storm door units are available in sizes to fit most exterior entry units, in a variety of sizes, colors, and designs.

Units for Takeoff

Folding doors and partitions are taken off by the square foot of opening they occupy. The takeoff description should note size (width × height), composition of the door and finishes, method of door operation (manual or mechanical), STC classifications, fire rating (if applicable), and any special locking hardware.

Pocket doors are taken off and listed by the piece (EA), according to size, composition, and width of the wall. The EA designation includes all components in the unit.

Steel and fiberglass insulating entry units are taken off by the piece and listed according to size, wall thickness, function (in-swing or out-swing), design or model, exterior casing type, and the preparation for the door (boring for lockset or deadbolt).

Storm door units are taken off and listed by the piece according to size, design, special options (e.g., insulated glazing, locksets), and color of finish. Size is often designated as the size of the entry unit.

Overhead doors are taken off and listed by the piece according to size (width x height), thickness of the door panels, composition and finish of the panels, door style or design, manufacturer's model number (if applicable), and special options such as glass lights, electrical operators, safety devices, and security mechanisms. Additional information, such as R-value or the amount of overhead clearance distance (from the head of the door to the ceiling above), may also be required for accurate pricing. Custom-sized overhead doors are taken off and qualified by the same units.

Other Takeoff Considerations

Folding partitions for commercial applications may be required to be fire-rated by building code. The contractor should review the specifications carefully for this requirement. Larger folding partitions may require special structural details to accommodate the weight of the partition itself. This is not usually part of the scope of the partition manufacturer or installer; clarification of the limits of the work is essential for accurate pricing.

Pocket doors require some level of assembly prior to installation. The amount of work depends on the manufacturer and the quality of the unit.

Insulated fiberglass entry units typically offer a special finishing kit to give the appearance of stained wood. The finishing kit is purchased from the manufacturer, as an option, and field-applied. The contractor should review the specifications for the required finish of the door.

Entrances and Storefronts

Entrances and storefronts are classified as metal framework enclosing fixed or operable windows and entrance doors. The metal framework is manufactured from aluminum alloy and extruded into many different shapes and sizes. Finishes of the extruded aluminum framework are anodized in a number of colors, or clear, to prevent oxidation from exposure. In addition to anodizing, durable coatings are available in a range of colors.

Insulated or plain glazing panels manufactured off site are installed within the channels of the extruded framework to provide a weathertight window or wall system. Specialty glazing can be used to obscure structural elements, or insulating nonglazed panels may be incorporated into the design. Entrance doors of the same construction with hinges or pivots can be manually operated or motorized, and come with panic and finish hardware. Operable windows may be used to provide ventilation.

The location and width of the entrance and storefront system can be found on the architectural floor plans. Additional drawings that show sections through the head, jamb, and sills of the various windows are usually provided on a separate sheet called *window details*. Elevation views are found on the exterior elevations drawings, which are part of the architectural drawings. Specifications should be carefully reviewed for the type of glass, frame system, manufacturer, door and window systems and hardware, laminated safety and tempered glass, samples, and shop drawings. Figure 11.6 is an example of exterior window elevations and the corresponding details for a storefront system.

Units for Takeoff

Each individual component of the glass and glazing system is taken off separately. The aluminum framework is taken off by the linear foot, including all vertical and horizontal mullions, jambs, heads, and sills. Extruded sections should be separated according to size, thickness of the material (gauge), finish, and shape.

Aluminum flashings at the sills, or head of the system, should be taken off by the LF. Flashings over 12″ in width may be converted to SF. Sealants at the perimeter of the unit are taken off by the LF.

Glass and insulated glass panels are taken off by the SF. The square footage refers to the area of the individual insulated glass panels that fit within an extruded frame system. These should be listed separately according to size (width x height), thickness of the insulating panel or pane of glass, type of glass (e.g., clear, tinted, spandrel), and special treatment of glass (e.g., tempered, laminated safety glass). Insulated nonglazed panels are taken off and listed by the SF according to the size, thickness, facing or finish of the panel, and the insulating material within the panel.

Windows are taken off by the piece (EA) according to size (width x height), function (e.g., casement, awning, hopper, sliding), hardware requirements, material and finish of the sash, and the glass type.

Doors are taken off by the piece (EA) according to size (width x height), thickness, stile and rail materials, hardware (such as hinges or pivots), panic devices, locking mechanisms, and type of glass.

In addition to the above takeoff procedure for fabrication quantities, sizes of the individual frames (that fit within a rough opening) are listed by the SF for estimating labor to install the frame at the job site.

Other Takeoff Considerations

Custom fabrication of the glass panels and the framework require shop drawings that detail and dimension the exact sizes of each fabrication. This requires field dimensioning and verification of the rough opening sizes. The cost of shop drawings is considered as part of the cost of the work, and should be accounted for in the takeoff. Shop drawings are listed as a lump sum (LS).

Wood and Plastic Windows

Wood and plastic windows are used to provide natural light and ventilation to the rooms of a building. They are furnished as complete, factory-assembled units including frame, sash, operating hardware, weatherstripping, and glazing. Wood windows are constructed of kiln-dried, clear, straight-grain woods, usually pine. The wood frame has been treated with a water-repellent preservative, and is available primed and either ready for field-applied paint, factory-painted, aluminum-clad, or vinyl-clad. Prefinished and clad windows are available in a variety of colors.

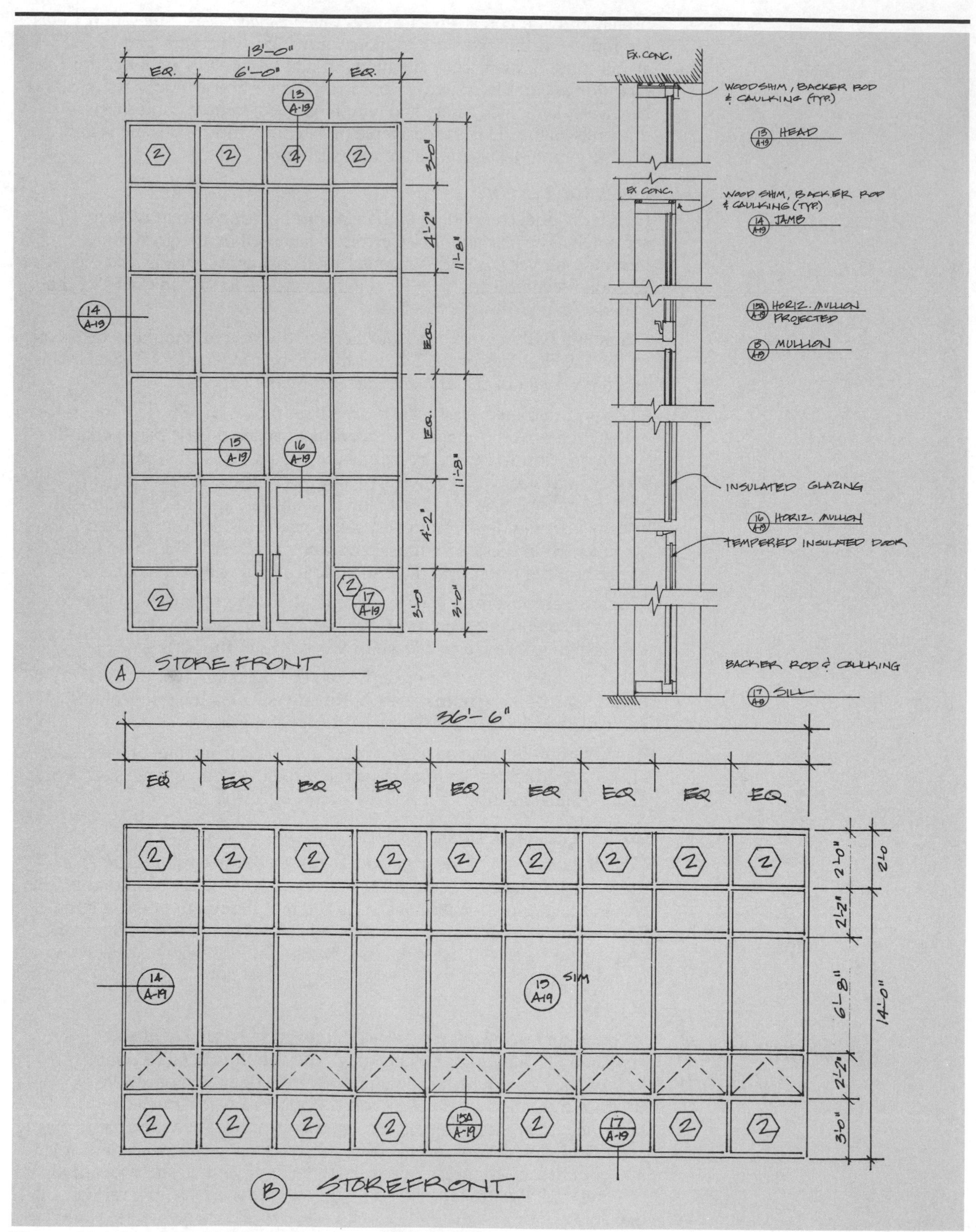

Figure 11.6 Storefront Window Elevations and Details

Plastic windows are constructed without wood. Jambs, heads, sills, and sashes are made of various grades of PVC, and are also available in a variety of colors, white being the most common. Insulated glass panels are set within plastic stiles and rails. Wood and plastic windows are available in literally hundreds of sizes and functions (see Figure 11.2). Combinations of various styles and sizes are often joined together in the factory or field to produce a desired appearance.

The locations of wood and plastic windows are shown on the architectural plan view drawings, and can be cross-referenced by the exterior elevations. Window Schedules (see Figure 11.4) should be studied for the specific manufacturer, model, and size of the individual or combined unit. Occasionally, special details or sections through the unit may be used for clarification.

The specifications should be reviewed for detailed information concerning the type of glass, finish of the window, and accessories such as screens or grilles.

Units for Takeoff

Wood and plastic windows are taken off by the piece (EA). Quantities should be noted according to manufacturer, model or size, function (e.g., double-hung, casement, awning, fixed sash), wall thickness, type of exterior finish, color of the finish, type of insulating glass, and exterior casing (for nonclad wood windows). Special options such as grilles, screens, or blinds can be taken off separately or included as part of the actual window takeoff.

Other Takeoff Considerations

For larger window units such as bays or bows, many manufacturers offer the option of making normally fixed sashes operable, such as the center sashes in a four- or five-lite bow window. This often constitutes an additional cost, and should be noted in the takeoff for accurate pricing.

Many manufacturers fabricate windows to custom specifications, which can be expensive and cause a considerable delay. To accurately price custom window fabrication, the contractor should contact the manufacturer and obtain a direct quote.

While most wood windows are manufactured for a normal wall thickness (2″ x 4″ construction), those constructed of thicker framing members, such as 2″ x 6″, require the use of finish wood pieces to bring the jambs, head, and sill out flush to the interior finish surface. These are called *extension jambs* and constitute an additional cost.

Finish Hardware

Finish hardware comprises a wide variety of items that are fitted to a door and frame to perform a specific function. The most common examples are hinges, lock sets, latch sets, closers, stops, deadbolts, thresholds, weatherstripping, and panic devices. Most finish hardware is available in a variety of finishes, designated by the U.S. Code Symbol Designation. The following is a list of the most common U.S. Code Symbol numbers and their respective finishes:

- US P–Primed paint coat
- US 3–Polished brass
- US 4–Satin brass
- US 9–Polished bronze

- US 10–Satin bronze
- US 10B–Satin bronze–oil-rubbed
- US 14–Polished nickel
- US 15–Satin nickel
- US 20–Statuary (light) bright bronze
- US 20D–Statuary (dark) bright bronze
- US 26–Polished chrome
- US 26D–Satin chrome
- US 28–Satin aluminum–anodized
- US 32–Polished stainless steel
- US 32A–Satin stainless steel

The contractor must study the architectural drawings with careful attention to the Door Schedule (see Figure 11.3). Door Schedules follow two basic formats for noting finish hardware. The first and most common method is to designate the particular hardware set for each door by number, as illustrated in the example at the beginning of this chapter. Each hardware set is defined in the specifications by its components. The second method employs a series of columns included in the Door Schedule, each of which is headed by a particular item of finish hardware. The row lists the door mark and the corresponding hardware items required with a number that defines the specific item in the specifications.

Units for Takeoff

Hardware items are taken off by counting the individual units and are listed as EA. Items should be separated according to the manufacturer, series, model, type, finish, and any other means of identification specified, such as Federal Specification Series Designation or ANSI (American National Standards Institute) series number.

Other Takeoff Considerations

A considerable portion of the cost of hardware is the finish. Some of the less expensive finishes, such as US 28, are readily available and may be a stock item. Others, such as US 32, are special orders and may require long lead times for delivery. The contractor should consult the finish hardware supplier for delivery schedules so that temporary devices may be included in the estimate if necessary.

Many construction projects require the use of temporary cores for the lock sets during construction, with the permanent cores installed at turnover. This should be noted in the specifications and may be an additional cost.

Most doors supplied to project sites have been "prepped" for the hardware off site. Occasionally, doors are mortised and bored for hardware on the site. The contractor must know how the doors will be supplied prior to pricing the hardware installation.

Mirrors

Mirrors are sheets of float glass, usually 1/8" or 1/4" in thickness, to which a reflective coating has been applied. The reflective coating is then sealed to protect against moisture and handling damage. Mirrors are made available in different shades by using a combination of different coatings with clear or tinted glass. The most common shades are gray, clear, and bronze. Mirrors that have been cut from larger pieces require dressing of the edges. The most common form of edgework is a plain polished edge, although beveled edges are sometimes specified for a particular application. Mirrors can be installed using adhesives applied to the back, or supported by special tracks and hardware. Trim pieces may also be required to cover vertical and horizontal seams or inside and outside corners. The trims are usually aluminum or plastic, with a reflective coating to match the mirror.

Mirrors are usually shown on the interior elevations of architectural drawings for wall applications, and on reflected ceiling plans for ceiling applications. The specifications should be reviewed for the type, thickness, and method of application, as well as the associated trim pieces needed.

Units for Takeoff

Mirror is taken off by the SF and listed according to size, type or color, thickness, edge treatment, and method of installation. The contractor may elect to take off the quantity of edge treatment separately, by measuring the perimeter of the individual pieces and listing by the LF. Special trim pieces are taken off by the LF and listed according to type (inside corner, outside corner, track, or seam). Adhesives are taken off by the quantity per SF and extended to the gallon or manufacturer's typical sales unit. Quantities per SF are determined by the individual product's coverage.

Other Takeoff Considerations

Cutting special shapes or boring holes in mirrors for special applications can be expensive and should be noted in the takeoff for accurate pricing.

Summary

The takeoff sheets in Figure 11.7 are the quantities for the work of Division 8—Doors and Windows for the sample project. Items of work for this division should be listed with as much information as necessary to accurately price both the materials and labor for installation.

A final note: Because of the enormous variety of doors and windows available, the contractor must review the drawings and specifications thoroughly for any information that will help to accurately price the material.

Division 8 Checklist

The following list can be used to ensure that all items for this division have been accounted for.

Hollow Metal Doors and Frames

- ☐ Hollow metal frames
- ☐ Hollow metal doors
- ☐ Special preparations
- ☐ Fire-rated doors and frames

Wood Doors and Frames

- ☐ Wood doors
- ☐ Hollow-core wood doors
- ☐ Solid-core wood doors
- ☐ Fire-rated wood doors
- ☐ Wood frames
- ☐ Prehung door units
- ☐ Bifold door units

Sliding Glass Doors

- ☐ Sliding glass doors
- ☐ Stationary glass panels
- ☐ Accessories

Special Doors

- ☐ Overhead garage doors
- ☐ Automatic door operators
- ☐ Folding doors
- ☐ Folding partitions
- ☐ Pocket doors
- ☐ Steel-insulated entry units
- ☐ Fiberglass-insulated entry units
- ☐ Sidelights
- ☐ Storm doors

Entrances and Storefronts

- ☐ Aluminum extrusion framework
- ☐ Glass
- ☐ Insulated glass
- ☐ Special glazing
- ☐ Spandrel glass
- ☐ Operable windows
- ☐ Doors
- ☐ Special door hardware
- ☐ Flashing
- ☐ Insulated panels

Wood and Plastic Windows

- ☐ Wood windows
- ☐ Clad wood windows
- ☐ Plastic windows
- ☐ Accessories
- ☐ Grilles
- ☐ Screens
- ☐ Extension jambs

Finish Hardware

- ☐ Lock sets
- ☐ Cylindrical
- ☐ Mortise
- ☐ Latch sets
- ☐ Closers
- ☐ Hinges
- ☐ Panic hardware
- ☐ Thresholds
- ☐ Weatherstripping
- ☐ Stops
- ☐ Push and kick plates

- ☐ Silencers
- ☐ Miscellaneous or special hardware

Mirrors

- ☐ Mirrors
- ☐ Edge treatments
- ☐ Trims and fasteners

Means Forms

QUANTITY SHEET

SHEET NO. 8-1/3

PROJECT: SAMPLE PROJECT — ESTIMATE NO. 1

LOCATION: — ARCHITECT: HOME PLANNERS — DATE:

TAKE OFF BY: WJD — EXTENSIONS BY: WJD — CHECKED BY: KF

DESCRIPTION	NO.	DIMENSIONS				UNIT		UNIT		UNIT		UNIT
DIVISION 8: DOORS & WINDOWS												
INTERIOR DOOR UNITS												
6-PANEL PINE PREHUNG INTERIOR DOOR UNITS W/ 4 5/8" JAMB THICKNESS												
2'-6" × 6'-8" × 1 3/8"	8				8	EA					8	EA
6-PANEL PINE INTERIOR BIFOLD UNITS W/ TRACK & HARDWARE												
2'-6" × 6'-8" × 1 3/8"	1				1	EA					1	EA
4'-0" × 6'-8" × 1 3/8"	5				5	EA					5	EA
6 PANEL PINE POCKET DOOR FOR 4 5/8" WALL THK. WITH TRACK & HARDWARE												
2'-4" × 6'-8" × 1 3/8"	1				1	EA					1	EA
EXTERIOR DOOR UNITS												
INSULATED STEEL DOOR UNIT (6 9/16" THICK JAMB)												
2'-8" × 6'-8" × 1 3/4"	1				1	EA					1	EA
3'0" × 6'-0" × 1 3/4" W/ 4 LITE TRANSOM 9"×9"	1				1	EA					1	EA

Figure 11.7

Means Forms

QUANTITY SHEET

SHEET NO. 8-2/3

PROJECT SAMPLE PROJECT

ESTIMATE NO. 1

LOCATION

ARCHITECT HOME PLANNERS

DATE

TAKE OFF BY WJD

EXTENSIONS BY: WJD

CHECKED BY: KF

DESCRIPTION	NO.	DIMENSIONS			UNIT	UNIT	UNIT	UNIT
6'0" x 7'-8" SLIDER UNITS W/TRACK HARDWARE, SCREEN & 21 LITE GRILLES	3				3 EA			3 EA
WINDOWS (ALL GLAZING TO BE DOUBLE INSULATED)								
56" x 73" FIXED GLASS UNIT W/ 30 LITE GRILLE @ D.R. & STUDY (6 9/16" JAMB)	2				2 EA			2 EA
20" x 36" DOUBLE HUNG UNITS W/ GRILLES & SCREEN	4				4 EA			4 EA
32" x 36" DOUBLE HUNG UNITS W/ GRILLES & SCREEN	3				3 EA			3 EA
32" x 24" DOUBLE HUNG UNITS W/ GRILLES & SCREEN	1				1 EA			1 EA
32" x 24" (2) DOUBLE HUNG W/ MULLION UNIT W/ GRILLES & SCREENS	4				4 EA			4 EA
HARDWARE								
PASSAGE SETS @ SWING DOOR (INT.)	6				6 EA			6 EA
LOCKSETS @ EXTERIOR UNITS (CYLINDRICAL)	2				2 EA			2 EA
PRIVACY SETS @ BATH UNITS	2				2 EA			2 EA

Figure 11.7 (continued)

Means Forms

QUANTITY SHEET

PROJECT SAMPLE PROJECT						SHEET NO. 8-3/3
						ESTIMATE NO. 1
LOCATION		ARCHITECT HOME PLANNERS				DATE
TAKE OFF BY WJD		EXTENSIONS BY: WJD				CHECKED BY: KF

DESCRIPTION	NO.	DIMENSIONS				UNIT		UNIT		UNIT		UNIT
SPECIAL DOOR												
8'-0" x 7'-0" 16 PANEL WOOD GARAGE DOOR WITH TRACK & HARDWARE	1				1	EA					1	EA
ELECTRIC GARAGE DOOR OPENER	1				1	EA					1	EA

Figure 11.7 (continued)

Chapter Twelve

FINISHES

(Division 9)

Chapter Twelve

FINISHES
(Division 9)

Division 9 includes a variety of interior work items, such as drywall and plaster construction, tile, acoustical ceilings, wood and resilient flooring, carpeting, and painting. Although Finishes do not typically represent a major portion of the construction costs, they are the most visible component of the structure to its owners/occupants, and have the greatest impact on the aesthetic value of the interior of the building.

The work of Division 9 is shown on the architectural drawings. Quantities are most often determined from the floor plans, interior elevations, and the reflected ceiling plan. Critical information for determining the components and specific applications of the finishes are often shown in sections and details that correspond to the various architectural drawings. For example, wall sections may be used to determine the thickness or height of the drywall or plaster system. Details of sections or elevations can provide information on the height of the transition from one finish material to another, or the substrate to which the finish is applied. Wall, ceiling, and floor finishes of individual rooms are found on the room finish schedule.

Room finish schedules are more frequently used in commercial construction, where the number of rooms and variety of finishes are numerous, rather than on residential drawings. The room finish schedule uses a column format to list the rooms by name or number, specifying the flooring and base material; wall finishes such as paint, wallcoverings, or tile; and often the ceiling height.

Division 9 of the specifications should indicate the specific materials used for each type of finish, the quality of workmanship, and special methods of application. The specifications should also be reviewed for the work of other trades that may affect the pricing of a particular finish.

Lath and Plaster

Lath is comprised of wood strips or perforated metal sheets that are used as a base for plaster. Most plaster used in residential/light commercial construction is applied using a thin coat, called *skim coat plaster*, applied over a gypsum composition base. The gypsum base is a special gypsum-core board faced with a multi-layer, specially treated paper designed to bond the veneer plaster to the base. Skim coat plaster can be applied in a one- or two-coat system for a strong, highly abrasion-resistant surface. Gypsum plaster base is manufactured in various sized sheets,

4 feet in width by 8, 9, 10, 12, or 14 feet in length. Manufacturers' standard thicknesses are 3/8", 1/2", and 5/8". Gypsum plaster base is applied by nailing or screwing to the framing or furring components. Plaster base is also manufactured for fire-resistive applications, as well as foil-backed to retard vapor transmission. Veneer plaster is available in 50- and 80-pound bags. Coverage is based on the individual product, specified thickness, and type of finish. Conventional three-coat plaster systems require that the plaster be mixed with sand, perlite, or vermiculite aggregate in varying proportions to obtain the desired mix.

Less common is metal or wire lath, sheets of perforated or stranded metal, available galvanized or rust-inhibitive painted, and either flat or dimpled (self-furring). The metal lath is nailed on wood framing, or tied with wire to metal supporting systems. It can also be used to reinforce plaster to other surfaces.

Additional materials used in plaster systems include special trim pieces such as corner bead, joint reinforcing tape, and screeds, sometimes referred to as plaster "grounds".

Units for Takeoff

Quantities are determined by measuring the surface area to be plastered. All plaster systems can be taken off and priced either by individual components or as a whole system. Skim coat plaster and gypsum base, when taken off as a system, are quantified by the square foot. The SF unit should include all items necessary for the complete installation, including reinforcing tape, trims, plaster, and gypsum base, as well as labor and equipment. Conventional or three-coat plaster systems quantified as a whole system are taken off by the SF and extended to the square yard.

Takeoff quantities for individual components are as follows. Gypsum base for skim coat plaster systems is taken off by the SF, and can be converted to the required number of sheets. Skim coat plaster is taken off by the SF of each coat and can be converted to bags of plaster. Trims such as corner bead and reinforcing tape are taken off by the linear foot and separated by the type and size.

Metal lath is taken off by the SF and converted to either SY or individual sheets required. Overlap of metal lath should be in accordance with the specifications. Gypsum plaster is taken off by the SF and can be converted to bags or weight (pounds or tons). Sand for mixing with plaster is determined by using the specifed sand-to-plaster ratio to arrive at a weight. For example, a 2:1 ratio requires 2 lbs. of sand for every lb. of plaster. Perlite or vermiculite aggregates are specified by cubic feet of aggregate per 100 lbs. of plaster. To determine the quantity of perlite or vermiculite aggregate needed, one first determines the total weight of plaster (in lbs.), then multiplies the CF of aggregate per lb. to arrive at the total CF required. Metal trims and grounds are taken off by the LF.

Other Takeoff Considerations

The amount of plaster for both skim coat and conventional multi-coat plaster is based on the thickness of each coat, the texture of the plaster, the surface of the substrate, and the quality of workmanship. These must all be taken into account to determine the quantity of plaster materials needed. All plaster should be separated according to thickness, base (lath) materials, fire rating (if required), and class of workmanship (first class or ordinary). Plastering should also be separated according to location of the work, such as walls, ceilings, high work requiring staging, large open areas with few obstructions, and confined areas where free movement is restricted. All will have a direct effect on productivity.

Procedures used to deduct for openings vary, but in general, window openings less than 4 SF are not deducted, while larger openings without plastered *reveals* or returns are deducted in full. Reveals at openings should be calculated at 1-1/2 times their actual area. Specialty items such as arches, pilasters, columns, rosettes, centerpieces, and special patterns are listed separately and priced on an individual basis. Screws or nails for lath are estimated by allowing approximately 1 fastener for every SF of lath material. Multiple layers will follow the same procedure. The size and type of fastener is determined by the thickness of the lath and the corresponding requirement in the specifications.

Gypsum Drywall

Gypsum wallboard, more commonly known as *drywall,* is manufactured by combining gypsum powder with other ingredients and water and sandwiching this material between two layers of treated paper—a smooth finish paper at the exposed surface and edges, and a backing paper at the back side of the sheet. Because of its mineral core, drywall does not support combustion. It is durable for most applications, and it has sound-insulating characteristics. Drywall is manufactured in 1/4″, 3/8″, 1/2″, and 5/8″ thicknesses, a width of 4′, and lengths of 8′, 9′, 10′, 12′, and 14′. Drywall is specially manufactured for a variety of specific applications, such as fire-resistant, water- and moisture-resistant, foil-backed for retarding vapor transmission, and prefinished. Specialty products, such as 3/4″-thick fire-resistive drywall, exterior gypsum sheathing for applications at exterior sidewall and ceilings, and 1″-thick shaft wall liner for use in area separation and cavity wall construction, are also available. Drywall sheets are installed on wood or metal framing systems with screws. The seams of abutting drywall sheets and screw holes are taped with a vinyl-based joint treatment compound and reinforcing tape. Joint taping is a multi-step process usually requiring three coats, including the tape coat. The tape is embedded in a coat of joint compound and allowed to dry. Subsequent coats are applied, with sanding between coats to remove the imperfections. The resulting finish conceals the joints, and should be smooth and ready for paint or other finishes.

Accessories for drywall construction, such as corner bead for outside corners, "J" bead, and "L" bead for the termination of drywall edges, are similar in appearance and function to those used in plaster work, and are taken off in the same way.

Drywall ceiling installations can be covered with a textured finish to achieve a simulated acoustical ceiling finish. Textured finishes are powder products with a polystyrene aggregate that is mixed with water or paint and sprayed or roller-applied to ceilings. Finishes are available in textures ranging from fine to super coarse, and are sold in dry bag form.

Units for Takeoff

Drywall can be taken off as a whole system, taped and finished, or by the individual components. Drywall taken off as a whole system should include all the components and should be quantified by the SF. Takeoff units for individual components are as follows. Drywall is taken off by the SF of surface area to be covered, with quantities separated according to type and thickness. SF quantities can be extended to the number of specific size sheets required. Multiple drywall layers for fire-rated assemblies should be listed in the takeoff according to their position in the assembly (i.e., first layer, second layer, finish layer) for accurate pricing of the taping and finishing work. Taping and finishing is taken off by the SF and should be listed according to the level of finishing required (i.e., tape coat only for first and second layers in multi-layer systems, and finish taping for the exposed layer). The quantity of joint compound required is determined by the SF area to be finished and the coverage per the manufacturer's sales unit. Reinforcing tape is taken off by the LF and converted to the quantity of rolls needed. Trims are taken off by the LF and listed according to type and size required.

Textured finishes for ceiling applications are taken off by the SF area to be covered, and can be converted to the manufacturer's standard sales unit depending on the coverage of the individual product. Quantities should be separated according to method of application, specifically spray or handwork.

Other Takeoff Considerations

The specifications should be reviewed for the type of drywall needed for different locations within the structure, such as water-resistant, fire-rated, and prefinished sheets. Architectural drawings often provide small cross-sectional views through the various types of walls used on the project. These are referred to as *wall types*, and should be studied carefully for the various components that comprise the particular wall.

Like lath work, drywall work should be separated according to location of the work, such as walls; ceilings; high work requiring staging; large open areas; and confined areas for accurate labor pricing. Listings of areas to receive textured finishes should follow the same procedure for accurate labor pricing.

Deducting openings in the surface area should be done using the same procedure as used for plastering work (see previous section). The quantity of screws for attaching drywall is calculated at approximately 1 screw per SF of area to be fastened. Screws for multiple-layer installations are calculated the same way.

Joint compound for taping and finishing work is available ready-mixed in various-sized containers, the most common of which is the five-gallon pail, as well as in dry powder form for on-site mixing. Special joint compounds with accelerators for faster drying between coats are also available and may be substituted for regular compound. Coverage of joint compound depends on the individual product and the number of external corners or trims to be "mudded." Numerous external corners or trims can reduce the coverage by as much as 30%.

Metal Studs and Furring

Metal stud framing and furring has become a popular choice for walls and ceilings on commercial projects beause they cannot support combustion. Discussion of metal framing and furring in this section is limited to nonload-bearing applications. For information on load-bearing light-gauge metal framing systems, refer to Chapter 8. Nonload-bearing metal framing and furring is manufactured from cold-rolled galvanized metal, and is available in a number of sizes, thicknesses (gauge), and shapes.

Various shapes include metal stud and track for wall and ceiling framing; "Zee" furring channel and "hat" furring channel for wall furring; resilient channels for acoustical control; and "C" channel for use in ceiling suspension systems.

Metal Stud and Track

The most common sizes of metal stud are 1-5/8", 2-1/2", 3-5/8", 4", and 6", with thickness of 20-25 gauge. Standard lengths range from 8′ to 16′, in 2′ increments. Longer lengths of larger sizes may be available. Metal stud partitions are framed in much the same way as wood partitions. Channel-shaped runners called *tracks* are positioned at the top and bottom of the wall and are similar to the wood plates in wood framing. The tracks are anchored to the floor and overhead structure with nails, screws, or, in the case of concrete or steel, a powder-actuated fastener. Metal studs are installed perpendicular to the track by fastening the flanges of the stud and track together with self-tapping screws. The studs are located within the track at the specified on-center spacing. Figure 12.1 shows a typical nonload-bearing metal stud partition assembly.

Metal furring members for installing gypsum-based boards on concrete or masonry wall surfaces are one of three main types:

- "Hat" furring channels are hat-shaped channels used for furring at walls and ceilings. The hat channel is installed at a specified on-center spacing by nailing, screwing, or powder-actuated fastening the flanges at the edge of the channel to the substrate.
- "Zee" furring channel is used in conjunction with rigid board insulation. Its two flanges are supported by a perforated web, differing in size to support rigid insulation of varying thicknesses. Zee furring is installed in the same manner as hat channel.
- Resilient channels are used to provide a separation between the gypsum panels and the framing members. They are used extensively on common walls and ceilings in multi-unit residential construction to dampen noise transmission between units.

Special steel framing components for use in cavity shaft wall and fire-rated construction are manufactured in 20-, 22-, 24-, and 25-gauge thicknesses and are galvanized to resist corrosion. They are assembled with

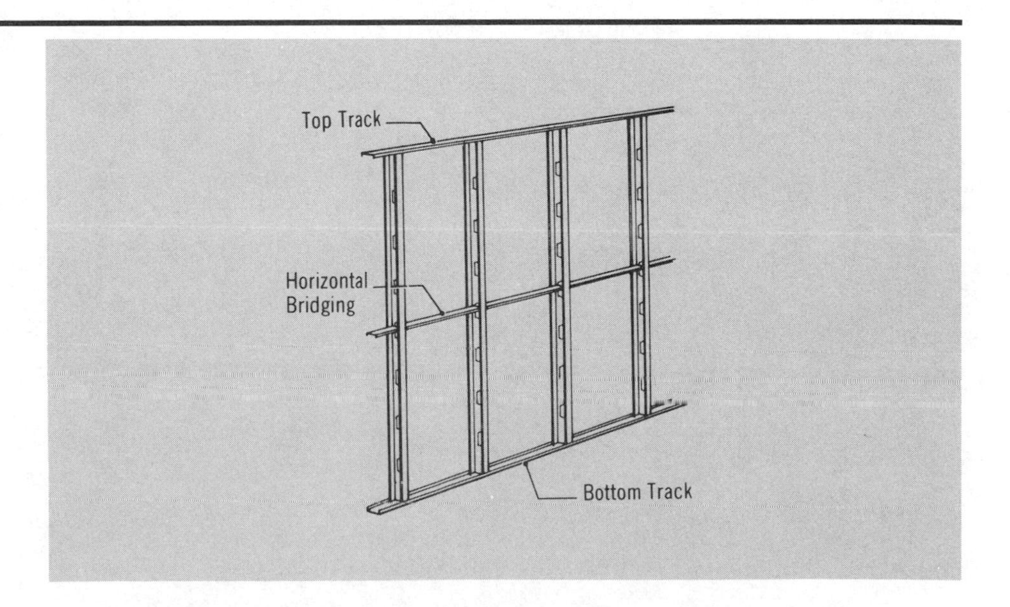

Figure 12.1 Nonload-Bearing Metal Stud Partition

components similar to regular metal stud framing, but are made to include support for 1″ shaft wall liner panels. They are available in 2-1/2″, 4″ and 6″ widths, and lengths from 8′-28′, depending on the shape and width specified. Some of the more common shapes encountered in shaft wall construction are C-H studs installed between abutting liner panels, E-studs used to cap panels at vertical intersections of walls, and J and C runners used as track at the top and bottom of cavity wall construction. Cavity wall framing materials are all nonload-bearing. Figure 12.2 illustrates the different types of studs and track used in cavity wall construction.

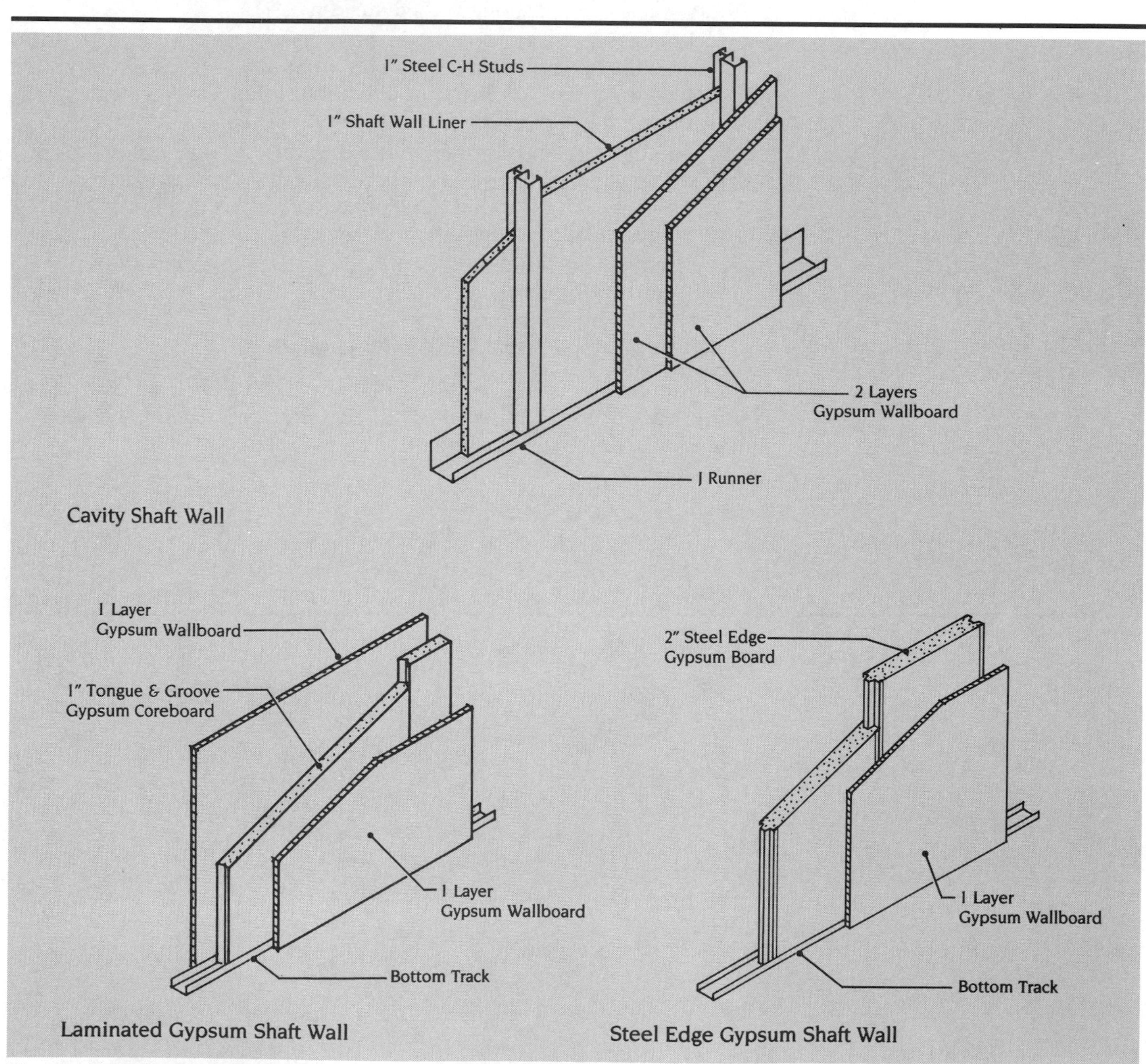

Figure 12.2 Studs and Track Used in Cavity Wall Construction

Other specialty members, such as "CRC" channels for use as main runners in ceiling suspension systems, are manufactured in 3/4", 1-1/2", and 2" sizes. The material is 16-gauge cold-rolled galvanized steel and is available in 16′ and 20′ lengths. It is typically suspended by means of wire tied to the structure above. CRC channels are spaced a maximum of 4′-0" on center longitudinally, with hat channel fastened transversely to the CRC channel at 12", 16", or 24" on center by wire or special clips.

Units for Takeoff

Metal stud and track are taken off by the LF and separated according to size, gauge, type, and length of studs. Quantities of studs and track are determined by dividing the length of the wall (in feet) by the on-center spacing (also in feet) of the studs, and adding one stud for the end.

Furring channels are taken off by the LF and separated according to type, size, and application (such as installation on concrete walls or wood studs, or wired to CRC channel).

CRC channels are taken off by the LF and separated according to size and location. To determine the LF quantity, the length of the ceiling (in feet) is divided by the on-center spacing (in feet). The resulting number is multiplied by the length of the individual pieces. Cavity wall framing members are taken off by the LF and separated according to type, size, and length of the individual member. The quantity of J and C runners are determined by measuring the length of the wall at the top and bottom. The number of C-H studs is determined by dividing the length of the wall (in feet) by the spacing (also in feet) and adding one stud for the end.

Other Takeoff Considerations

Determining the quantity of framing screws is based more on experience than on any given formula. While it is prudent to check the specifications for type and size of screws, the contractor will generally assume that for every LF or SF measure of area a certain number of pounds of screws will be needed, based on industry standards. When the final purchase is made the exact size and quantity will be determined. The type and size of screws should be designated in the specifications, which will also give specific requirements on related work in other sections, such as the installation of insulation within the walls.

Additional studs may also be needed for regular metal framing at door and window openings. Although there are no structural headers required, a piece of track is used to terminate the gypsum surface at the head of the window, and additional studs are needed at the jambs of doors and windows. This requirement should be considered for cavity wall framing members as well.

Powder-actuated fasteners are typically expensive and should be calculated by dividing the total length of track (top and bottom) and the total length of furring channels by the on-center spacing of the fasteners, as dictated by the specifications.

Tile

Tile is manufactured from clay, porcelain, or stone. It is available in a wide variety of sizes, shapes, colors, textures, patterns and thicknesses. Tile provides a hard, durable, and virtually maintenance-free surface for both interior and exterior use. It is a good choice where wear and tear is a concern.

In addition to standard tile pieces, special shapes or trim pieces are also manufactured to match the field tile. Some of the more common wall trims are bullnose, cove base, inside and outside corner pieces, custom transition shapes, and accessories such as toilet paper holders, towel bars, and soap dishes.

In addition to the actual type, size, and quality of the individual tile materials, the method used to install, or set, tile can have a large effect on the overall cost. Tile can be set with water-resistant premixed adhesives, which is the most common method for wall and floor tiles in residential bath applications. The adhesive is used directly from the shipping container and trowelled onto the substrate.

Another common method is the use of dry-set Portland cement mortar, commonly referred to as *thin set mortar*. Thin set mortar is available in a variety of different-sized bags, and is mixed with water to obtain a "toothpaste-like" consistency. It is trowelled onto the surface, and the tile is installed in the setting bed while it is still plastic. Specialty types of thin-set mortars for flexibility, exterior applications, and epoxy-based thin-sets are also available and in wide use.

Tile is available in individual pieces, or in back-mounted sheets for faster installation. Tile is set with small spaces between the individual pieces to give the appearance of a "grid" around the tile work. Size of the spacing, called the *grout joint*, depends on the type, size, location, and specific design of the tile. Some ceramic tile is manufactured with preset spacing "dimples;" others are spaced by the tile setter. Once the tile has set, the joints between the individual pieces are grouted to provide a continuous surface.

Grout is manufactured in many colors and textures, such as unsanded for wall tile applications and sanded for floor grout. Grout is a Portland cement-based mortar mixed with water or latex bonding additives and applied with a float trowel. The excess is cleaned off by sponge or other means to leave the surface clean.

The contractor should review the architectural drawings for the location of various tile work. Room finish schedules are also used in determining the location, amount and type of tile to be set. Floor tile quantities are determined from the architectural floor plans. Wall tile quantities can be taken off from the interior elevations, or from the perimeter of the walls shown on the floor plans. Special details showing sections through walls are used to determine the vertical limits of the wall tile, the substrate to be used, and the base or cap trim pieces. The technical specifications describe the type, size, and specifics concerning the product, as well as the setting method to be used.

For the purpose of takeoff, tile work can be classifed into two basic categories: ceramic tile and marble or stone tile.

Ceramic Wall and Floor Tile

Ceramic wall and floor tile is manufactured from clay, porcelain, or similar materials, and baked in a kiln to a permanent hardness. Ceramic tile is available glazed or unglazed. Glazed tile has an impervious glassy facial finish available in a multitude of colors. Special trim pieces, such as bullnose tile, are used as transition pieces to terminate the tile work at adjacent surfaces. Inside and outside corner pieces are used to change the direction of the bullnose tile, or at the intersection of tiled surfaces. Special trim pieces that are used at the intersection of the wall and floor tile are called *coves* or *bases*, and are available with inside and outside corners. Other trim pieces for countertop edging and swimming pool coping are also manufactured. Ceramic accessories, such as toilet paper holders, soap dishes, and towel bars, are also manufactured to match the main, or *field* tile.

Ceramic *mosaic tile* is smaller, with a face area of less than 6 square inches. Mosaic tiles are mounted pre-spaced on backing sheets to facilitate setting. They are manufactured glazed or unglazed, and in various colors, patterns, and designs. Ceramic mosaic tile is used primarily for floor or decorative wall applications. Figure 12.3 shows the typical locations and uses of field and trim ceramic tile.

An extremely hard, durable form of ceramic tile called *quarry tile* is manufactured for use primarily as a floor tile. It is long wearing and has characteristics that resemble brick pavers. Quarry tile is also available with abrasive chips embedded in the wearing or facial surface for non-skid applications. Trim pieces similar to ceramic wall tile trims are also available to match the field tile.

Units for Takeoff

Field tile for ceramic and quarry tile applications is taken off by the SF. Takeoff quantities should be separated according to type, size, color, surface finish (glazed or unglazed), and location (floor or wall). Accent tile of a different color or size within the field tile should be taken off and listed separately. Special tiles with logos, designs or surface features should be taken off and listed separately, usually by the piece (EA). Ceramic mosaic tiles are taken off by the SF and listed separately. Ceramic and quarry trim pieces, such as bullnose and base, are taken off by the LF and can be

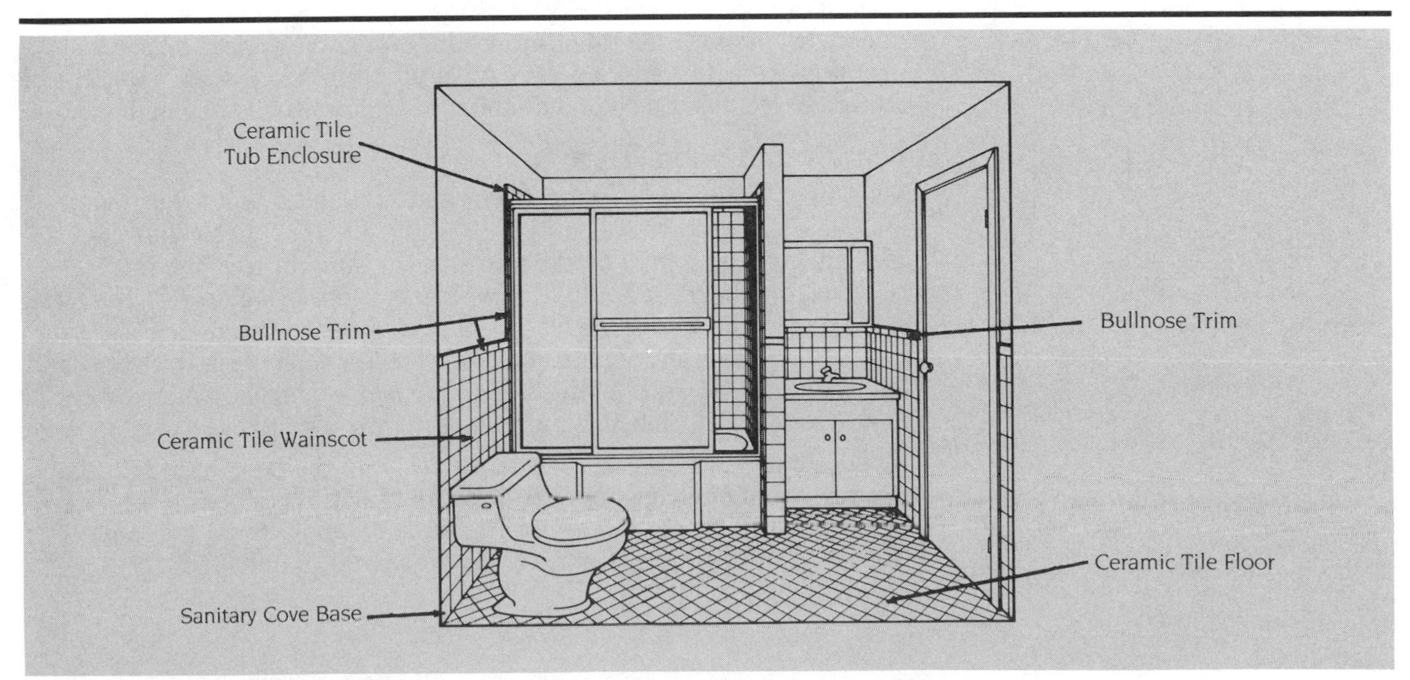

Figure 12.3 Uses of Ceramic Tile

converted to the number of individual pieces required. Inside and outside corner pieces are taken off by the piece and listed separately. Deductions for window and door openings should be in full, but the trim pieces that may be required to finish the perimeter of the opening must be included in the takeoff.

The quantity of setting materials, such as thin-set or adhesive, is calculated by the SF of tile to be set; SF quantities are then converted to sales units based on the manufacturer's coverage for the particular product. Required bonding additives are determined by the manufacturer's ratio of liquid to the dry material.

Calculating the grout for tile joints is done using the same procedure for setting materials. The SF area and the size of the grout joint (width and depth) are required to accurately determine the quantity of grout needed. The grouting and cleaning phase can be listed in the takeoff by the SF.

Other Takeoff Considerations

Special designs within the field of wall or floor tile should be noted in the takeoff, as they require additional time for layout.

Most ceramic tile cutting is done by a hand tile cutter. Occasionally harder or more dense types of tile and special-shaped cuts require the use of an electric saw with a water-cooled diamond blade. Projects with a large number of cuts should be noted, as this can affect the productivity of the tile setter as well as the wear and tear on the saw.

A time lapse is required between setting the tile and grouting the joints to allow the setting materials to cure. Special setting materials that allow grouting within hours of installation are available; but the extra cost should be considered. Delays for curing time can adversely affect the cost of tile work on small areas, such as bathrooms.

Because of the characteristic weight and bulk of tile, shipping and handling costs can be quite high. The contractor should secure quotes from tile suppliers F.O.B. job site.

Special cleaning solutions such as muriatic acid may be necessary to adequately remove the dried grout film on the tile surface after initial grouting operations are complete. As most specifications require that the tile surface be left clean, a second cleaning operation may be required.

Marble and Stone Tiles

Tiles manufactured from natural rock and stone, such as marble, granite and slate, are used as wall and floor finish surfaces. Marble and granite tiles are cut from larger blocks into smaller, uniform tiles, the most common of which is 12″ x 12″. Thicknesses vary from 1/4″ to 3/4″. The facial surface of marble and stone tiles are factory-polished to a high gloss. Slate tile used for floor and wall applications are less uniform in thickness, and vary between 1/4″ and 1/2″ thick. Sizes are in 3″ multiples, ranging from 6″ x 6″ to 12″ x 12″. Slate tiles are packaged by square footage.

Most setting materials used in ceramic tile work can be used in setting stone tiles, although specialty products for certain types of stone tiles may be required. Trim pieces are not commonly manufactured for stone tiles; consequently, the external corners may require field mitering.

Grouting joints between marble and stone tiles is accomplished in the same manner as in ceramic tile work. Cleaning grouted surfaces is done mainly with water and light detergents, in contrast to the use of acid in some ceramic work.

Units for Takeoff

Stone tiles are taken off by the SF, and can be extended to the quantity of individual pieces required. Marble, granite and slate tile quantities should be separated according to size, thickness, type (color, species, manufacturer, lot number, etc.), location (wall or floor), and method of setting. Mitering of external corners, and field polishing should be taken off by the LF and listed separately according to the operation. Specialty pieces such as marble thresholds are taken off by the individual piece and listed according to size and type. Deductions for openings are in full. Determining quantities of setting materials and grout is the same as for ceramic tile installations, and can be expressed in the takeoff as a SF quantity.

Other Takeoff Considerations

Stone tile cuts are done by an electric saw with a water-cooled diamond blade. Projects with large quantities of cuts or miters should be noted in anticipation of reduced productivity.

Special setting materials for use with certain colors of marble can be expensive and may also reduce productivity because of their lack of workability. This should be taken into account in the takeoff.

Special materials for sealing the surface of stone tiles may be specified. The quantity of sealant is determined by the SF of tile, converted to the manufacturer's sales unit based on coverage.

Acoustical Ceiling Systems

Acoustical ceiling systems are designed to absorb and generally reduce sound transmission within a space. Most acoustical ceiling tile is composed of mineral fibers and is available in a variety of colors, patterns, and textures. For takeoff purposes, acoustical ceilings can be classified by application in three groups: acoustical ceiling tiles attached directly to ceiling substrates such as wood or gypsum; suspended acoustical ceiling tiles with a concealed spline; and acoustical tile installed in an exposed suspension system.

Acoustical tiles are manufactured in size increments of 12" in each dimension, with 12" x 12", 24" x 24", and 24" x 48" being the most common. Other, nonacoustic, ceiling tiles can also be installed in a suspension system. Decorative thin metal panels, either embossed or plain, and in chrome or brass finishes, are available in some sizes. Vinyl-coated gypsum panels are manufactured for ceilings that require washable surfaces, such as kitchens and food processing areas.

Suspension systems usually consist of aluminum or steel main runners of light, intermediate, or heavy-duty construction, spaced two, three, four, or five feet on center, with snap-in cross tees in one- to five-foot lengths. Main runners are hung from the supporting structure with tie wire at the required spacing. Small metal angles, called *wall angles*, are attached to the perimeter wall or vertical surface at the ceiling height to complete the grid. Suspension systems are available in colors or metallic plating.

The contractor should refer to the architectural drawings, with special attention to the reflected ceiling plan for ceiling tile requirements. Quantities are determined by measuring the length and width of the room to receive the ceiling. Building cross-sections should show the "length of the hang" of suspended systems. Specifications should indicate the manufacturer, model, color, fire rating, and any other specifics that will affect the cost of the materials. Room finish schedules should also be studied to determine the various ceiling heights and the location of different types of tile.

Units for Takeoff

Acoustical ceiling systems are taken off by the SF. Quantities of the individual components—acoustical tile, main runners, cross tees, wall angle, and accessories—may be taken off separately for accurate materials pricing. If this method is used, tiles are taken off by the SF and converted to the number of pieces needed. Quantities should be separated according to size, type, color, and special characteristics, such as reveal or square edge and fire ratings.

Suspension systems can be converted from SF of tile areas to the LF of wall angle and main runner. Cross tees, hanging wire, and hold-down clips are taken off by the piece. Individual components should be listed according to size, color or finish, manufacturer, model, and fire-rated characteristics.

Other Takeoff Considerations

In determining SF quantities, the contractor should separate large, unobstructed areas from smaller, confined areas, as this will affect productivity.

The length the ceiling will hang from the supporting structure, and the supporting structure materials, should also be noted as it will have a direct effect on the cost of installation. For example, hanging from open web bar joist requires less labor than drilling anchors into precast concrete plank.

Special staging required for workers to reach the ceiling height should also be noted in the takeoff.

Quantities of materials, as well as cost of installation, are directly affected by the specified layout. Laying out tile to center on sprinkler heads or other items projecting from the ceiling, as well as equal size "cuts" along the perimeter, may entail a greater waste allowance in the material quantity, which should be added into the takeoff.

Since many ceiling tiles are directional—that is, the pattern on the exposed face of the tile has a direction—cut pieces from one side may not always be usable on the opposite side. In such cases, extra quantities should be included (to the nearest tile width or length) for waste allowance.

Ceilings with sloped soffits or *splays* should be noted separately in the takeoff, by the SF or LF, as they will affect installation productivity.

Special fire-rated systems should also be taken off and noted separately.

Reveal edge tiles that have been cut to fit a grid opening must be grooved at the cut edge to replicate the reveal along this edge. This is called a *kerf* and allows the cut edge to fit within the grid system.

Wood Flooring

Wood flooring is available in a variety of styles, grades, species, and patterns for use as interior finish flooring. Wood flooring is manufactured in solid and laminated planks, or solid and laminated parquet. Laminated flooring can be defined as a high-grade wood veneer laminated over a lesser-grade base material. Laminated wood flooring is sold prefinished. Solid wood flooring is available prefinished or unfinished, and is classified by grade or quality. Commonly sold hardwood floor materials are milled from oak, ash, maple, beech, walnut, cherry and mahogany. Other exotic hardwoods include teak, ebony, zebrawood, and rosewood. Popular softwood flooring materials include fir, pine, cedar, and spruce.

Prefinished flooring materials range in thickness from 3/8″ for parquet to 3/4″ for strip or plank flooring. Strip or plank flooring materials are sold in several different milling formats and combinations, including tongue-and-groove and matched, and square-edge and jointed with square edges and splines. The installation of tongue-and-groove flooring usually requires blind nailing, while square-edge flooring is face-nailed. Strip and plank flooring is usually installed over wood subflooring attached to joists. A layer of building paper is installed between the subfloor and the strip flooring to act as a cushion and reduce noise.

Parquet flooring is manufactured in prefabricated panels of various sizes, which are milled with square edges, and tongue-and-groove. The panels are available with backing materials to protect the flooring from moisture, provide insulation, reduce noise, and add comfort to the walking area. Adhesives are typically used to attach panels to the subfloor of wood, concrete, plywood, or other sound subfloor material. Both strip and parquet flooring can be installed in a variety of designs or patterns, and strip plank flooring can be installed with plugs at the end of each piece to simulate a method of installation.

Unfinished wood flooring involves repeated machine sanding with increasingly finer-grit sand paper, followed by the application of a finish coating. Finish coatings can vary from polyurethanes to tung oil, and may require more than one application.

Architectural drawings show the locations and areas of wood flooring. Special attention should be given to floor plans and room finish schedules as well as the technical specifications for the grade, species, type, finish, and method of installation for the wood flooring. Figure 12.4 shows types of wood flooring.

Units for Takeoff

Wood flooring is taken off and listed by the SF, according to type (plank, strip, parquet), species of wood, grade, finished or unfinished, and method of application. Large, open areas should be separated from smaller areas such as stair landings and closets, where productivity will decrease.

Area in SF and the required nailing pattern determine nail quantities for face or blind nailing.

To calculate adhesive quantities for applying parquet flooring, convert the SF area to gallons based on the manufacturer's recommended coverage. Calculate underlayment paper quantity by converting the SF area to rolls. Include allowances for side and end laps.

Sanding and finishing unfinished strip flooring is quantified by determining the SF area of installed flooring. Finish materials are converted from SF to the typical sales unit (one- or five-gallon containers), based on

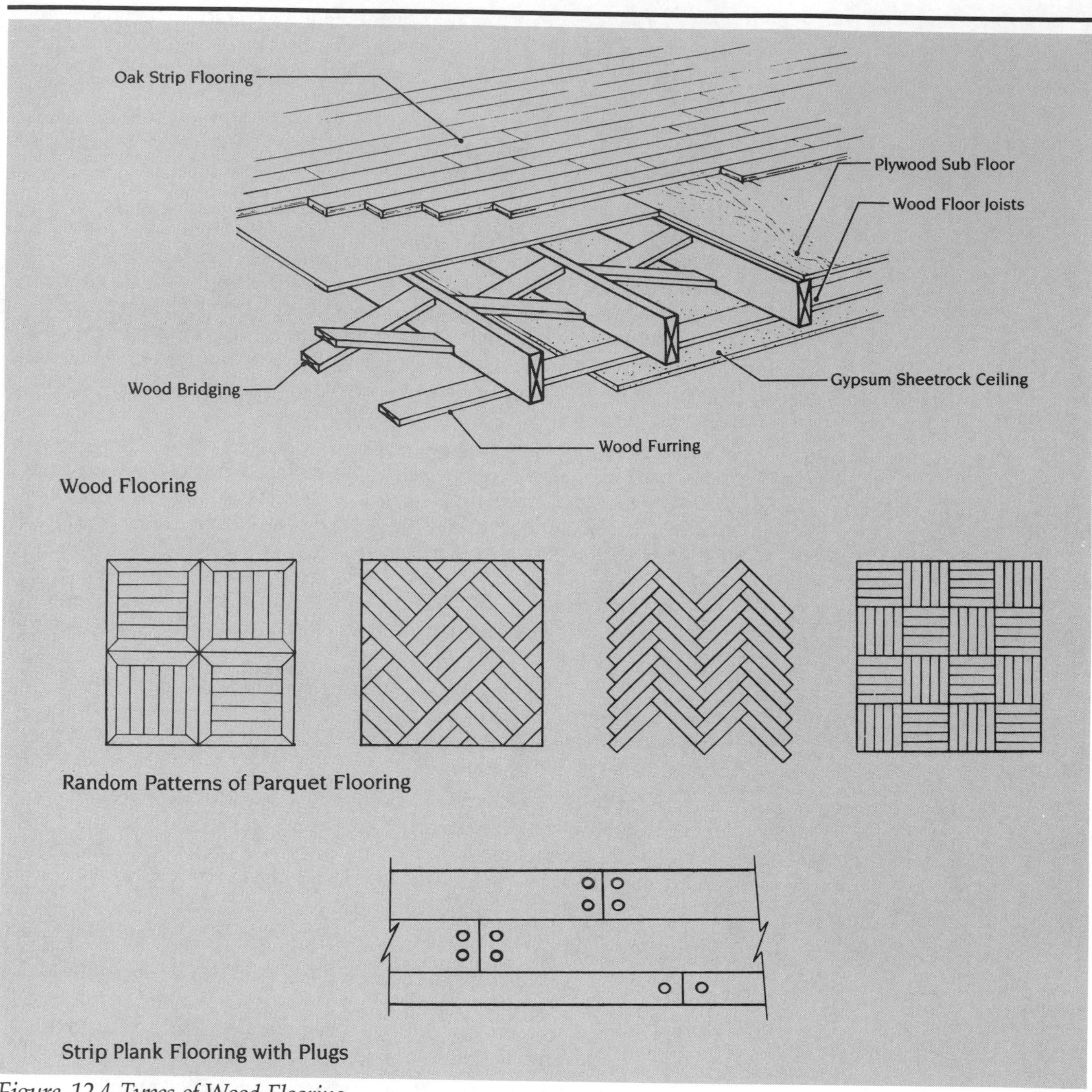

Figure 12.4 Types of Wood Flooring

the manufacturer's recommended coverage. The procedure is repeated for additional coats of finish or light sanding between coats.

Other Takeoff Considerations

Transition pieces, such as wood thresholds, between different types of flooring should be noted and taken off separately by the LF. For transition pieces not in the same plane as the floor, additional labor for hand finishing may be included.

Areas requiring special designs and patterns should be noted in the takeoff, to allow for reduced productivity and additional waste materials. Waste on prefinished flooring depends on the amount of cutting, the patterns, and the individual product. In general, larger open areas have less waste per SF than smaller areas where cutting is more frequent. Waste on unfinished strip flooring materials is also directly related to the grade and species of wood being used. Higher quality flooring materials have less waste resulting from natural imperfections that are found in lesser-quality wood materials.

Determining the quantity of stain needed for wood flooring follows the same procedure as for finish coatings.

Resilient Flooring

The category of resilient flooring includes materials such as asphalt tiles, vinyl composition tiles, cork tiles, rubber tiles, and sheet goods composed of rubber, vinyl, and polyvinyl chloride. Resilient flooring is designed for situations where durability, low maintenance, and flooring longevity are of primary consideration. All of the materials are manufactured with or without resilient backing, and are available in a wide variety of sizes, colors, patterns, textures, and styles. Rubber and vinyl accessories for resilient flooring include base, thresholds, transition strips, stair treads and risers, and nosings.

Resilient flooring and accessories are installed with adhesives, although some composition tiles are manufactured with a self-adhering backing.

Architectural drawings, specifically floor plans, contain information on quantities of resilient flooring materials and accessories. Refer to room finish schedules and the applicable technical specification section for the location, limits, product information, and method of installation of the various resilient flooring types.

Units for Takeoff

Resilient flooring is taken off by the SF. Sheet goods are extended from the SF to the SY. Takeoff quantities should be listed according to type, size, thickness (gauge), method of installation, and location.

Accessories such as base, transition strips, and thresholds are taken off by the LF, while stair treads, riser covers, and nosings are taken off by the LF or by the piece.

Quantities of adhesives for the installation of resilient flooring are calculated by converting the SF of flooring to be installed to units (most often gallons) of the individual product. Coverage depends on the manufacturer's recommended square footage over the specific subfloor.

Other Takeoff Considerations

Allowances for waste depend on the layout required. Most specifications call for the resilient tiles to be placed beginning at the center of the floor area so that cuts along opposite walls will be the same size. The contractor may want to increase the measurement of room dimensions to the nearest full tile on each side. Because of the thinness and flexibility of resilient goods, defects in the subflooring easily "telegraph" through the flooring. Consequently, the subfloor materials and the quality of surface preparation are very important. Floor preparation, even for new construction, should be included in the takeoff. This is expressed in SF of area to be prepped.

Carpeting

Carpeting is used as a flooring material for residential and commercial construction because it provides a comfortable, sound-absorbing, and attractive finish surface. Carpeting comes in a variety of different materials, including nylon, wool, acrylic, polyester, and rayon. Carpet is specified by weight, pile thickness, and density. Special carpets for commercial applications, such as fire-rated materials and anti-static carpet for use with computers, are also available. Commercial-grade carpeting is manufactured in rolls and tiles; residential carpeting is available mainly in rolls. Size of rolls will vary with the type, quality, model, and manufacturer of the specific carpeting. Carpet rolls are generally available in 9-, 12-, 15-, and 18-foot widths; 12- and 15-foot widths are the most common. Carpet tiles are available in 18" x 18" and 24" x 24" sizes.

Carpeting is installed over cushion material called *padding*, which is manufactured from many different products, including animal hair, rubberized fibers, cellular rubber, and polypropylene. Some types of carpeting—mainly commercial carpets and carpet tiles—are manufactured with an integral cushion backing.

Accessories for carpeting include metal or rubber edging at the intersection of carpet and other types of flooring, as well as specialty carpeting for stair runners.

Most carpeting is installed using one of two methods: directly gluing the carpet to the floor, or using pad underlayments and perimeter fastening.

The contractor should refer to the architectural drawings, especially Room Finish Schedules and floor plans, for location and limits of the carpeting. Specifications should be reviewed for the type, style, manufacturer, model, color, weight, and method of installation of the carpet.

Units for Takeoff

For the purpose of estimating, carpeting can be divided into three main categories: carpet installed with a separate pad; pre-backed carpet glued directly to the substrate; and carpet tiles.

All carpeting is taken off by the SF and extended to the SY, with the exception of carpet tiles, which are extended to each piece for accurate material pricing. Quantities should be separated according to manufacturer, model, type, style, weight, color, and method of installation. Padding materials are taken off by the SF; certain types may be extended to the SY. Quantities of pad are listed according to type, composition, manufacturer, and model. Fasteners for the perimeter of carpet with separate pad are taken off by the LF of abutting vertical surfaces (walls).

Adhesives, referred to as *mastic,* are calculated by converting the SF area to sales units (usually 1- or 5-gallon containers) based on the manufacturer's recommended coverage.

Accessories, such as edge strip, are taken off by the LF. Carpet runners for stairs are taken off by the LF and width of the runner. Quantities of stair runner are determined by adding the total SF of all risers and treads (including overhang of the treads) in the flight of stairs.

Other Takeoff Considerations

Designs that incorporate different types of carpeting within one room (e.g., different color border along the perimeter of the room) should be taken off by the SY of each carpet and listed separately.

When determining the actual quantity of carpet required for one room, consideration should be given to the available width of the particular carpet. For example, a room that is 10′-3″ x 20′-3″ has an area of 23 SY, but if the carpet is available in 12′-wide rolls, a piece 12′ x 21′ or 28 SY would have to be purchased. This difference represents a 22% waste if the cut-off piece could not be used elsewhere. Alternate methods that employ the use of seams may be acceptable, and may reduce the amount of waste. The contractor should refer to the specifications for rules governing the use of seams.

Carpeting with special designs or repeating patterns requires careful calculation to allow for matching the pattern. Carpeting that simulates an Oriental rug is a good example of this.

Painting

Painting and finishing of materials or surfaces that are formerly unfinished include a wide variety of items, both at the interior and the exterior of the structure. Painting and finishing are required to protect interior and exterior surfaces against wear and deterioration, and to provide a decorative appearance to the work. Painting and finishing work includes minor preparation to the surface, such as light sanding, filling of nail holes, and the removal of oils or dust. More involved preparation for painting or finishing includes applying wood conditioner or priming raw surfaces.

Painting and finishing materials include latex or water-based paints and stains, alkyd or oil-based paints and stains, epoxy coatings that are catalyzed to produce durable chemical-resistant finishes, and urethanes that are polymers with drying oils.

Methods for applying paints and stains include brushing, rolling, and spraying.

Architectural drawings should be examined for the various items that will require painting or finishing. Room finish schedules should be studied for the location and limits of the interior painting work on wall, ceiling, and floor surfaces. Exterior and interior elevations should delineate the areas of painting and finishing, and should be cross-referenced with the door and window schedules for the quantity of windows and doors to be painted or finished. Drawings showing sections or details of installed casework, trim, shelving, and miscellaneous items that may require field painting should also be reviewed.

Roof plans showing rooftop equipment requiring painting should be studied. Mechanical plans should be reviewed for piping in exposed areas that require identification painting, such as water, gas, fire-protection, or chemical or process piping.

Technical specifications should be studied carefully for the type of products to be used, method of application, number of coats, preparation work, colors or finishes, and the items to be painted.

For the purpose of quantity takeoff and estimating, painting and finishing work can be classified in two ways: exterior painting and interior painting.

Exterior Painting

Exterior work includes the painting or staining of exterior trims and soffits, siding, shutters, columns, doors and frames, windows, decks, porches, and rooftop equipment.

Units for Takeoff

Exterior trims and soffits less than 12" in width are taken off by the LF. Trims larger than 12" in width can be extended to the SF of area to be painted. Siding is also taken off by the SF, by measuring the actual surface area to be painted. Shutters, columns, doors and frames, and windows are taken off by each piece. Descriptions in the takeoff should include the size and number of sides to be painted on items such as shutters and columns. Decks and porches are taken off by the SF, and railings and balusters are taken off by the LF and piece (EA), respectively. Exterior stairs are taken off by the individual components: treads and risers are by the piece (EA) and stringers, skirts, or cheek boards are by the LF.

Rooftop equipment is taken off by the individual piece (EA), with notation of size and type of equipment.

All work should be taken off and listed separately according to type of work (painting or staining), item (siding, trim, door, etc.), product, number of coats, color, and method of application.

Other Takeoff Considerations

Exterior work often involves the use of ladders or special staging to gain access to high areas. Productivity decreases as a result of frequent moves of ladders or staging. This should be noted for accurate labor pricing. Priming of all surfaces of trim prior to installation, commonly referred to as *backpriming*, should be noted separately, and should be taken off by the SF of surface area to be painted. This provides more accurate estimates of required quantities of paint.

Wood moldings installed on other wood trims that are to be painted different colors should be noted in the takeoff. This type of painting reduces productivity because of the added time required for "cutting-in".

Interior Painting

Interior work includes the painting or staining of drywall or plaster walls and ceilings, doors and frames, windows, wood trims such as baseboards, casings at doors and windows, chair rails, wainscoting, cornices, stair parts, wood shelving, casework, and concrete floors.

Units for Takeoff

Painting of interior walls and ceilings is taken off by the SF of surface area to be painted. Walls are calculated by multiplying the perimeter of the room by the height of the walls. Openings larger than 4 SF are deducted in full; openings smaller than 4 SF are negligible. Ceilings are calculated by multiplying the length by the width of the surface. Walls and ceilings should be listed separately in the takeoff.

Interior wood trims, such as baseboards, cornices, chair rails, casings, railings, and stair cheek or skirt boards are taken off by the LF and should be listed separately. Other stair parts such as balusters, treads, risers, and newel posts are taken off by the piece (EA) and listed separately. Windows and accessories such as grilles are taken off by the piece (EA). Interior doors and frames are taken off by the piece (EA).

Casework, cabinetry, and shelving are taken off by the SF or the individual piece (EA), based on what best represents the work.

Wainscoting or wood ceilings are taken off by the SF. Descriptions of the work, such as raised panel wainscoting or tongue-and-groove plank ceiling, should be noted for accurate pricing.

Concrete floors are taken off by the SF and are computed by multiplying the length by the width of the space.

Painting of piping is taken off by the LF, and should be listed according to the diameter of the pipe. Solvents for cleaning the oils and coatings off piping should be included as part of the preparatory work.

All interior painting should be listed separately according to type of work or item to be painted, type of coating (paint or stain), number of coats, color, product, and method of application.

Other Takeoff Considerations

Special preparations for interior work, such as acid etching of concrete surfaces, minor sanding, or filling of nail holes, should be noted on the takeoff.

Special staging required for high areas, such as cathedral ceilings, should be noted for accurate labor pricing.

Most specifications provide that the painter do minor touch-up work after the installation of the work of other trades. This is most apparent at the interior of a building where worker activity is high. Since touch-up work is labor intensive (as the materials are usually left over from the main work), allowances for touch-up are made by the labor-hour. There is no accurate way to predict the amount of touch-up that will be required. Estimates of labor-hours should be derived from historical data on similar work.

Figure 12.5 is a guide for determining SF of surface area for a variety of interior and exterior painted surfaces.

Wallcoverings

Wallcovering refers to a wide selection of materials that are applied as a decorative treatment to the surface of interior walls. Wallcoverings are manufactured, printed, or woven, in fabric, paper, vinyl, leather, suede, cork, wood veneers, and foils. Wallcoverings are available in different weights, backings, and quality.

Wallcoverings are installed with adhesives, sometimes referred to as *paste*. Certain wallcoverings require the use of special pastes that are compatible with the wallcovering.

Preparation of the wall surface prior to installation is necessary to ensure proper bonding of the wallcovering. Preparation includes minor sanding and repairing of wall defects, and applying *sizing*, or primer for proper adhesion.

Architectural drawings should be reviewed with special attention to floor plans for determining room perimeters, and room finish schedules for determining the location of wallcoverings within specific rooms, as well as the ceiling heights in the respective rooms. Interior elevations can be used

Each area should be compared to a flat wall surface by increasing the area with a predetermined percentage on a square-foot basis as follows:

Item	Type	Measure
* Balustrades:		1 Side x 4
* Blinds:	Plain	Actual area x 2
	Slotted	Actual area x 4
* Cabinets:		Front area x 5
* Clapboards and Drop Siding:		Actual area x 1.1
* Cornices:	1 Story	Actual area x 2
	2 Story	Actual area x 3
	1 Story Ornamental	Actual area x 4
	2 Story Ornamental	Actual area x 6
* Doors:	Flush	150% per side
	Two Panel	175% per side
	Four Panel	200% per side
	Six Panel	225% per side
* Door Trim:		LF + 50% per side
* Fences:	Chain Link	1 side x 3 for both sides
	Picket	1 side x 4 for both sides
* Gratings:		1 side x 4/6
* Grilles:	Plain	1 side x 200%
	Lattice	Area x 2 per side
	Moldings Under 12″ Wide	1 SF/LF
* Open Trusses:		Length x Depth x 2.5
* Pipes:	Up to 4″	1 SF per LF
	4″ to 8″	2 SF per LF
	8″ to 12″	3 SF per LF
	12″ to 16″	4 SF per LF
	Hangers Extra	
* Radiator:		Face area x 7
* Shingle Siding:		Area x 1.5
* Stairs:		Number of risers x 8 widths
* Tie Rods:		2 SF per LF
* Wainscoting, Paneled:		Actual area x 2
* Walls and Ceilings:		Length x Width, no deducts for less than 100 SF
* Sanding and Puttying:	Quality Work	Actual area x 2
	Average Work	Actual area x 50%
	Industrial	Actual area x 25%
* Downspouts and Gutters:		Actual area x 2
* Window Sash:		1 LF of part = 1 SF

Figure 12.5 Square Foot of Surface Area

as a means of cross referencing the limits and location of wallcoverings. Technical specifications should be studied for the individual product, adhesive, and surface preparation.

Units for Takeoff

Wallcoverings are taken off by the SF. Square footage is calculated by multiplying the linear footage of the wall by the height of the wall. Deductions are taken in full for windows and doors as well as baseboards and wainscoting.

The SF area is converted to rolls, where a single roll contains approximately 36 SF. This determines the basis for calculating the number of rolls required. Wallcoverings are generally manufactured in double or triple roll units called *bolts*. Commercial wallcoverings are manufactured in widths from 21" to 54" and in lengths up to 100-yard bolts. For an accurate quantity of the material needed for an individual product, the linear footage of walls to be covered is divided by the width of the wall covering (in feet), which results in the number of "strips". The number of "strips" per bolt can be determined by dividing the length per bolt by the height of the hang. Adjustments in the number of strips per bolt will be required based on the repeat of the pattern (if applicable). The number of strips required (rounded up to the nearest whole number) is divided by the number of strips per bolt (whole number) to calculate the total number of bolts required.

Quantities of adhesive or paste are calculated by dividing the total SF area to be covered by the manufacturer's recommended coverage per sales unit (typically gallon). Coverage may be expressed in rolls.

Area of surface preparation is taken off by the SF, and should be listed according to the level of surface preparation required.

Quantities of different wallcoverings should be listed according to type, width, size of bolt or roll, pattern repeat, and type of adhesive required.

Other Takeoff Considerations

Commercial wallcoverings are often required to be fire-treated so as not to support combustion. This can greatly affect the cost of the product.

Waste on wallcoverings can range from a low of approximately 10% for wallcoverings with no matching pattern to as much as 60% for wallcoverings with intricate pattern repeats.

Summary

The takeoff sheets in Figure 12.6 are the quantities for the work of Division 9—Finishes for the sample project. The separation of blueboard and plaster quantities into walls and ceilings in addition to individual rooms will be helpful when doing the takeoff of other Division 9 items such as painting and wallcovering. These quantities can be referred to as a means of checking painting and wallcovering quantities. The takeoff quantities for the painting of exterior and interior trims can be checked against the actual quantities of trim found in Division 6 takeoff sheets.

Note: Many specifications require that the contractor, as part of the work, provide the owner with a percentage of materials used in the finishes for future repair, replacement, or touch-up. The most common examples are carpet tiles; ceramic, quarry, and marble tiles; resilient flooring tiles and base; parquet flooring; acoustical ceiling tiles; paints or stains; and wallcoverings. The contractor should be aware of this requirement and include the cost of additional materials and delivery to the owner as part of the cost of the work.

Division 9 Checklist

The following list can be used to ensure that all items for this division have been accounted for.

Lath and Plaster

- ☐ Gypsum and metal lath
- ☐ Skim coat plaster
- ☐ Conventional three-coat plaster
- ☐ Fire-rated lath and plaster
- ☐ Screws, nails, and trim accessories

Gypsum Drywall

- ☐ Fire-rated drywall
- ☐ Moisture-resistant drywall
- ☐ Foil-backed drywall
- ☐ Taping and finishing
- ☐ Fire taping
- ☐ Spray textures for ceilings
- ☐ Screws and accessories

Metal Stud Framing and Furring

- ☐ Metal stud walls
- ☐ Metal stud ceilings and soffits
- ☐ Suspended ceiling systems for drywall
- ☐ Resilient channel furring
- ☐ Wall and ceiling furring
- ☐ Cavity and area separation wall construction
- ☐ Powder-actuated fasteners and framing screws

Tile

- ☐ Ceramic wall and floor tile
- ☐ Ceramic trim tile
- ☐ Stone tiles
- ☐ Thin-sets and adhesives
- ☐ Grouting and cleaning
- ☐ Ceramic bath accessories

Acoustical Ceiling Systems

- ☐ Suspended grid system
- ☐ Acoustical tiles
- ☐ Fire-rated ceiling systems
- ☐ Special ceiling tiles (vinyl-coated gypsum, metal tiles)
- ☐ Splays or sloped soffits
- ☐ Integrated ceiling systems
- ☐ Kerfing reveal edge tile
- ☐ Special staging

Wood Flooring

- ☐ Prefinished wood floors
- ☐ Hardwood strip floors
- ☐ Parquet floors
- ☐ Sanding and finishing flooring
- ☐ Nails and adhesives
- ☐ Staining floors
- ☐ Transition pieces

Resilient Flooring

- ☐ Resilient tiles
- ☐ Resilient sheet flooring
- ☐ Resilient base
- ☐ Stair treads, risers
- ☐ Nosings and transition pieces
- ☐ Adhesives

Carpeting

- ☐ Carpet and pad
- ☐ Direct-glue carpet
- ☐ Carpet tiles
- ☐ Adhesives and "tackless"
- ☐ Accessories

Painting

- ☐ Exterior painting
- ☐ Exterior trims and soffits
- ☐ Siding
- ☐ Doors, frames, and windows
- ☐ Backpriming
- ☐ Staging
- ☐ Interior Painting
- ☐ Walls
- ☐ Ceilings

Trims

- ☐ Doors, frames, and windows
- ☐ Casework
- ☐ Painted floors
- ☐ Piping
- ☐ Staging
- ☐ Surface preparation

Wallcoverings

- ☐ Fire treatment
- ☐ Sizing and surface preparation
- ☐ Adhesives

Means Forms

QUANTITY SHEET

PROJECT: SAMPLE PROJECT

SHEET NO. 9-1/8

ESTIMATE NO. 1

LOCATION

ARCHITECT: HOME PLANNERS

DATE

TAKE OFF BY: WJD

EXTENSIONS BY: WJD

CHECKED BY: KF

DESCRIPTION	NO.	DIMENSIONS				UNIT		UNIT
DIVISION 9: FINISHES								
BOARD & SKIM COAT PLASTER								
1/2" BLUE BOARD BASE W/ SKIM COAT PLASTER FINISH @ WALLS								
@ STUDY	1	59	9.12		539	SF		
DEDUCT FOR OPENINGS	1	6	7.67		(46	SF)		
	2	2	6		(24	SF)		
	1	4.5	6		(27	SF)		
	1	2.5	6.67		(17	SF)		
					425	SF	425	SF
@ LIVING ROOM	1	65	9.12		593	SF		
DEDUCT FOR OPENINGS	1	4	6.67		(27	SF)		
	1	3	3		(9	SF)		
	3	2.67	6		(48	SF)		
					507	SF	507	SF
@ HALL, FOYER, BSMT. STAIR, POWDER ROOM	1	120	9.12		1095	SF		
DEDUCTIONS FOR OPNGS.	1	3	6.67		(20	SF)		
	5	2.5	6.67		(84	SF)		
	2	2.33	6.67		(31	SF)		
	1	5	6.67		(34	SF)		
	1	2.67	6.67		(18	SF)		
					908	SF	908	SF
@ DINING ROOM	1	44	9.12		402	SF		
DEDUCTION FOR OPNG.	1	5	6.67		(34	SF)		
	2	2	6		(24	SF)		
	1	4.5	6		(27	SF)		
					365	SF	365	SF
@ KITCHEN	1	58	9.12		529	SF		
DEDUCTION FOR OPNGS.	2	6	7.67		(92	SF)		
	2	2.67	6.67		(36	SF)		
ADD FOR SOFFIT OVER CABINETS	1	20	2		40	SF		
					441	SF	441	SF

Figure 12.6

Means Forms

QUANTITY SHEET

PROJECT SAMPLE PROJECT | SHEET NO. 9-2/8 | ESTIMATE NO. 1

LOCATION | ARCHITECT HOME PLANNERS | DATE

TAKE OFF BY WJD | EXTENSIONS BY: WJD | CHECKED BY: KF

DESCRIPTION	NO.	DIMENSIONS				UNIT		UNIT		UNIT		UNIT
@ UPSTAIRS HALL	1	56	8.12		454	SF						
DEDUCTIONS @ OPNGS	5	2.5	6.67		(84	SF)						
	1	7	7	.5	(25	SF)						
					345	SF					345	SF
@ BEDROOMS, BATHS,												
& CLOSETS	1	172	8.12		1396	SF						
	1	74	8.12		600	SF						
DEDUCTIONS FOR OPNG	5	2.5	6.67		(84	SF)						
	8	4	6.67		(214	SF)						
ADD FOR SOFFIT OVER												
SHOWER & TUB	2	5	1		10	SF						
					1708	SF					1708	SF
5/8" TYPE 'X' BLUEBOARD												
@ GARAGE WALL TO HSE	1	21	9		189	SF						
DEDUCTION OPNG	2	2.67	6.67		(36	SF)						
					153	SF					153	SF
@ CEILINGS												
@ STUDY	1	13.5	16		216	SF						
	1	2	7.5		15	SF						
@ DINING RM	1	10.67	10		107	SF						
	1	2	7.5		15	SF						
@ LIVING RM	1	20	13		260	SF						
@ HALL, FOYER &												
POWDER ROOM	1	3.5	10.67		38	SF						
	1	15.5	3.33		52	SF						
	1	4	3.83		16	SF						
	1	3.25	9.4		31	SF						
	1	4	4		16	SF						
@ BSMT. STAIR	1	14.5	3		44	SF						
@ KITCHEN	1	19	13		247	SF						
KITCHEN SOFFIT	1	8.5	3		26	SF						
	1	13	1		13	SF						
@ UPSTAIRS HALL	1	19	3.1		59	SF						
	1	2	5		10	SF						
	1	4.1	2		8	SF						
@ 2ND FLR. STAIR	1	3	12.5		38	SF						
					1212	SF					1212	SF

Figure 12.6 (continued)

Means Forms

QUANTITY SHEET

PROJECT: SAMPLE PROJECT — SHEET NO. 9-3/8

ESTIMATE NO. 1

LOCATION — ARCHITECT: HOME PLANNERS — DATE

TAKE OFF BY: WJD — EXTENSIONS BY: WJD — CHECKED BY: KF

DESCRIPTION	NO.	DIMENSIONS				UNIT		UNIT		UNIT		UNIT
CEILINGS (CONT)												
@ BATH	1	8.67	5		44	SF						
	1	8.5	5		43	SF						
@ BEDROOM, CLOSETS	1	14	11.83		166	SF						
	1	4.5	8.5		39	SF						
	2	12.83	9.67		249	SF						
	2	4.67	2.1		20	SF						
					561	SF					561	SF
5/8" TYPE 'X' BLUEBOARD @ GARAGE CEILING	1	13.7	20.33		279	SF					279	SF
15% MATERIAL WASTE FOR 1/2" BOARD	6625 X .15				993	SF					993	SF
SCREWS FOR BLUEBOARD	6625 X 1/SF				6625						6625	EA
CORNER BEAD @ EXTERNAL CORNERS	20	10'			200	LF					200	LF
CERAMIC TILE												
1/2" TILE BACKER BOARD @ TUB & SHOWER	1	7	10.5									
SURROUND	1	5.5	10.5		131	SF					131	SF
4 1/4" x 4 1/4" CERAMIC TILE @ TUB SURROUND FULL HEIGHT SET W/ MASTIC	1	5.5	10.5		58	SF					58	SF
2"x6" BULLNOSE @ TUB	2	7			14	LF					14	LF
4 1/4"x4 1/4" CERAMIC TILE @ SHOWER SET W/ THIN SET — WALLS	1	7	10.5		74	SF					74	SF
FLOOR	1	2.67	5		14	SF					14	SF
CEILING	1	2.67	5		14	SF					14	SF

Figure 12.6 (continued)

Means Forms

QUANTITY SHEET

PROJECT	SAMPLE PROJECT
LOCATION	
TAKE OFF BY	WJD
ARCHITECT	HOME PLANNERS
EXTENSIONS BY:	WJD
SHEET NO.	9-4/8
ESTIMATE NO.	1
DATE	
CHECKED BY:	KF

DESCRIPTION	NO.	DIMENSIONS			UNIT	UNIT	UNIT	UNIT
2"x6" BULLNOSE @								
SHOWER PERIMETER	1	19			19 LF			19 LF
2"x2" CERAMIC TILE								
FLOOR SET w/THIN								
SET @ BATH FLRS.								
@ POWDER R/M	1	7.33	3.25		24 SF			
@ COMMON BATH	1	5	6		30 SF			
@ MASTER BATH	1	5.5	5		28 SF			
					82 SF			82 LF
CERAMIC TILE BASE								
@ BATH ROOMS								
@ POWDER ROOM	1	18			18 LF			
@ COMMON BATH	1	11			11 LF			
@ MASTER BATH	1	20.5			21 LF			
					50 LF			50 LF
GROUT & CLEAN								
CERAMIC TILE								
@ FLOORS	1	96			96 SF			
@ WALLS/CEILING	1	196			196 SF			
					292 SF			292 SF
CERAMIC SOAP DISH								
@ TUB/SHOWER	2	EA			2 EA			2 EA
CERAMIC TOWEL BAR								
@ TUB SHOWER (24")	2	EA			2 EA			2 EA
16 GA COPPER SHOWER								
PAN @ SHOWER								
2'-8" x 5'-0" w/ 6" SIDES	1	EA			1 EA			1 EA
CERAMIC TILE @ FOYER								
& HALL	1	10.67	3.5		38 SF			
	1	3.33	15.5		52 SF			
	2	4	3.25		26 SF			116 SF
GROUT & CLEAN FOYER/	1	116			116 SF			116 SF
HALL TILE								

Figure 12.6 (continued)

Means Forms

QUANTITY SHEET

PROJECT: SAMPLE PROJECT — SHEET NO. 9-5/8

ESTIMATE NO. 1

LOCATION — ARCHITECT: HOME PLANNERS — DATE

TAKE OFF BY: WJD — EXTENSIONS BY: WJD — CHECKED BY: KF

DESCRIPTION	NO.	DIMENSIONS				UNIT		UNIT		UNIT		UNIT
RESILIENT TILE												
PATCH & PREP KITCHEN FLOOR FOR RESILIENT TILE	1	18.5	10		185	SF						
	1	7	4.5		32	SF						
					217	SF					217	SF
12"X 12" RESILIENT FLOOR TILE SET IN MASTIC @ KITCHEN	1	217			217	SF					217	SF
WOOD FLOORING												
No 1 RED OAK 25/32" STRIP FLOORING SET & FINISHED W/ 2 COATS POLY.												
@ STUDY	1	13.5	16		216	SF						
@ D.R	1	12	10.67		129	SF						
@ L.R	1	20	13		260	SF						
DEDUCT FOR HEARTH	1	(6	1.5)		(9	SF)						
					596	SF					596	SF
ADD 15% FOR MATERIAL WASTE	596	15			90	SF					90	SF
UNDERLAYMENT PAPER	1.15	596			686	SF					686	SF
CARPET & PAD												
CARPET & PAD @ BEDROOMS, HALL, DRSG. AREA UPSTAIRS	1	19	3.1		59	SF						
	2	2	5		20	SF						
	1	1	3		3	SF						
	1	4.5	8.5		39	SF						
	1	14	11.83		166	SF						
					287	SF					32	SF

Figure 12.6 (continued)

Means Forms

QUANTITY SHEET

PROJECT: SAMPLE PROJECT

SHEET NO. 9- 6/8

ESTIMATE NO. 1

LOCATION:

ARCHITECT: HOME PLANNERS

DATE:

TAKE OFF BY: WJD

EXTENSIONS BY: WJD

CHECKED BY: KF

DESCRIPTION	NO.	DIMENSIONS				UNIT		UNIT		UNIT		UNIT
CARPET & PAD (CONT.)	2	9.67	12.83		249	SF						
	2	4.67	25		24	SF						
					273	SF					31	SY
INTERIOR PAINTING												
CEILING PAINT @ PLASTER CEILINGS (1 COAT)	1	2052			2052	SF					2052	SF
PRIMER @ PLASTERED WALLS (1 COAT)	1	4573			4573	SF					4573	SF
LAYTEX FINISH PAINT @ WALLS (2 COATS)	1	4573			4573	SF					4573	SF
SAND FILL, & PRIME, PAINT (2 COATS) @ STANDING & RUNNING TRIM												
@ CEILING	1	275			275	LF					275	LF
@ FLOOR	1	424			424	LF					424	LF
@ CHAIR RAIL	1	28			28	LF					28	LF
@ PANEL MOLDING	1	75			75	LF					75	LF
@ KITCHEN BEAMS												
- 5X5 1/2	1	36			36	LF					36	LF
- 8 3/4 X 9 1/4	1	13			13	LF					13	LF
- PER. 9 1/4 BEAM	1	64			64	LF					64	LF
- DOOR CASINGS	1	637			637	LF					637	LF
- WINDOW CASING & STOPS	1	269			269	LF					269	LF
- DOOR UNITS	17	EA			17	EA					17	EA
- SLIDERS	3	EA			3	EA					3	EA
- WINDOWS												
- FIXED UNITS	2	EA			2	EA					2	EA
20X36 D.H	4	EA			4	EA					4	EA
32X36 D.H.	3	EA			3	EA					3	EA
32X24 D.H.	1	EA			1	EA					1	EA
(2) 32X24 D.H	4	EA			4	EA					4	EA

Figure 12.6 (continued)

Means Forms

QUANTITY SHEET

PROJECT SAMPLE PROJECT | SHEET NO. 9-7/8
ESTIMATE NO. 1
LOCATION | ARCHITECT | DATE
TAKE OFF BY WJD | EXTENSIONS BY: WJD | CHECKED BY: KF

DESCRIPTION	NO.	DIMENSIONS				UNIT	UNIT	UNIT		UNIT
STAIN & FINISH (2 COATS) POLYURETHANE @ STAIR PARTS -										
- RAILS	1	16			16	LF			16	LF
- ROSETTES	2	EA			2	EA			2	EA
- BALUSTERS	16	EA			16	EA			16	EA
- TREADS	15	EA			15	EA			15	EA
- RISERS	16	EA			16	EA			16	EA
- CHEEK/SKIRT	1	24			24	LF			24	LF
SAND, PRIME, FINISH PAINT (2 COATS) EDGE BANDED SHELVING (BOTH SIDES)	2	74			148	SF			148	SF
STAIN & FINISH MANTELS										
- @ KITCHEN	1	EA			1	EA			1	EA
- @ LIVING RM	1	EA			1	EA			1	EA
EXTERIOR PAINTING										
PRIMER @ BEVEL SIDING (1 COAT)	1	2051			2051	SF			2051	SF
TWO COATS LAYTEX FINISH PAINT @ BEVEL SIDING	1	2051			2051	SF			2051	SF
PRIMER (1 COAT) @ STANDING + RUNNING TRIM @ EXT.										
- 1X6	1	582			582	LF			582	LF
- 1X8	1	28			28	LF			28	LF
- 1X10	1	46			46	LF			46	LF
- 1X12	1	71			71	LF			71	LF
- 5/4 X 2 1/4"	1	180			180	LF			180	LF
- 5/4" X 4"	1	10			10	LF			10	LF
- 5/4 X 6"	1	476			476	LF			476	LF
- 5/4 X 8"	1	257			257	LF			257	LF

Figure 12.6 (continued)

Means Forms

QUANTITY SHEET

SHEET NO. 9-8/8

PROJECT SAMPLE PROJECT

ESTIMATE NO. 1

LOCATION

ARCHITECT HOME PLANNERS

DATE

TAKE OFF BY WJD

EXTENSIONS BY: WJD

CHECKED BY: KF

DESCRIPTION	NO.	DIMENSIONS				UNIT		UNIT		UNIT		UNIT
TRIM (CONT.)												
– 5/4" x 10"	1	94			94	LF					94	LF
– 5/4" x 12"	1	31			31	LF					31	LF
– 3/4" x 4 1/4 CROWN	1	11			11	LF					11	LF
3/8" PLYWOOD SOFFITS –	1	301			301	SF						
FINISH PAINT (2 COATS) @ STANDING/RUNNING TRIM –												
– 1x6	1	582			582	LF					582	LF
– 1x8	1	28			28	LF					28	LF
– 1x10	1	46			46	LF					46	LF
– 1x12	1	71			71	LF					71	LF
– 5/4" x 2 1/4"	1	180			180	LF					180	LF
– 5/4 x 4"	1	10			10	LF					10	LF
– 5/4 x 6	1	476			476	LF					476	LF
– 5/4 x 8	1	257			257	LF					257	LF
– 5/4 x 10	1	94			94	LF					94	LF
– 5/4 x 12	1	31			31	LF					31	LF
– 3/4" x 4 1/4" CROWN	1	11			11	LF					11	LF
3/8" PLYWOOD SOFFIT	1	301			301	SF					301	SF
PRIME (1 COAT) & FINISH PAINT (2 COAT) WOOD SHUTTERS	10	EA			10	EA					10	EA

Figure 12.6 (continued)

Chapter Thirteen

SPECIALTIES

(Division 10)

Chapter Thirteen

SPECIALTIES
(Division 10)

Division 10 includes items that are manufactured off site and shipped prefinished and often preassembled for quick and easy on-site installation. This division includes chalkboards and tackboards, directories and signs, lockers, fire extinguishers, and toilet partitions and accessories. There are residential items included in this division as well, such as freestanding or prefabricated fireplaces, toilet and bath accessories, woodburning stoves, and cupolas. Specialties are usually installed at the end of the project.

Because Division 10 is somewhat of a "catch-all" category, containing a wide variety of items that don't quite fit anywhere else, the contractor should review all drawings in the bid set for locations of Specialties items. Ground-set flagpoles, for example, may not be shown anywhere except on the site plans. Architectural drawings sometimes offer only a general note that refers to the specialty item.

Technical specifications for Division 10 should be studied carefully, as this is often the most important information for accurate pricing of the specialty items. Specialties, by definition, are special or unique, and often the designer has carefully researched and specified an exact item. Many designers provide all pertinent information for the product, including the manufacturer or vendor's name, address, and telephone number, in the technical specification. This is a good indication that substitutions will not be accepted.

Units for Takeoff

Specialties in general are taken off by the individual piece (EA). For accurate pricing, items should be listed in the takeoff according to defining characteristics such as manufacturer model or series number, finish or color, and other physical characteristics such as size, length, and weight. For items that will be purchased and installed by the contractor, specific notes in the takeoff should include the level of site assembly required, special equipment needed for installation (such as a crane for flagpoles), manufacturer's recommendations for installation, and any general information that will facilitate accurate installation pricing.

Other Takeoff Considerations

Many of the items in Division 10 are designated *furnished and installed*. The contractor should solicit bids on this basis for items unfamiliar to his or her crew.

The cost for shipping and handling of Specialties may also need to be considered, especially for large, cumbersome, or heavy items.

The contractor should pay particular attention to the work of other specification sections that may affect the installation of the specialty item. For example, wood blocking for toilet partitions, structural steel shapes to support operable partitions, and concrete bases for lockers are not considered part of the work of Division 10, but must be included in the estimate.

Summary

The takeoff sheets in Figure 13.1 are the quantities for the work of Division 10—Specialties for the sample project. Residential construction, generally speaking, does not include many items of Division 10. The most common ones are toilet and bath accessories, mailboxes, and prefabricated fireplaces.

Division 10 Checklist

The following list can be used to ensure that all items for this division have been accounted for.

Specialties

- ☐ Chalkboards
- ☐ Toilet partitions
- ☐ Shower compartments
- ☐ Metal wall louvers
- ☐ Wall and corner guards
- ☐ Prefabricated fireplaces
- ☐ Fireplace accessories
- ☐ Woodburning stoves
- ☐ Flagpoles
- ☐ Cupolas
- ☐ Canopies
- ☐ Mail chutes
- ☐ Mail boxes
- ☐ Portable partitions
- ☐ Medicine cabinets
- ☐ Coat racks and wardrobes

Means Forms

QUANTITY SHEET

PROJECT SAMPLE PROJECT

SHEET NO. 10-1/1

ESTIMATE NO. /

LOCATION

ARCHITECT HOME PLANNERS

DATE

TAKE OFF BY WJD

EXTENSIONS BY: WJD

CHECKED BY: KF

DESCRIPTION	NO.	DIMENSIONS				UNIT		UNIT		UNIT		UNIT
DIVISION 10: SPECIALTIES												
TOILET ACCESSORIES												
TOILET PAPER HOLDERS	3	EA			3	EA					3	EA
TOWEL BARS (24")	3	EA			3	EA					3	EA
16"×36" MEDICINE CABINET (RECESSED)	3	EA			3	EA					3	EA
18"×36"×1/4" PLATE GLASS MIRROR W/ FRAME @ VANITIES	3	EA			3	EA					3	EA
SLIDING GLASS SHOWER ENCLOSURE W/ TRACK & HARDWARE 5'0" WIDE	2	EA			2	EA					2	EA

Figure 13.1

Chapter Fourteen

EQUIPMENT

(Division 11)

Chapter Fourteen

EQUIPMENT
(Division 11)

Division 11 includes the specialized apparatus and fixtures, often permanently installed, that allow the building to function as it was designed. Examples are kitchen appliances; central vacuum cleaning systems; and banking, library, medical, food service, and darkroom equipment.

The contractor's responsibility for equipment should be spelled out in the specifications. Equipment is often purchased directly by the owner, furnished and installed. Sometimes the owner purchases and arranges delivery of equipment, and installation is done by the contractor. For a new home, most appliances are the contractor's responsibility both for furnishing and installation.

Equipment is not always defined on the drawings with the clarity and thoroughness of other aspects of work. Vague notations such as "Furnished by owner, installed by general contractor" and "N.I.C." (not in contract) may be the only indications regarding contractor responsibility.

The contractor should review all plans in the bid set for correlation and coordination with the equipment specified. Specifications should be studied thoroughly to identify the need for related work (of other sections) that may be connected with the equipment. Examples include the hookup of equipment to building services such as water, waste, gas, electrical, or telecommunications. This is often indicated under the appropriate specification section in Division 15, Mechanical, or Division 16, Electrical. Other related work may include special structural elements for support, or wood blocking for attachment.

The items discussed in this chapter may be included in the contractor's scope of work.

Vacuum Cleaning Systems

Vacuum cleaning systems generally refer to permanently piped outlets at various locations with a centralized collection system. This type of system is most often encountered in residential and, occasionally, light commercial construction. It consists of thin-wall PVC or ABS piping, 2" in diameter, installed within the framework of the walls, floors, and ceilings. Individual lengths of pipe are connected by a variety of plastic fittings, such as 90-degree bends, 45-degree bends, couplings, and wyes, with PVC cement. *Ports*, or outlets, are attached to floor or wall framing members at various locations during the rough-in phase, and are later fitted with finish outlet covers. Piping is typically cut by hand with a hacksaw. Installation of piping requires the boring of wood members for access between floors. A two-lead low voltage wire is connected to each outlet cover that includes a set of contacts for activating the collection device. The collection device is a single- or two-stage electrical motor centrally located in a basement, garage, or maintenance area.

Units for Takeoff

Piping for vacuum cleaning systems is taken off by the linear foot and is extended to the length of pipe required. The pipe is typically sold in ten-foot lengths. Fittings and outlets are taken off by the individual piece (EA) and listed according to type and function. Low-voltage wire is taken off by the LF and converted to the roll size, most often 50′ or 100′ in length. The collection unit is taken off by the piece and listed according to manufacturer, model, weight, and power requirements.

Other Takeoff Considerations

The installation method and surface material on which the collection unit will be mounted (e.g., mounted on concrete walls or wood frame, or suspended from the structure above) may also be noted for a more accurate installation pricing.

The contractor should also ensure that the electrical specifications have made provisions for power and control wiring, and that accessories such as the hose, power head, and attachments are included as part of the work.

Residential Appliances

Residential appliances include items normally found in kitchens and laundry rooms. Specific items include refrigerators, freezers, trash compactors, ranges, range exhaust hoods, dishwashers, microwave ovens, washers, and clothes dryers.

Residential architectural drawings are limited, at best, in defining appliances, and generally show only locations and arrangements of appliances. Refer to the technical specifications for the manufacturer, model number, color, and type of appliance to be included in the takeoff.

Occasionally appliances are specified as an allowance item to be selected during construction. Allowances may be noted in Division 11, or in Division 1 under "Allowances."

Units for Takeoff

Residential appliances are taken off by the individual piece (EA) and listed separately according to the type of appliance, manufacturer, model number, color, and method of operation (gas or electrical, if applicable). Included in the takeoff should be provisions for shipping, handling, and installation. These may be listed as a lump sum (LS).

Other Takeoff Considerations

If allowances are to be included as part of the bid, the contractor should clearly indicate what they include. For appliances supplied by the owner, the contractor may need to include the cost of handling and "setting in place" for installation by the respective trades. Special consideration for handling is particularly important in the case of multiple-dwelling units, such as apartments or condominiums, where many appliances are needed.

A primary responsibility of the contractor is ensuring that individual subcontractors have included hook-up and utility services for the appliances. Specific queries include venting for the range and/or range hood; gas and/or electric for clothes dryers and ranges; waste, water, and electric for clothes washers and dishwashers; and water for automatic ice machines in refrigerators.

Note: For specialized equipment, a bid should be obtained from a qualified contractor specializing in the specific type of work. The general contractor's primary responsibility is to ensure that all items (both the equipment package and the related work that coincides with the equipment) has been provided for in the takeoff and estimate. This requires a thorough understanding of the specifications, and the scope of work of each party.

Summary

The takeoff sheets in Figure 14.1 are for the quantities of work of Division 11—Equipment for the sample project. Residential projects have very little in the way of specialized equipment, other than the kitchen and laundry appliances.

Division 11 Checklist

The following list can be used to ensure that all items for this division have been accounted for.

Equipment

- ☐ Vacuum cleaning systems
- ☐ Bank and vault equipment
- ☐ Ecclesiastical equipment
- ☐ Educational equipment
- ☐ Athletic equipment
- ☐ Food service equipment
- ☐ Library equipment
- ☐ Laboratory equipment
- ☐ Medical equipment
- ☐ Loading dock equipment
- ☐ Detention equipment
- ☐ Theater and stage equipment
- ☐ Residential equipment

Means Forms

QUANTITY SHEET

PROJECT: SAMPLE PROJECT	SHEET NO. 11-1/1
	ESTIMATE NO. 1
LOCATION	ARCHITECT: HOME PLANNERS
	DATE
TAKE OFF BY: WJD	EXTENSIONS BY: WJD
	CHECKED BY: KF

DESCRIPTION	NO.	DIMENSIONS			UNIT	UNIT	UNIT	UNIT
DIVISION 11: EQUIPMENT								
KITCHEN APPLIANCES								
24" DISHWASHER	1	EA			1 EA			1 EA
30" ELECTRIC RANGE W/OVEN	1	EA			1 EA			1 EA
36"x 6'-6" MAX. REFRIGERATOR	1	EA			1 EA			1 EA
UNCRATE & SET IN PLACE	3	EA			3 EA			3 EA
CLOTHES WASHER	1	EA	SUPPLIED BY OWNER					
CLOTHES DRYER	1	EA	SUPPLIED BY OWNER					

Figure 14.1

Chapter Fifteen

FURNISHINGS

(Division 12)

Chapter Fifteen

FURNISHINGS
(Division 12)

Furnishings include furniture, artwork, window treatments (blinds, draperies, etc.), floor mats, rugs, seating, plants, and file cabinet systems. Furnishings are sometimes supplied by the owner under a separate contract (to include installation), but may require coordination and scheduling by the contractor. Some items may be supplied by the owner, with installation by the contractor.

The contractor should review the bid documents to determine the items he or she is responsible for, paying special attention to Division 12 for the actual scope of the furnishings. Occasionally, for commercial projects, designers provide a furnishings or fixture plan as part of the architectural drawings to show the locations of items such as chairs, tables, bar stools, office furniture, floor mats and rugs, and artwork. These plans are usually included only when the contractor is responsible for installation.

Units for Takeoff

For furnishings to be supplied and installed by the contractor, items should be taken off by the individual piece (EA) and listed according to type, manufacturer, model number, color, size, and any identifying characteristics. One example is fire-treated fabric for window treatments or seating.

Items to be supplied by the owner but handled and installed by the contractor should be taken off by the piece (EA) and listed according to type, size, weight, location (e.g., first floor, third floor), and level of assembly required. This information should be noted for accurate pricing of handling labor.

Other Takeoff Considerations

Some furnishings installed by the contractor require assembly. Examples include fastening table tops to pedestal bases, installing seat cushions on built-in benches, installing peg legs on couches or office furniture, and assembling office desks.

Artwork, such as framed paintings, murals, or three-dimensional items, may also need to be fastened in place. Most commercial installations, especially with public traffic, such as restaurants, require the use of security fasteners to discourage vandalism or theft. These often involve considerably more installation time than simple picture hanging.

Many items in Division 12 are shipped in protective packing such as boxes, crates, and plastic wrappings. Removal and disposal of packaging debris and light cleaning of the items are often the contractor's responsibility, and should be included as part of the cost of installation.

Specifications should also be reviewed for projected delivery dates of long-lead items. Often special or custom furnishings are ordered in advance at the start of construction. Delivery dates that conflict with the construction schedule may require that items be stored until the appropriate phase in the construction.

Summary

For this particular sample project there is no work from Division 12–Furnishings specifically included in the Contract Documents.

Division 12 Checklist

The following list can be used to ensure that all items for this division have been accounted for.

Furnishings

- ☐ Artwork
- ☐ Murals
- ☐ Window treatments
- ☐ Office furniture
- ☐ Tables
- ☐ Seating
- ☐ Rugs and mats

Chapter Sixteen

Special Construction

(Division 13)

Chapter Sixteen

SPECIAL CONSTRUCTION
(Division 13)

Division 13 includes a wide variety of specialty subsystems such as clean rooms, walk-in refrigeration units, sports courts, swimming pools, and vaults. Items in this division are usually purchased by the owner and installed under a separate contract by a specialty contractor.

Bid documents should be reviewed thoroughly for any information on Special Construction items. Plans may refer to the item or system using only a general description and notes such as "N.I.C." (not in contract) or "Furnished and installed under separate contract."

Units for Takeoff

Because most of the items included in Division 13 require specialized estimating experience, normal takeoff procedures do not apply. A review of the bid documents will reveal related work of other specification sections (particularly mechanical, electrical, structural, and temporary work) that may be required for Special Construction items.

The contractor should also review the exact scope of the Special Construction work with the designated specialty contractor to determine what, if any, coordination is needed. It should be noted that manufacturers of Division 13 systems often require that only trained personnel install the item to sustain material and performance warranties.

Summary

For this particular sample project there is no work from Division 13—Special Construction specifically included in the Contract Documents.

Division 13 Checklist

The following list can be used to ensure that all items in this division have been accounted for.

Special Construction

- ☐ Air-supported structures
- ☐ Integrated ceilings
- ☐ Athletic courts or rooms
- ☐ Cold storage rooms
- ☐ Saunas and steam baths
- ☐ Radiation protection
- ☐ Acoustical enclosures

- ☐ Greenhouses
- ☐ Swimming pools
- ☐ Pre-engineered structures
- ☐ Ice rinks
- ☐ Clean rooms
- ☐ Observatory
- ☐ Vaults

Chapter Seventeen

CONVEYING SYSTEMS

(Division 14)

Chapter 17

CONVEYING SYSTEMS

(Division 14)

Division 14 is comprised of specialized transportation systems that move people or objects within a building or between different buildings. It includes items or subsystems such as elevators, chair lifts for disabled people, escalators, dumbwaiters, material handling systems, and moving walks and ramps.

Items or systems specified in Division 14 are often purchased directly by the owner, who also arranges installation under a separate contract. In some cases, projects will require that the contractor assume responsibility for the conveying systems.

The contractor should review all types of drawings in the bid set for information concerning conveying systems. Locations of elevators, escalators, dumbwaiters, and so on are typically shown on architectural drawings, especially floor plans. Building sections through elevator shafts should provide travel distance, number of stops, type of elevator or lift, and general operation of the system. Special details or sections through parts of the system may also be shown. Careful attention should be given to general notes about the limits of the contractor's responsibility, such as "N.I.C." (not in contract) and "Furnished and installed under separate contract."

It is not unusual for a designer to provide only general information regarding conveying systems, and depend on the manufacturer to provide drawings and specifications for the actual installation. The contractor should make careful inquiries concerning who will provide drawings and when they will be available.

Units for Takeoff

Because the conveying systems found in Division 14 require specialized estimating experience, normal takeoff and estimating procedures do not apply. The contractor must instead study the specifications to become familiar with the type of system and its method of operation. A checklist is helpful for classifying the type and operation of the system as a means of qualifying the subcontractors' proposals. Figure 17.1 is an example.

Division 14 specifications should also be studied to determine the exact scope of the work, as well as any related work of other trades, in preparation for the conveying systems. The most common examples of related work are masonry walls for elevator or dumbwaiter shafts, electrical power to elevator motors, elevator hoistway enclosures at the roof, and concrete work for elevator pits and escalators.

Other Takeoff Considerations

Other considerations may include the type of construction. Installations in new work tend to be less costly than in renovated applications. In addition, service agreements for the maintenance of the system for a specified period may be part of the contract requirements. This should be noted in the checklist.

When doing a takeoff for an elevator or dumbwaiter, doors must be included for each floor at which the elevator or dumbwaiter stops.

Summary

For this particular sample project there is no work from Division 14–Conveying Systems specifically included in the Contract Documents.

Division 14 Checklist

The following list can be used to ensure that all items for this division have been accounted for.

Conveying Systems

- ☐ Elevators
- ☐ Material handling systems
- ☐ Pneumatic tube systems
- ☐ Vertical conveyors
- ☐ Handicap

Dumbwaiters

__ Capacity	__ Size
__ Overall height	__ Number of stops
__ Speed	__ Finish

Elevators

__ Hydraulic or electric	__ Car size
__ Capacity	__ Number of stops
__ Overall height or floors	__ Door type
__ Speed	__ Machinery location
__ Interior finish	__ Number required
__ Signals	__ Cab finish
__ Geared or gearless	

Moving Stairs and Walks

__ Capacity	__ Speed
__ Story height	__ Floors
__ Machinery location	__ Finish
__ Size	__ Horizontal or inclined
__ Number required	__ Special requirements

Figure 17.1 Checklist for Qualifying Subcontractor's Proposals

Chapter Eighteen

MECHANICAL

(Division 15)

Chapter Eighteen

MECHANICAL

(Division 15)

Division 15 includes fire protection systems; plumbing systems; and heating, ventilating, and air conditioning systems commonly referred to as HVAC. The work of Division 15 is typically performed by trades or firms with specialized training in the particular discipline, often requiring licenses and permits separate from those of the general contractor. To take off and estimate Division 15 work requires a working knowledge of the particular trade or system and, often, specialized education and training not normally within the realm of the general contractor's experience.

The work of Division 15 is typically shown on the mechanical drawings (see "Mechanical Drawings" in Chapter 1). It is labeled with prefixes: *M* for mechanical, *P* for plumbing, *FP* for fire protection, and *H* or *HVAC* for heating, ventilating and air conditioning. All drawings in the bid set should be carefully reviewed for related work that may be shown on other drawings, as well as for detailed information on dimensions and measurements for room sizes, floor-to-floor heights, location of services entering the building, and the coordination of other work within the particular area. Mechanical drawings that employ the use of schedules are helpful in determining types and quantities of materials for takeoff (see "Schedules" in Chapter 1). Specialized details such as riser diagrams that show the configuration and components of piping systems are often included in the mechanical drawings (see "Mechanical Drawings" in Chapter 1). Riser diagrams are for the graphic representation of information only, and are not drawn to scale.

Division 15 specification sections should be reviewed thoroughly for the products, methods and techniques of installation, as well as the related work of other trades that would affect the pricing of Division 15 work. A review of Division 1 of the specifications may be necessary to determine what, if any, special requirements may be included in the work.

For the purpose of this text, a discussion of the basic procedures and methods for takeoff will be limited to the mechanical systems normally encountered in residential and light commercial construction.

Fire Protection Systems

Many of the residences and light commercial structures under construction today require the use of fire protection systems. The three main components of a fire protection system are detection, alarm, and suppression. The system may also require the cooperation of other systems not typically specified under Division 15. The suppression of fire may be accomplished by the use of fire standpipe systems, automatic sprinkler systems, or a combination of the two.

A *fire standpipe* is a system of piping for firefighting purposes, comprised of a water supply that serves one or more hose stations. Hose stations are placed at various locations within the structure, based on design and local code requirements. There are two types of standpipe systems. In a *wet standpipe* system, all piping is filled with water under pressure and is discharged upon the opening of any hose valve. A *dry standpipe* system is filled with air to prevent freeze-up in unheated areas. Water enters the piping either automatically by a reduction in air pressure within the system, or by manual activation at the individual hose station.

Automatic sprinkler systems are a series of piping, valves, and sprinkler heads for the automatic distribution of water or chemicals to suppress a fire. The water or chemical, such as halon, carbon dioxide, or high expansion foam, is dispersed through special discharge fittings called *sprinkler heads* located according to an engineered design. Either the sprinkler head itself or an independent device detects a fire and releases the water or chemical through the head to suppress the fire.

There are several different classifications of automatic sprinkler systems designed for specific firefighting applications. The more common are the wet pipe system, the dry pipe system, the pre-action system, the deluge system, the halon system, and the fire cycle system. All systems deliver the water and chemicals through a series of piping, heads, valves, and equipment.

The contractor should study the fire protection drawings for locations of sprinkler heads, the various sizes and material of the piping, and the location of valves, fittings, and appurtenances. Special equipment such as compressors for dry systems, backflow preventers, and alarm bells may be shown on riser diagrams or details. The specifications should be studied for the materials to be used, the method of installation, and the required compliances of the various governing agencies.

Units for Takeoff

Riser and distribution piping for standpipe and automatic sprinkler systems are taken off by the LF, and are listed according to type of pipe, size (diameter), and method of connection (grooved joint or threaded).

Fittings are taken off by counting and are listed by the piece (EA) according to type (elbow, tee, reducer, etc.), size, method of connection, and material composition of the fitting (steel, iron, galvanized iron, PVC, etc.).

Valves and special appurtenances (such as gauges, couplings, flanges, water motor alarms and bells, siamese connections, backflow preventers, control panels or devices, and storage cylinders) are taken off by the individual piece (EA) and listed according to type, manufacturer, model, size, use or application, and any other identifying information that is necessary for accurate pricing.

Sprinkler heads or discharge nozzles are taken off by the piece (EA) and listed according to type (pendent, upright, sidewall), temperature range, manufacturer's and model. Special trim pieces called *escutcheons*, which are used to provide a finished appearance around the sprinkler head, are taken off by the piece (EA) and listed according to type, finish, model, and application.

Other Takeoff Considerations

Takeoff quantities should be listed separately according to type and classification of system. In addition to materials, special items such as shop drawings, permits, fees, and special staging or rigging equipment for the installation of larger diameter pipes and valves should be noted for accurate pricing.

In most cases, the sprinkler contractor's work begins at the interior of the building where the fire protection water service line enters the structure. Site piping and related excavation and backfill are typically the responsibility of the site contractor. This may vary according to local practices or codes. The contractor should review carefully the specifications regarding the exact scope of work.

Testing of water pressure, plan review with local government agencies, and review/approval processes by insurance underwriters may also be required by the contract documents, and should be noted in the takeoff.

Plumbing

All buildings that will be occupied require some type of plumbing, from simple toilets in warehouses, to sophisticated plumbing systems in hospitals and restaurants, to elaborate bathrooms in upscale residences. As with fire protection, plumbing work requires special knowledge and training of both the estimator and the tradesperson. Plumbing work requires the tradesperson to be licensed, and a separate permit and inspection process.

The contractor must carefully review and be familiar with all the plans of the bid set, including the plumbing drawings. Plumbing drawings with plan views and riser diagrams are used in determining piping quantities. Fixture and Equipment Schedules are helpful in determining quantities and types of plumbing fixtures and equipment to be furnished and/or installed. Other drawings within the bid set, such as special equipment plans (e.g., kitchen equipment plans for restaurants), should be studied for equipment that is furnished by others and installed under the plumbing contract. The contractor should review the graphic symbols and abbreviations legend (see Appendix A) provided on the plumbing plan for a clear understanding of the work. Figure 18.1 is a plumbing plan showing waste, vent, and water piping.

Figure 18.2a illustrates an example of a riser diagram for soil and vent piping for the preceding plumbing plan.

Figure 18.2b is the riser diagram for the hot and cold water piping for the plan in Figure 18.1.

Special drawings with enlarged plan views and details for clarification of certain aspects of work are common, as in the example of the domestic water heater shown in Figure 18.3.

Division 15 of the specifications should be reviewed for product information and acceptable methods of installation, as well as the related work in other sections of the specifications.

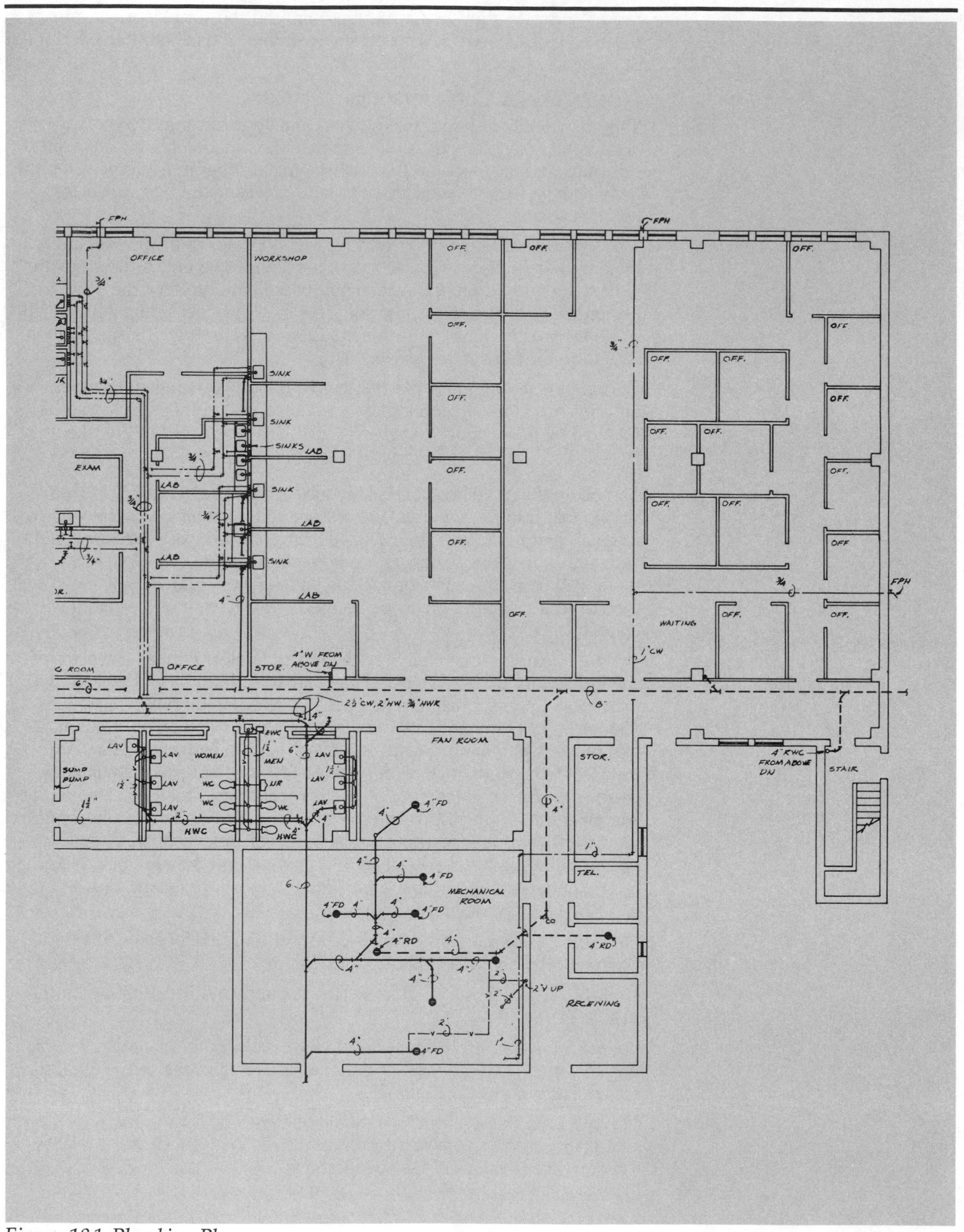

Figure 18.1 Plumbing Plan

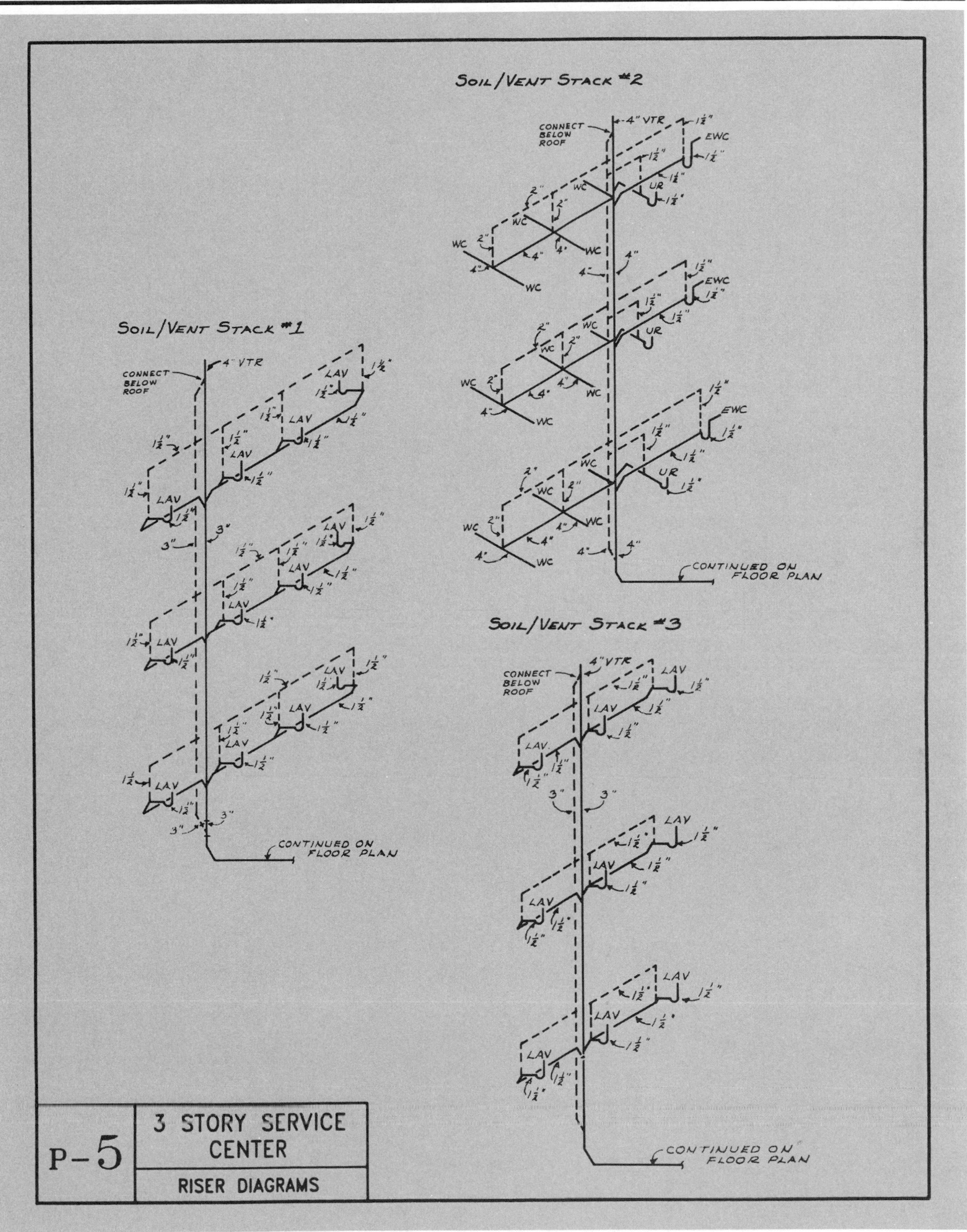

Figure 18.2a Riser Diagrams for Soil/Vent Piping

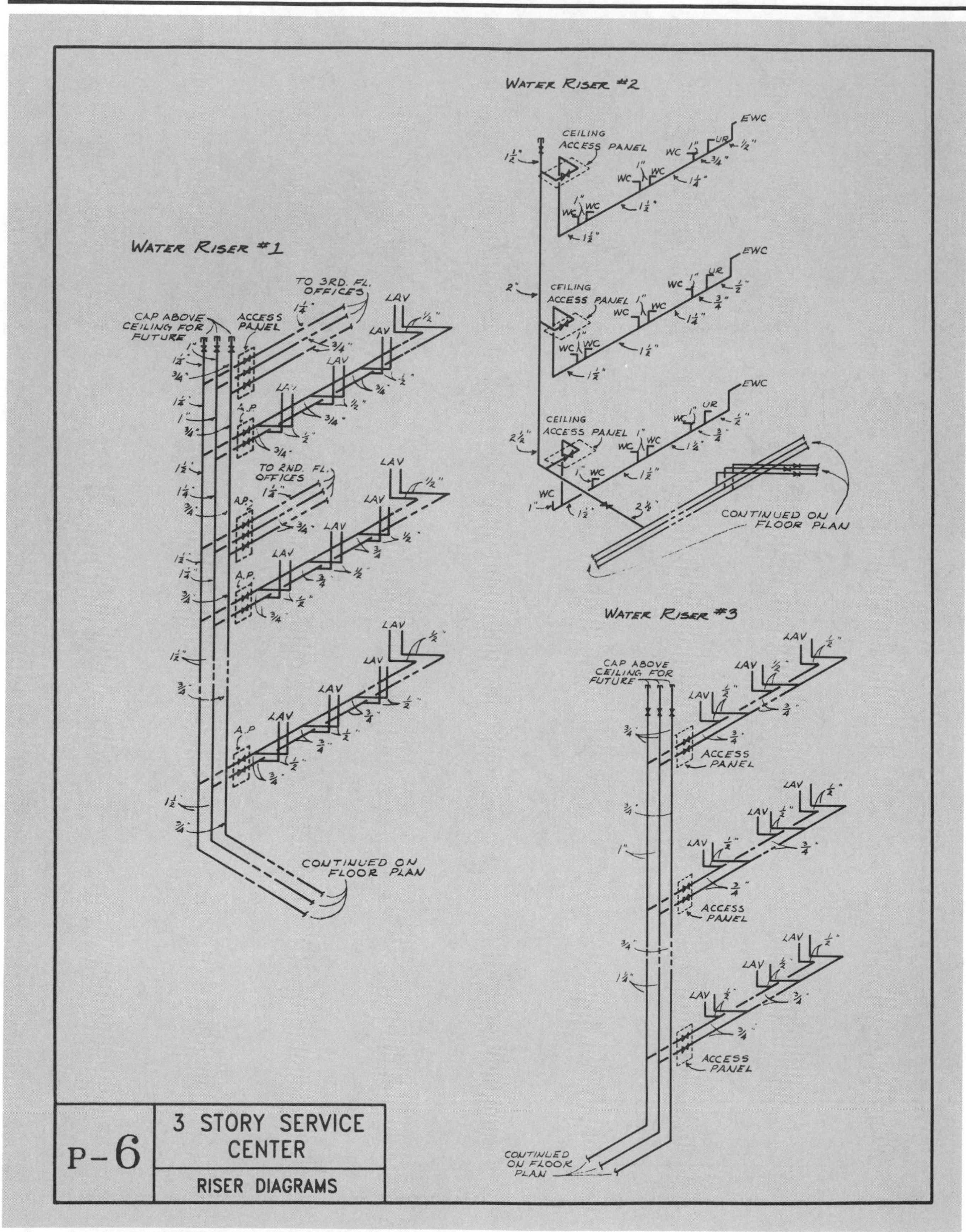

Figure 18.2b Riser Diagrams for Water Piping

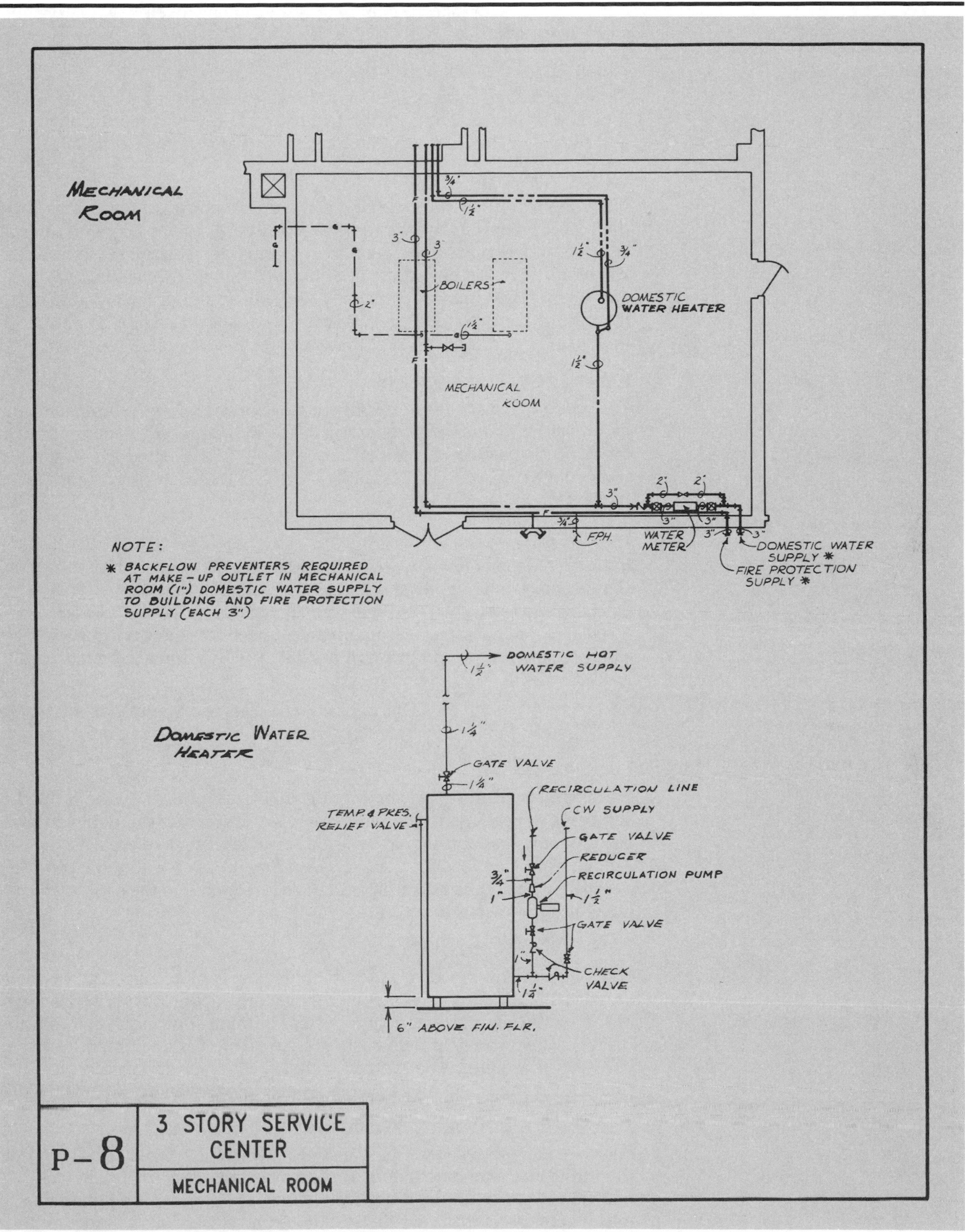

Figure 18.3 Mechanical Room Plan

For the purpose of takeoff, plumbing work can be broken down into eight subsystems, which are described in the following pages.

Plumbing Fixtures

Plumbing fixtures include such items as water closets, urinals, shower stalls, tubs, and lavatories. Trims for the fixtures, such as valves, faucets, shower and tub valves, trip levers, and drains, are included in the fixture takeoff.

Units for Takeoff

Plumbing fixtures are taken off by counting the individual pieces and listing them as EA. Trims follow the same procedure. Quantities for fixtures may be used in determining the quantities of trims. For example, each lavatory will require a faucet and drain; each tub will require a shower/tub valve and a trip lever drain; each water closet will require a toilet seat. Each fixture and corresponding trim is listed according to type, manufacturer, model, color or finish, and special features such as handicap compliance.

Other Takeoff Considerations

To reduce the chance for error, the contractor may choose to separate fixture and trim quantities according to specific bathrooms or toilets, or by the floor for multi-level commercial restrooms. This method allows the contractor to complete the takeoff for each bathroom or toilet before proceeding to the next one.

Equipment

Equipment for plumbing work includes such items as water heaters, water storage tanks, interior grease interceptors, and sump pumps. In addition to equipment furnished and installed under the plumbing contract, the installation or "hook-up" of equipment supplied by others, such as garbage disposals, gas ranges and ovens, refrigerators with automatic ice-makers, and dishwashers, must be included.

Special devices such as washing machine outlets and venting kits for water heaters should also be included in the takeoff.

Units for Takeoff

Equipment is taken off by counting the individual piece (EA) and is listed according to type, manufacturer, model, size, and capacity. Equipment that requires only hook-up should be listed separately by the piece with the same qualifications as noted above. Special devices needed for the complete installation should be noted separately with reference to the equipment for which it is required.

Other Takeoff Considerations

Some equipment provided under the plumbing contract may require the work of other trades, such as power wiring of water heaters, power wiring of draft inducers, and special flues for gas-burning appliances such as water heaters. Penetrations through roofing systems and flashing of roof vents may also be required. The contractor should study the specifications carefully for the exact scope of work related to equipment.

Below-Grade Sanitary Waste and Vent Piping

Below-grade sanitary waste piping carries the waste, typically sewage, from the individual plumbing fixture and above-grade waste piping in the house to the drain and out of the building. Below-grade vent piping ties into vertical risers that vent the system to the exterior of the structure. A variety of fittings and devices is used, in addition to the piping itself, to special functions and change direction of the piping. Typical fittings and devices include couplings, elbows, tees, tee-wyes, clean-outs, traps, and floor drains, to mention just a few. Fittings, devices, and piping are available in a number of different materials, from PVC to cast iron. The contractor should review the specifications for the type of material for the specific application.

Plumbing plans are used for determining quantities, size and type of piping, fittings, and devices that will be used in the below-grade application.

Units for Takeoff

Fittings and devices are taken off by counting the individual piece (EA) and listing them separately according to type (elbow, tee, clean-out, etc.), material composition (PVC, service-weight cast-iron, extra heavy-weight cast iron, etc.), method of connection to piping (lead and oakum, hubless, neoprene joint, or PVC cement), and size (diameter).

Piping is taken off by the LF, and should be listed separately according to type, size, and method of installation. Linear feet of individual types and sizes of pipe can be converted to actual lengths of pipe required.

If the cast-iron soil pipe is of the lead-joint design, it will be necessary to determine the total amount of lead, oakum (jute packing), and gas (propane) necessary to complete the installation. This is done after the fittings and piping takeoff has been completed so that the total number of joints is known. Cast-iron soil pipe is manufactured in 5′ and 10′ lengths; for lead, the 5′ length is used. The amount of lead used per joint is determined by the diameter of the pipe. Figure 18.4 lists the amount of lead (in pounds) required per joint to caulk cast-iron pipe.

The following example shows how to determine quantities of lead, oakum, and gas for cast-iron soil pipe, assuming the following quantities of service-weight soil pipe and fittings have been taken off:

LEAD REQUIRED TO CAULK CAST IRON SOIL PIPE JOINTS

Pipe & Fitting Diams. Inches	Lead Ring Depth Inches	Service Weight		Extra Heavy Weight	
		Cu. Ins.	Wt. Lbs.	Cu. Ins.	Wt. Lbs.
2	1	2.81	1.15	2.91	1.19
3	1	3.90	1.60	4.17	1.71
4	1	4.98	2.04	5.25	2.15
5	1	6.06	2.49	6.24	2.56
6	1	7.15	2.93	7.42	3.04
8	1.25	15.06	6.17	15.49	6.35
10	1.25	18.90	7.75	19.34	7.93
12	1.25	25.53	10.47	26.02	10.67
15	1.5	43.09	17.67	43.38	17.8

Figure 18.4

20 LF—4" service-weight soil pipe
3 EA—4" 1/8 bends
1 EA—6" x 4" wye

The procedure is as follows:

(Note that each hub, rather than the opening, is the basis for the number of joints.)

20 LF of 4" (5' length) = four 4" joints × 2.04 lbs./joint	*= 8.16 lbs.*
3 EA—4" 1/8 bends = 3 × (1—4" joint) = 3 × 2.04	*= 6.12 lbs.*
1 EA—6" x 4" wye = 1 × (1—6" joint + 1—4" joint)	
= (1 × 2.93) + (1 × 2.04)	*= 4.97 lbs.*
Total lead	*=19.25 lbs.*

Oakum is typically estimated at one-tenth the weight of lead. Therefore: 19.25 divided by 10 = 1.92 lbs. of oakum.

Gas consumption is approximated at one (instopropane) cylinder per 200 lbs. of lead.

If the cast-iron soil pipe is of the neoprene joint clamp type (hubless), or if the waste piping is PVC, the contractor must count the number of joints to arrive at the quantity of clamps, gaskets, or couplings required.

Other Takeoff Considerations

Below-grade sanitary waste piping may include excavation and backfill for the placement of the piping. Excavation and backfill work is typically the responsibility of the site or general contractor. The specifications and General Conditions should be reviewed to determine the exact scope of work to avoid costly duplication or omission.

Above-Grade Sanitary Waste and Vent Piping

Above-grade sanitary waste and vent piping allows for the flow of waste above grade (such as upper floors of the building) to the below-grade system where it will exit the structure. Vent piping will allow the escape of gases generated by the waste through the upper level of the structure, typically the roof.

The piping and fittings used below grade will be similar to those used above grade and are determined by the specifications and the plans.

Units for Takeoff

Procedures for takeoff follow those of the below-grade system. Fittings are counted and listed as EA according to identifying types and sizes. Piping is taken off by the LF and listed according to type, size, and method of installation. The procedures for calculating lead, oakum, gas and/or gaskets, clamps, and couplings are the same as for the below-grade system.

Additional items for the support of piping, such as pipe hangers and supports, are taken off by the individual piece and listed as EA, according to size, application, and type. Because pipe hangers and supports are omitted on the drawings for clarity, it is necessary to complete the piping takeoff for horizontal run and riser to determine the quantity of each. The contractor must refer to the specifications for the required intervals of hangers and supports to determine the number needed.

Other Takeoff Considerations

The takeoff procedure for above-grade waste and vent piping should be altered slightly to accommodate the fact that above-grade piping is in two planes: horizontal runs and branches in the horizontal plane, and risers and drops to fixtures and equipment in the vertical plane. Riser diagrams provide dimensions showing floor-to-floor heights. With this information, the length of risers and the approximate length of drop pieces to fixtures can be calculated. Horizontal piping and risers are taken off separately to reduce the chance for error in referencing between multiple drawings.

Special staging or rigging may be required to install some of the heavier cast-iron pipe in above-grade applications where it is supported from the structure above.

Below-Grade Storm System Piping

The below-grade storm drainage system is the lowest part of the piping system that receives clear water from roof leaders (on flat roof buildings), cooling or condensate water, or other clear water within the structure. This system conveys the drainage to the building's storm sewer or drain by gravity.

Units for Takeoff

The takeoff procedure and units are the same as those for the below-grade sanitary waste and vent piping system. Although similar, the storm system takeoff should be kept separate from the sanitary system.

Above-Grade Storm System Piping

Above-grade storm system piping consists of roof drains, leaders, and horizontal offsets that will tie into the below-grade storm system at the floor level.

Units for Takeoff

The takeoff procedure and units are the same as those for the above-grade sanitary waste and vent piping. Again, although similar, these two systems should be kept separate in the takeoff.

Other Takeoff Considerations

The contractor should keep horizontal pipe offsets for above-grade storm systems (that will be above finished ceilings) separate on the takeoff, since these sections of pipe are normally insulated to prevent condensation.

Hot and Cold Water Piping

Hot and cold water piping is a part of virtually every plumbing job. It includes piping, fittings, valves, control devices, and all the related appurtenances for conveying water to plumbing fixtures and equipment. This is often referred to as *domestic* water piping, and excludes piping for fire protection systems. The cold-water supply typically starts at the point where the water service enters the building. From this point the water is distributed to the various fixtures and equipment within the structure. The hot-water supply starts at the hot-water generating equipment, usually a water heater, and from there begins its distribution to the fixtures.

Standard piping materials used for the distribution of domestic water are types L and K copper tubing and, in some limited applications, brass, galvanized steel pipe, and PVC. The most common method of joining copper pipe and fittings is by solder joint, or rolled-groove pressure fittings for copper tubing, threaded fittings for steel and brass pipe, and cement joint fittings for PVC. Fittings include elbows, tees, 45- and 22-1/2-degree bends, couplings, and reducing fittings. Control devices include a variety of valves, such as check, globe, gate, ball, and butterfly. Other special devices include such items as backflow preventers, relief valves, pressure-reducing valves, shock absorbers, vacuum breakers, and frostproof hose bibs.

Units for Takeoff

Fittings, valves, and control devices are taken off by counting the individual piece (EA) and listing them according to type, material composition, size, and application. Water piping is taken off by the LF, following a similar procedure to above-grade sanitary piping. The contractor starts by taking off mains and branches, then risers and drops. The LF quantities can be converted to individual lengths of pipe. All piping quantities should be listed according to type, grade, and size (diameter).

Pipe hangers and supports are taken off by dividing the total LF of water piping in each size category by the specified intervals, as noted in the specifications or by code requirements.

Solder, flux, and gas are the materials used for joining copper water pipe. Copper pipe and fittings are joined by soft (non-lead) solder. Solder, flux, and gas are difficult items to estimate, but by using the chart in Figure 18.5, the contractor can arrive at a relatively accurate amount of each. The number of joints required for each size fitting and device must be counted to determine a total number of joints in each size category.

Other Takeoff Considerations

Many designs (as well as local plumbing and energy codes) require the use of pipe insulation to retard heat loss and prevent condensation. Pipe insulation is available in a wide variety of sizes and compositions for different applications. Insulation is manufactured in both rigid and flexible forms, with or without fittings.

Once the water piping and fittings have been taken off, the contractor can calculate the quantity of insulation for piping and the valves and fittings. Pipe insulation is taken off by the LF; insulation for individual fittings is taken off by the piece (EA). Insulation should be listed in the takeoff according to the diameter and length of the pipe it will cover and the type of insulation.

ESTIMATED POUNDS OF SOFT SOLDER REQUIRED TO MAKE 100 JOINTS*

Size	3/8"	1/2"	3/4"	1"	1 1/4"	1 1/2"	2"
Pounds	.5	.75	1.0	1.4	1.7	1.9	2.4
Size	2 1/2"	3"	3 1/2"	4"	5"	6"	8"
Pounds	3.2	3.9	4.5	5.5	8.0	15.0	32.0

*Two oz. of flux will be required for each pound of solder. One tank of PRESTO gas will be required for every 500 joints.

Figure 18.5

Access panels installed in the finish surface of the wall or ceiling may also be included as part of the work. These are taken off according to size, type, manufacturer, model, and location, and are listed by the piece (EA).

Natural Gas System Piping

Piping for the distribution of natural gas within the structure is often considered part of the plumbing work. Natural gas piping starts at the entrance of the gas service to the building, and is distributed to the various gas-fueled appliances within the building, such as water heaters, furnaces, boilers, ranges, clothes dryers, and rooftop HVAC units. Piping materials are typically black steel pipe with malleable iron-threaded fittings. Fittings for gas piping are similar to those of other piping systems and include elbows, bends, unions, and tees. Valves for the control of the flow of gas within the pipe are called *gas cocks*, and are typically brass.

Mechanical plans and Division 15 specifications should be studied for the location, size, and type of pipe, and for the appliances to be connected.

Units for Takeoff

The procedure for taking off piping is similar to that of above-grade sanitary waste and vent piping. Fittings and valves are taken off by the individual piece (EA) and listed according to type and size (diameter). Piping is taken off by the LF, according to type and size. Since gas pipe and fittings are joined by means of a threaded connection, lead and oakum or solder are not required. Pipe hangers and supports are determined using the same procedure as for water piping.

Other Takeoff Considerations

Special devices for regulating the pressure of gas supplied to an appliance may also be required at certain appliances (such as ranges or commercial ovens) and must be included in the takeoff. Flexible gas connectors for connecting movable appliances to a stationary gas supply may also be required and included in the takeoff.

Some local codes may require a separate permit for gas work. This usually constitutes an additional fee over and above the plumbing portion of the work.

Heating, Ventilating, and Air Conditioning

Heating, ventilating, and air conditioning systems, commonly referred to as HVAC, include the various systems that provide heating, cooling, and fresh air to the occupied space of the building or residence. One of the most common methods involves the use of a gas- or oil-fired furnace that supplies warm air to the respective spaces or rooms through a series of supply and return air ductwork. This same system of ductwork can be used to supply cooled air in summer. Separate systems can also be installed.

Alternate methods of heating employ gas- or oil-fired boilers that force hot water through a system of radiant baseboard installed in individual rooms.

Mechanical plans should be studied carefully for the layout, locations, and sizes of ductwork and fin tube radiation baseboard. Special mechanical plans and details that illustrate the components of boilers and rooftop HVAC units may also be included. Division 15 specifications also list the specific materials, manufacturer and model of the heating and cooling units, and the various appurtenances that are required for a complete system.

For purposes of takeoff, HVAC work can be divided into two general categories: ducted systems for the distribution of heated or cooled air, and radiant heating systems for the distribution of forced hot water.

Both takeoff and estimating of HVAC systems, like other Division 15 work, require a specialized knowledge of the individual system, which is not normally within the realm of the general contractor's experience.

Ducted Systems

Ducted systems include a wide variety of heating and cooling systems that employ the use of metal ductwork for the supply and return of conditioned air to and from spaces within a building. The heated or cooled air can be supplied from a single self-contained unit or from separate components at various locations within or outside the structure. Ductwork and equipment shown on drawings have been engineered to suit the specific application based on the design criteria. In much the same fashion as piping for water and waste systems, ductwork is installed through the structure in a series of main trunks and branches to specific areas as required. Different "fittings" or transition pieces allow the round or rectangular-shaped ducts to change direction, circumvent obstacles, or reduce in size, as required by the particular application.

Special devices that control the flow of air within the ductwork are called *dampers*. Round or rectangular outlets that diffuse the air delivered to the space are called *diffusers*; they are typically located at the ceiling level. Similar devices that have a grille and damper for regulating air flow at the device are called *registers*. Louvered or perforated panels at the inlet to return air ducts are called *grilles*.

Controls that regulate the temperature of the space, called *thermostats*, signal to the furnace the need for more or less heat. Figures 18.6a and 18.6b illustrate gas- and oil-fired warm air ducted systems, respectively.

Units for Takeoff

Ductwork is taken off by the LF and listed according to type, size, and application (supply or return). In addition to horizontal mains, vertical risers and drops are also necessary for the distribution of air between multiple floors. Flexible ducts for short runs to diffusers are taken off by the LF.

Since the supply ductwork and fittings are usually insulated, it is helpful to separate the quantities of each on the takeoff. The various fittings, transition pieces, reducers, collars for the connection of flexible ducts, and dampers are taken off by the individual piece (EA) and listed according to type, size, and application. Devices installed in the finished space, such as registers, grilles, thermostats, and diffusers, are taken off by the individual piece (EA) and listed according to type, size, manufacturer, model, and finish. Once the ductwork portion of the takeoff has been completed, the contractor can calculate the quantity of insulation needed. To do so, the total surface area of the various-sized ducts and fittings must be determined. The easiest method considers the fittings as ductwork, and measures through the fittings when doing a takeoff from the plan. For example:

> *Assuming a takeoff quantity of 43′ of 12″ x 16″ supply duct to be insulated:*
> *Surface area = (12″ + 16″) × 2 = 4.67 SF per LF of duct*
> *= 43 LF × 4.67 SF/LF = 200.8 SF*

Sound lining in the interior of the duct is taken off and listed by the SF using the same procedure as for duct insulation.

Equipment takeoff for furnaces, air-conditioning condensing units, evaporators, electric coils, and heat pumps is done by the individual piece

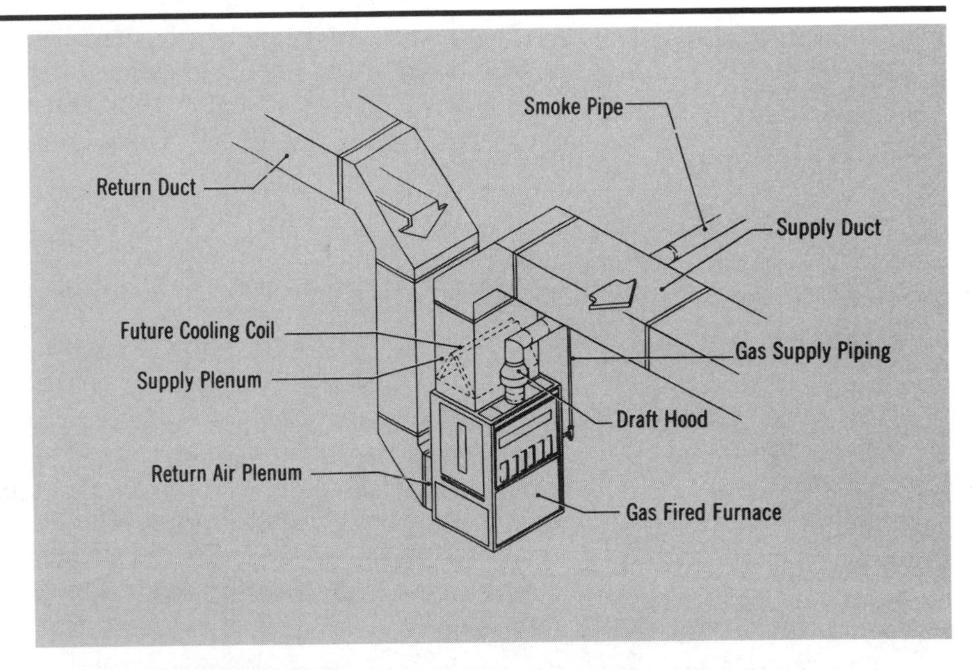

Figure 18.6a Gas-Fired Warm Air System

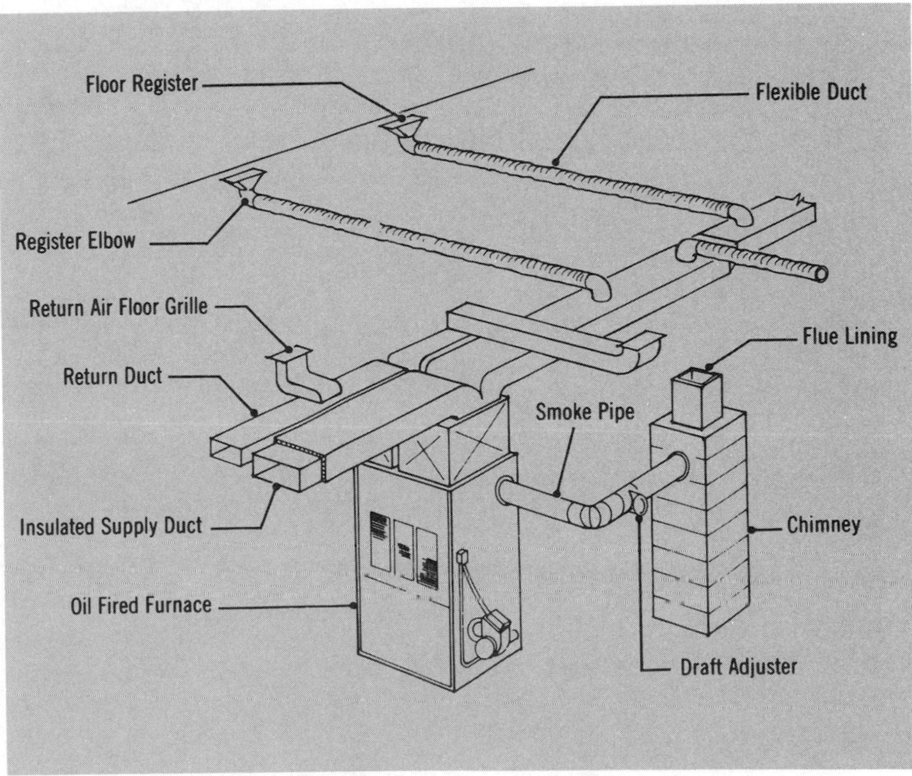

Figure 18.6b Oil-Fired Warm Air System

(EA) and listed according to type, size or capacity, manufacturer, model or series, and any other special identifying criteria for accurate pricing. Components required to complete the system may include flues for the furnace, control wiring for the thermostat, testing and balancing, and filters. These items should be taken off individually and can be listed as each (EA), lump sum (LS), or whatever units best represent the scope of work.

Other Takeoff Considerations

Drawings and specifications should be carefully reviewed to determine the exact scope of HVAC work. Items such as power wiring to the furnace, installation of a condensing unit or compressor, furnishing and installation of oil tanks for oil-fired systems, and gas piping for gas-fired furnaces are not typically part of the HVAC contractor's work.

For commercial projects with rooftop equipment, hoisting or crane services will be required to set the unit on the roof. The contractor should consult the specifications to determine who will provide the crane or hoisting of the equipment. This information should be noted in the takeoff.

Testing and balancing of completed duct systems have become mandatory for most commercial projects. This is done by an independent contractor at the expense of the HVAC contractor. Reports are typically required to confirm that the design criteria has been met.

Sheet metal ductwork, fabricated from galvanized steel sheets, is often converted to weight (lbs.) for the pricing of the raw material (sheets). The calculation for this conversion takes into account the gauge (thickness) and weight per SF of the material being used. The chart shown in Figure 18.7

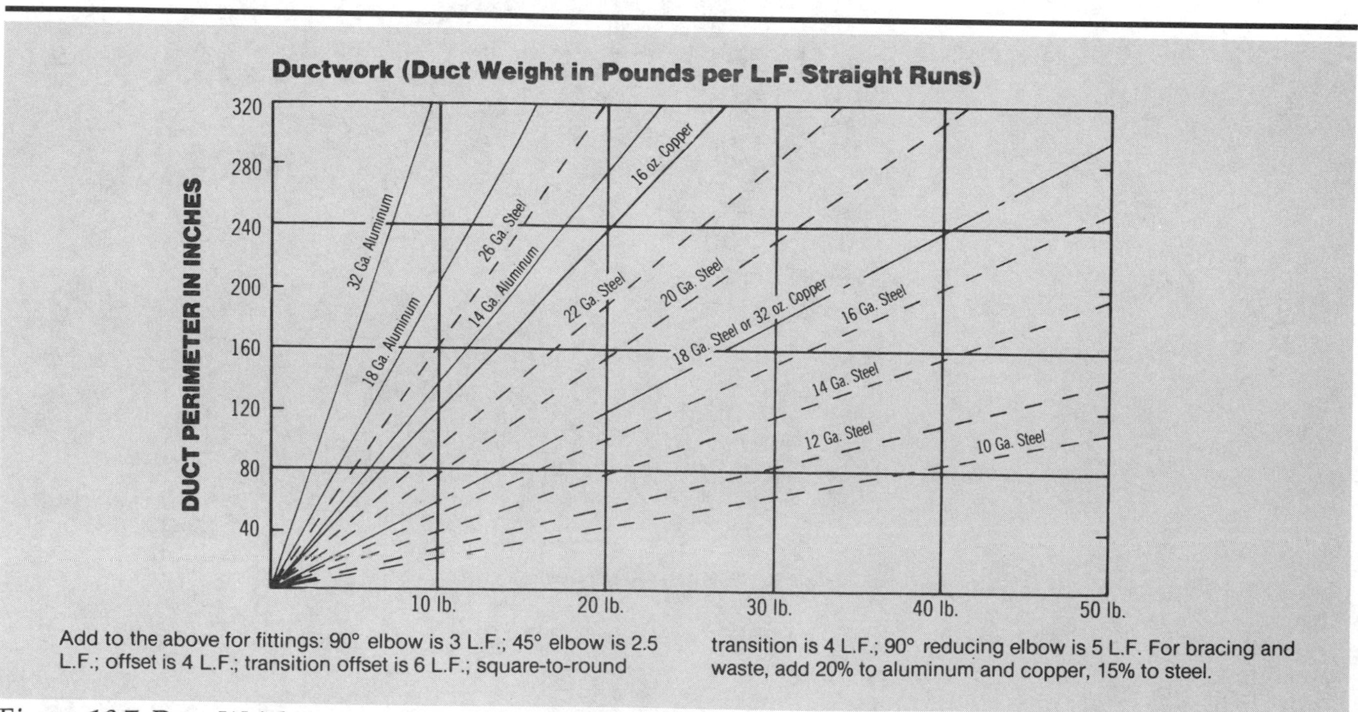

Add to the above for fittings: 90° elbow is 3 L.F.; 45° elbow is 2.5 L.F.; offset is 4 L.F.; transition offset is 6 L.F.; square-to-round transition is 4 L.F.; 90° reducing elbow is 5 L.F. For bracing and waste, add 20% to aluminum and copper, 15% to steel.

Figure 18.7 Duct Weights

can be used to obtain the weight per LF of ductwork for various types of sheet metal.

The contractor's price per pound for metal ductwork should include an allowance of 10% – 15% for waste, slips, and hangers.

Forced Hot-Water Systems (Hydronic Heating Systems)

An alternate method of heating employs the use of oil- or gas-fired hot-water (or steam) boilers. Heating boilers are manufactured in cast-iron, steel, or copper, and are available pre-assembled (packaged) or in sections for field assembly.

Simply stated, water is heated by the boiler and "forced" through piping (with "fins" for dissipating heat) by the use of a circulating pump. As the control device (usually a thermostat) calls for heat, the boiler heats the water, and the circulator forces the water through a closed-loop system until the temperature has been satisfied. Figure 18.8 illustrates a forced hot-water heating system.

The same basic principle is applied to other forms of radiant heating where loops of piping are installed within the floor, ceiling, or walls. A common example is the use of "looped" copper or steel piping embedded within the concrete floor for slab-on-grade structures. The heat is distributed evenly over the entire floor area.

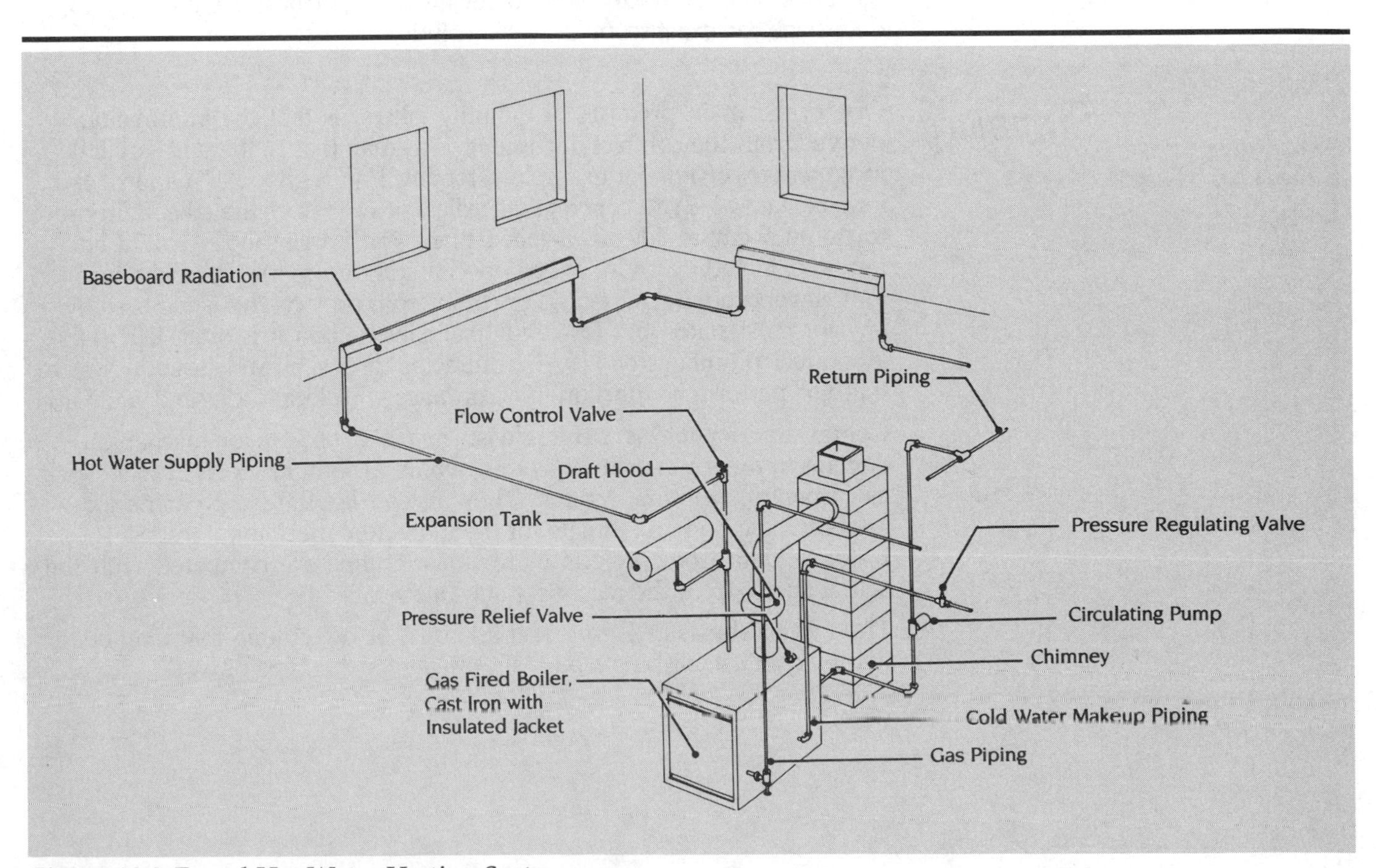

Figure 18.8 Forced Hot Water Heating System

In addition to the boiler itself, appurtenances required to complete the system include expansion tanks, pressure-relief and pressure-regulating valves, zone valves, circulators, pipe and fittings, flow control valves, oil burners (for oil-fired systems), operating controls, and fin tube radiant baseboard.

Units for Takeoff

Fin tube radiant baseboard is taken off by the LF, as is cast-iron baseboard. Quantities of each should be kept separate and listed according to type, manufacturer, model or rating (Btu output), and finish of the protective enclosure. The boiler itself is taken off by the individual unit and quantified as EA. Boilers should be listed according to size (Btu rating), type of fuel used, construction (steel, cast-iron), manufacturer, model or series, and level of assembly required. Appurtenances such as expansion tanks, circulators, zone valves, pressure valves, draft hoods, flues, and oil tanks are taken off by the piece (EA). They should be listed according to manufacturer, model, function or type, size, and any other identifying features specified.

Piping to and from radiant baseboard is taken off by the LF and listed according to size and material (steel, copper, etc.). Fittings and valves for the piping are taken off by the individual piece (EA) according to type, size, material, and method of joining. This same procedure is used for taking off looped radiant systems.

Other Takeoff Considerations

Related work of other trades may be necessary for a complete system. This work could include power wiring, gas piping, control wiring, and piping of oil tanks. The contractor should review the specifications carefully to determine the exact scope of work included.

Summary

One of the main premises of quantity takeoff is that the information derived from the contract documents is factual; that is, there is very little judgment involved. For example, a footing 1′-0″ high x 2′-0″ wide x 54′-0″ long contains 4 CY of concrete regardless of who does the takeoff. In order to do an accurate takeoff, detailed plans and specifications would be required, including an individual mechanical design for plumbing, heating, and air conditioning systems. For commercial projects this is required by law for most states, but for residential construction it is often left to the individual plumbing or HVAC contractor to design and install a system that will perform its function in accordance with local codes and standards.

Contractors do not, as a rule, do an itemized takeoff for mechanical systems in the same manner as they would for other aspects of work such as carpentry, painting, roofing. They instead list the basic criteria for establishing a budget estimate for the individual mechanical trades. Then historical data from projects with similar criteria are compared with the actual quotes from the respective subcontractors.

The takeoff sheets in Figure 18.9 list some of the criteria that may be necessary in establishing a budget estimate for this type of work.

Division 15 Checklist

The following checklist can be used to ensure that all items for this division have been accounted for.

Fire Protection Systems

- ☐ Fire standpipe systems
- ☐ Automatic sprinkler systems
- ☐ Shop drawings and insurance approval
- ☐ Permits

Plumbing

- ☐ Plumbing fixtures
- ☐ Equipment
- ☐ Hook-up of equipment provided by others
- ☐ Below-grade sanitary waste and vent piping
- ☐ Above-grade sanitary waste and vent piping
- ☐ Below-grade storm system piping
- ☐ Above-grade storm system piping
- ☐ Domestic hot and cold water piping
- ☐ Natural gas piping
- ☐ Specialty piping
- ☐ Pipe insulation
- ☐ Hangers and pipe supports
- ☐ Lead, oakum, gas, PVC, cement, gaskets, clamps
- ☐ Solder, flux, and gas
- ☐ Permit

Heating, Ventilating, and Air Conditioning

- ☐ Ductwork
- ☐ Ductwork, insulation, and sound lining
- ☐ Grilles, registers, diffusers, and dampers
- ☐ Heat-generating equipment
- ☐ Condensing units and evaporators
- ☐ Refrigerant lines
- ☐ Thermostats and controls
- ☐ Control wiring
- ☐ Testing and balancing
- ☐ Filters
- ☐ Cranes for hoisting rooftop equipment
- ☐ Hot-water boilers
- ☐ Radiant baseboard
- ☐ Radiant loop systems (coils)
- ☐ Piping for boilers
- ☐ Valves, circulators, expansion tanks, and appurtenances
- ☐ Related work of other trades

Means Forms

QUANTITY SHEET

SHEET NO. 15-1/5

PROJECT SAMPLE PROJECT

ESTIMATE NO. 1

LOCATION

ARCHITECT HOME PLANNERS

DATE

TAKE OFF BY WJD

EXTENSIONS BY: WJD

CHECKED BY: KF

DESCRIPTION	NO.	DIMENSIONS				UNIT		UNIT		UNIT		UNIT
DIVISION 15: MECHANICAL												
PLUMBING												
PLUMBING FIXTURES												
1.6 GAL WATER SAVER WATER CLOSETS AT EA BATH IN WHITE INCLUDING TRIM	3	EA			3	EA					3	EA
18" ROUND LAV SELF RIMMING DROP-IN (WHITE)	3	EA			3	EA					3	EA
SINGLE LEVER CHROME LAV FAUCET W/ POP-UP DRAIN KIT.	3	EA			3	EA					3	EA
32" x 60" ENAMELED CAST IRON TUB (WHITE)	1	EA			1	EA					1	EA
TRIP LEVER BATH DRAIN (CHROME)	1	EA			1	EA					1	EA
TUB/SHOWER VALVE (IN CHROME)	1	EA			1	EA					1	EA
W/ TUB SPOUT	1	EA			1	EA					1	EA
DRAIN TRIM @ SHOWER (CHROME)	1	EA			1	EA					1	EA
30" x 22" DOUBLE BASIN STAINLESS STEEL KITCHEN SINK	1	EA			1	EA					1	EA
SINGLE LEVER KITCHEN FAUCET W/ SPRAYER	1	EA			1	EA					1	EA

Figure 18.9

Means Forms

QUANTITY SHEET

PROJECT SAMPLE PROJECT

SHEET NO. 15-2/5

ESTIMATE NO. 1

LOCATION

ARCHITECT HOME PLANNERS

DATE

TAKE OFF BY WJD

EXTENSIONS BY: WJD

CHECKED BY: KF

DESCRIPTION	NO.	DIMENSIONS				UNIT		UNIT		UNIT		UNIT
STRAINER DRAIN @ KITCHEN SINK	2	EA			2	EA					2	EA
LAUNDRY TUB WITH FAUCET @ BASEMENT												
EXTERIOR HOSE CONNECTIONS	3	EA			3	EA					3	EA
75 GAL. GAS FIRED WATER HEATER	1	EA			1	EA					1	EA
HOOK-UP DISH WASHER (PROVIDED BY OTHERS)	1	EA			1	EA					1	EA
WATER METER @ SERVICE ENTRANCE	1	EA			1	EA					1	EA
INSULATION OF HOT & COLD WATER PIPING	1	LS			1	LS.					1	LS
PLUMBING PERMIT	1	EA			1	EA					1	EA
PROVIDE WASTE & WATER FOR WASHING MACHINE	1	EA			1	EA					1	EA

Figure 18.9 (continued)

Means Forms

QUANTITY SHEET

SHEET NO. 15-3/5

PROJECT SAMPLE PROJECT

ESTIMATE NO. 1

LOCATION

ARCHITECT HOME PLANNERS

DATE

TAKE OFF BY WJD

EXTENSIONS BY: WJD

CHECKED BY: KF

DESCRIPTION	NO.	DIMENSIONS				UNIT		UNIT		UNIT		UNIT
GAS PIPING												
GAS PIPING FOR												
BOILER	1	EA			1	EA					1	EA
WATER HEATER	1	EA			1	EA					1	EA
CLOTHES DRYER	1	EA			1	EA					1	EA

Figure 18.9 (continued)

Means Forms

QUANTITY SHEET

PROJECT: SAMPLE PROJECT

SHEET NO. 15-4/5

ESTIMATE NO. 1

LOCATION:

ARCHITECT: HOME PLANNERS

DATE:

TAKE OFF BY: WJD.

EXTENSIONS BY: WJD

CHECKED BY: KF

DESCRIPTION	NO.	DIMENSIONS				UNIT		UNIT		UNIT		UNIT
HEATING												
GAS-FIRED FORCED HOT WATER BOILER	1	EA			1	EA					1	EA
CIRCULATOR	1	EA			1	EA					1	EA
EXPANSION TANK & MISC. VALVES & APPURTENANCES @ BOILER	1	L.S.			1	L.S.					1	L.S.
ZONE VALVES FOR THREE ZONE SYSTEM												
- MASTER BEDRM. & MASTER BATH	1	ZONE			1	EA					1	EA
- UPSTAIR BEDROOM & COMMON BATH	1	ZONE			1	EA					1	EA
- FIRST FLOOR	1	ZONE			1	EA					1	EA
TOTAL HEATED LIVING AREA REQUIRING FIN TUBE BASEBOARD												
- 1ST FLR W/ 9'-0" CEILING	1	1230			1230	SF						
- 2ND FLR. W/ 8'-0" CEILING	1	744			744	SF						
					1974	SF					1974	SF
THERMOSTATS @ LIVING SPACES	3	EA			3	EA					3	EA

Figure 18.9 (continued)

Means Forms

QUANTITY SHEET

SHEET NO. 15-5/5

PROJECT SAMPLE PROJECT

ESTIMATE NO. 1

LOCATION

ARCHITECT HOME PLANNERS

DATE

TAKE OFF BY WJD

EXTENSIONS BY: WJD

CHECKED BY: KF

DESCRIPTION	NO.	DIMENSIONS				UNIT		UNIT		UNIT		UNIT
AIR-CONDITIONING SYSTEM												
TOTAL COOLED LIVING SPACE												
– 1ST FLR W/ 9'-0" CEILING					1230	SF						
– 2ND FLR W/ 8'-0" CEILING					744	SF						
					1974	SF					1974	SF
EXTERIOR PAD MOUNTED COMPRESSOR	1	EA			1	EA					1	EA
EVAPORATOR COIL & AIR HANDLING UNIT LOCATED IN ATTIC	1	EA			1	EA					1	EA
DUCTWORK & INSULATION	1	LS			1	LS					1	LS
GRILLES, REGISTERS & DIFFUSERS	1	LS			1	LS					1	LS
THERMOSTAT	1	EA			1	EA					1	EA
REFRIGERANT PIPING FROM COMPRESSOR TO AIR-HANDLING UNIT	1	LS			1	LS					1	LS

Figure 18.9 (continued)

Chapter Nineteen

ELECTRICAL

(Division 16)

Chapter Nineteen

ELECTRICAL
(Division 16)

Division 16 includes the work of various subsystems that distribute electricity for power and lighting, including more specialized wiring systems such as fire and security alarms, communications, electric heating, and low-voltage wiring. The work is performed by individuals or firms with specific training in electrical work, which requires licenses and permits separate from those obtained by the general contractor. To take off and estimate electrical work, one must have a working knowledge of the particular system in addition to academic training, neither of which are normally within the realm of the general contractor's experience. This chapter covers the procedures for takeoff of electrical systems generally encountered in light commercial and residential construction.

The work of Division 16 is typically shown on the electrical drawings labeled with the prefix *E*. (See Appendix A for a set of electrical symbols and abbreviations.) All drawings within the bid set should be reviewed for electrical work that may be shown on other drawings, such as site lighting, utilities plans, or the hook-up of equipment provided by others that might be shown on a fixture or equipment plan. For the electrical layout for the sample project, refer to the first and second floor plans in Appendix B. Mechanical drawings should be reviewed for the related work in other sections, such as power and control wiring for mechanical systems and wiring of detection and alarm systems for fire protection systems.

Architectural drawings should be reviewed and used during the takeoff as a source for information including dimensions, room sizes, floor-to-floor heights, location of services entering the building, and coordination of other work within the area.

Electrical drawings often use schedules that are helpful in determining the type and quantity of materials for takeoff (see "Schedules" in Chapter 1). Schedules are commonly used for lighting fixtures, panels, and feeders. Specialized details, such as the electrical riser diagram, illustrate the various components of the system and their configurations. As noted in Chapter 18, riser diagrams are for the graphic representation of information only, and are not drawn to scale.

Division 16 specifications should be reviewed thoroughly for the products, methods and techniques of installation, and the related work of other trades that must be included. A review of Division 1 of the specifications is necessary to determine what, if any, special requirements are to be included in the takeoff. A prime example may be the provision and maintenance of temporary power and lighting service during the work.

This chapter covers the procedures and methods for takeoff for electrical systems normally encountered in light commercial and residential construction. The takeoff and estimating of electrical work can be divided into the major categories that are discussed in the following sections. The order in which the categories are discussed is not intended to suggest the order in which takeoff should proceed; the starting point should be determined by the individual performing the takeoff.

Raceways

Raceways are channels constructed to house and protect electrical conductors. They include conduits, wireways, surface metal raceways, and underfloor duct. As part of the raceway system, fittings are needed to change direction, connect, and support the various types of raceway runs.

The most common type of raceway is conduit, including aluminum, rigid galvanized steel, steel intermediate conduit (IMC), rigid plastic-coated steel, PVC, and electrical metallic tubing (EMT). Conduits can be wall-mounted, suspended overhead, encased in concrete, or buried.

Units for Takeoff

Raceways in general are taken off by the linear foot and classified according to type, size, and application. Individual fittings for such items as wireways, underfloor ducts, surface metal raceways, and the larger diameter conduits are taken off by the piece (EA) and listed according to type, size, and material. Fittings for smaller-diameter conduits can accounted for by adding a percentage to the total conduit materials. Percentages will vary with the complexity of the run.

Other Takeoff Considerations

In most instances, fittings are not shown on drawings for standard conduit installations. Many of the applications of particular fittings are dictated by local electrical codes or the individual project requirements. The contractor should be familiar with the governing codes.

The takeoff of conduit should be divided into three categories: *power distribution*, *branch power*, and *branch lighting*. Power distribution includes the installation of the main conductors that will supply power to the various panels. Branch power and branch lighting refer to the branching of the panels to provide power and lighting on various locations in the project. Using these distinctions, all conduit quantities need not be taken off at one time. Quantities can instead be determined system by system.

Raceways installed higher than 15′ above the floor should be noted separately because of the reduced productivity for this type of installation.

Conductors and Grounding

A *conductor* is a wire or metal bar with a low resistance to the flow of electricity. *Grounding* is accomplished by a conductor connected between electrical equipment, or between a circuit and the earth. Wire is the most common material used to conduct current from the electrical source to electrical use. Wire is made of either copper or aluminum conductors with an insulating jacket, and is available in a variety of voltage ratings and insulating materials. Wire is installed within raceways, such as conduit or *flexible metallic conduit* (sometimes referred to as *Greenfield* or *flex*). Flexible metallic conduit is a single strip of aluminum or galvanized steel, spiral-wound and interlocked to provide a circular cross section of high strength and flexibility for the protection of the wire within. Other products similar to flex are covered with a liquid-tight plastic covering; they are used where protection from liquids is required.

Other types of conductors include armored cable (BX & MC), a fabricated assembly of cable with a metal enclosure similar in appearance to flex. Nonmetallic sheathed cable (Romex) is factory-constructed of two, three, or four insulated conductors enclosed in an outer sheath of plastic or fibrous material. It is available with or without a bare ground wire made of copper or aluminum conductors.

Special wires such as those used in low-voltage control wiring, signals, and telecommunications are also available for the performance required by the specific application.

In addition to the wire itself, special connectors or terminations at the end of each wire are required to complete the application. Various fasteners such as staples, clips, and flex fittings are also necessary.

Units for Takeoff

Wire, flex, and cables are taken off by the LF and are typically extended to 100 LF (CLF). For wire installed within conduits or flex, the total quantity of wire is determined by multiplying the number of conductors by the LF of conduit or flex, and dividing by 100 to arrive at CLF. All wire and cables should be listed in the takeoff according to type, size (rating), conductor material, and application (feeders, branch power, branch lighting).

Special fittings for the connection of wire or cables are taken off by the individual piece (EA) and listed according to type, size, application, and method of connection.

In addition to grounding conductors, accessory items such as ground rods, clamps, and exothermic weld metal are taken off by the individual piece (EA) and listed according to type, size, and application.

Other Takeoff Considerations

An allowance of 10% for waste and connections is usually acceptable, but may be increased for lengths of wire or cables with numerous interruptions such as intermediate connections or splices.

As a matter of procedure, wire should be taken off and separated according to application (feeders and service entrance, branch power, and branch lighting). Care is essential during takeoff to allow for sufficient lengths for connections, especially in larger feeders where incorrect footages may be costly. The contractor should be familiar with the governing codes concerning terminations to ensure that all items have been included.

Wiring Devices and Boxes

A box is used in electrical wiring at each junction point, outlet, or switch. Boxes provide access to electrical connections and serve as a mounting for fixtures or switches. They may also be used as pull points for wire in long runs or conduits. A wiring device can be defined as a mechanism that controls but does not consume electricity, such as a switch or receptacle.

Boxes often require accessories such as plaster rings, covers, and various fasteners to support them. Boxes are constructed of galvanized or coated steel, or high-density plastic.

Wiring devices, in addition to receptacles and switches, include pilot lights, relays, low-voltage transformers, and a variety of specialized controls and finish wall plates.

Units for Takeoff

Outlet boxes, pull or junction boxes, receptacles, switches, wall plates, relays, and wiring devices in general are taken off by the individual piece (EA). They should include the necessary accessories for the completed application. For example, outlet boxes should include plaster rings and extensions (if required); pull boxes should include covers; receptacles and switches should include plates; and so forth.

The various items should be listed according to type, size, composition, capacity or application, and color (if applicable).

Other Takeoff Considerations

The contractor should review the specifications to determine the exact scope of work concerning control devices, such as relays and low-voltage transformers, to be used in conjunction with the work of related trades. Special devices may be provided by other trades and installed and wired under the electrical specifications. Examples include relays for heating or cooling units, flow switches for automatic sprinkler systems, and temperature-sensing controls for heating applications.

Starters, Boards, and Switches

This section includes panelboards, starters for motors, control stations, circuit breakers, safety switches and disconnects, fuses, and meter centers and meter sockets.

Units for Takeoff

Control stations are taken off by the individual unit (EA) and listed according to type, manufacturer, classification, and application.

Circuit breakers are taken off by the individual piece (EA) and listed according to manufacturer, type (number of poles), capacity (rating), voltage, method of installation (plug-in or bolt-on), and classification (NEMA).

Panelboards are taken off by the individual unit (EA) and listed according to size, type, voltage, and manufacturer. Some standard board and breaker assemblies are available as pre-assembled units, such as *load centers* used in residential construction. A load center is a type of panel with breakers that is sold as a unit.

Starters are taken off by the individual piece (EA) and listed according to size, voltage, NEMA enclosure, and type.

Safety switches and disconnects are taken off by the individual unit (EA) and listed according to size, type (duty), number of poles, voltage, NEMA classification, and ampere rating.

Fuses are taken off by the individual piece (EA) and listed according to amp, voltage, and type or class.

Meter centers (unit for multiple sockets) and meter sockets are taken off by the individual unit (EA) and listed according to size and type for meter sockets, and by bus capacity, number of meter sockets, and type of enclosure for meter centers.

Other Takeoff Considerations

In addition to the above-mentioned components, special fasteners may be required to install the equipment. Plywood boards or concrete pads for mounting electrical equipment, or anchors to attach items to concrete or masonry surfaces, must be included as part of the individual takeoff. Refer to the specifications to clarify the exact scope of work. Since starters for motors are frequently furnished as part of the mechanical package, the contractor should verify that they are not furnished by others to avoid costly duplication or omission.

One fuse should be counted for each line (or phase) being protected.

Lighting

Lighting is fundamental to electrical construction. Varieties of lighting include interior and exterior, surface-mounted and recessed, emergency and exit fixtures, track lighting, and the lamps for the various fixtures. An enormous variety of light fixtures are manufactured to suit every application. Lamps for fixtures include incandescent, fluorescent, mercury vapor, metal halide, and high-pressure sodium.

Units for Takeoff

Light fixtures are counted and listed on the takeoff by the individual unit (EA). The number of lamps is determined per fixture and listed by individual lamp (EA).

Light fixtures should be listed in the takeoff according to the manufacturer, model, type, color or finish, location (wall, ceiling, room), and interior or exterior application. The level of assembly required for such items as ceiling fans or chandeliers should also be noted, if applicable.

Emergency and exit lighting should be kept separate from general light fixtures and are quantified by the individual piece (EA).

Other Takeoff Considerations

Interior and exterior light fixtures should be taken off and listed separately. Most exterior light fixtures are either wall- or pole-mounted, and larger fixtures may require the use of a crane or boom truck for installation. Large interior fixtures such as chandeliers may also require some type of rigging or hoisting, as well as some type of structural support for installation.

Special Systems

Special systems include fire alarms, cable and closed-circuit TV, intercoms, electric heating, energy management, and security systems. Each is meant to be a separate system, functioning independently of other electrical systems or in conjunction with other building systems.

Units for Takeoff

The takeoff procedure for special systems should follow that of other electrical systems previously outlined. Wiring should be taken off by the LF and listed according to identifying characteristics and application. Devices should be taken off by the individual piece (EA) and listed according to type, application, manufacturer, model, and function. The takeoff should include all necessary boxes, covers, plates, connectors, and equipment necessary for a complete system.

Other Takeoff Considerations

As many specialty systems are designed to perform a specific series of functions, specifications often require testing of the completed system to prove compliance with the design criteria. Testing may also be required by an independent firm at the expense of the contractor. Again, review of the specifications will help determine the scope and responsibility testing.

For buildings already in use, tie-ins to existing systems may have to be done during nights and weekends to avoid disruption of the building's normal operations. This should be noted in the takeoff for accurate pricing of labor.

Equipment Hook-ups

Much of the equipment within a building or residence requires power to operate. Appliances and equipment that cannot be simply plugged in must be "hard wired." This work is typically included as part of the electrical contractor's work. Examples include electric ranges and ovens, heating or cooling units, dishwashers, garbage disposals, and commercial equipment for specialty operations such as restaurants or manufacturing plants. This equipment is typically supplied and set in place by the equipment dealer.

The drawings and specifications should indicate the exact scope of required equipment hook-ups.

Units for Takeoff

Appliances or equipment that require hook-ups are taken off by the individual piece (EA) and listed according to type and method of installation (such as flexible-connected, piped, overhead, or in-floor), and any other special requirements.

Other Takeoff Considerations

Special requirements such as control devices or disconnect switches may be needed to meet governing codes and may not always be noted in the specifications. The contractor performing the takeoff should be familiar with the applicable codes governing the electrical installation of the particular equipment.

Miscellaneous

Other items to include in the takeoff may be harder to classify and are best categorized in a miscellaneous group. Some of the more common ones are permits and utility company tie-in fees, cutting and drilling for electrical access, and temporary power and lighting.

Units for Takeoff

Each miscellaneous item should be taken off and listed separately. Permits and utility company fees for tie-ins are taken off per occurrence, and can be listed as a lump sum (LS). Fees for electrical permits are typically based on the individual job and are paid when a permit is applied for or issued. Utility companies typically charge a set fee to tie in service or a transformer, per occurrence.

Cutting and drilling through wood or other light material is often considered part of the normal scope of work and therefore does not constitute a separate takeoff item. Cutting or coring through masonry or concrete may involve special equipment or additional time, especially for large quantities. Holes cut or cored through masonry or concrete are taken off by the piece (EA) and listed according to size (diameter or length x width), type of material, thickness of the material, and equipment needed.

Temporary lighting and power are often necessary during the construction process and are typically defined in Division 1—General Requirements. They may also be noted in Division 16—Electrical under "Related Work." Since temporary power panels and lights are reusable, the materials portion of the cost may include only lamps and wire, and can be taken off and listed as a lump sum (LS). The installation and removal of temporary power and lighting facilities are often quantified by labor-hours.

Other Takeoff Considerations

Minor excavation and backfilling for the installation conduits or nonmetallic sheathed cables may also be required. Excavation and backfill are typically part of the general or site contractor's work. The contractor should review the specifications for the exact scope of work concerning excavation and backfill for electrical work.

Summary

The same basic procedure for taking off quantities for Division 15—Mechanical (see "Summary" in Chapter 18) is used in the takeoff of Division 16—Electrical. Often residential designers include the layout of lighting, switches, and receptacles on the architectural drawings. This is used in establishing the cost of the electrical portion of the work. In addition, other power-using equipment such as furnaces, water heaters, air conditioning units and appliances are needed to appropriately determine the criteria for establishing the electrical budget.

The takeoff sheets in Figure 19.1 list some of the necessary criteria.

Division 16 Checklist

The following checklist can be used to ensure that all items for this division have been accounted for.

Raceways

- ☐ Conduits
- ☐ Wireways
- ☐ Surface wireways
- ☐ Fittings

Conductors and Grounding

- ☐ Wire
- ☐ Flexible metallic conduits
- ☐ Liquid-tight covering
- ☐ Armored cable
- ☐ Nonmetallic sheathed cable
- ☐ Special wire
- ☐ Fittings and connectors

Wiring Devices and Boxes

- ☐ Switches and receptacles
- ☐ Outlet boxes and pull boxes
- ☐ Finish plates
- ☐ Relays and controls
- ☐ Low-voltage transformers

Starters, Boards, and Switches

- ☐ Control stations

- ☐ Circuit breakers
- ☐ Panelboards and load centers
- ☐ Motor starters
- ☐ Safety switches and disconnects
- ☐ Meter centers and meter sockets
- ☐ Fuses

Lighting

- ☐ Lamps
- ☐ Interior light fixtures
- ☐ Exterior light fixtures
- ☐ Emergency and exit light systems
- ☐ Owner-supplied fixtures

Special Systems

- ☐ Telecommunications systems
- ☐ Electric heating
- ☐ Energy management systems
- ☐ Security systems
- ☐ Fire alarm and detection systems
- ☐ Music and paging systems
- ☐ Door bell

Equipment Hook-ups

Miscellaneous

- ☐ Cutting, drilling, and coring
- ☐ Permits
- ☐ Utility company fees or charges
- ☐ Temporary lighting and power
- ☐ Excavation and backfill

Means Forms

QUANTITY SHEET

PROJECT SAMPLE PROJECT

SHEET NO. 16-1/2

ESTIMATE NO. 1

LOCATION

ARCHITECT HOME PLANNERS

DATE

TAKE OFF BY WJD

EXTENSIONS BY: WJD

CHECKED BY: KF

DESCRIPTION	NO.	DIMENSIONS	UNIT	UNIT
DIVISION 16 – ELECTRICAL				
SURFACE MOUNTED LIGHT FIXTURES				
– INTERIOR	10	EA	10 EA	10 EA
– EXTERIOR	5	EA	5 EA	5 EA
– PORCELAIN KEYLESS @ BASEMENT, GARAGE & UNFINISHED SPACE	9	EA	9 EA	9 EA
RECEPTACLES				
– INTERIOR	49	EA	49 EA	49 EA
– EXTERIOR	2	EA	2 EA	2 EA
– GFI	4	EA	4 EA	4 EA
SWITCHES				
– SINGLE	23	EA	23 EA	23 EA
– 3 WAY	10	EA	10 EA	10 EA
ELECTRICAL WIRING OF				
– DISHWASHER	1	EA	1 EA	1 EA
– BOILER	1	EA	1 EA	1 EA
– WATER HEATER	1	EA	1 EA	1 EA
– RANGE	1	EA	1 EA	1 EA
– A/C UNITS	2	EA	2 EA	2 EA
– GARBAGE DISP.	1	EA	1 EA	1 EA
SMOKE DETECTORS	4	EA	4 EA	4 EA
HEAT DETECTOR (IN GARAGE)	1	EA	1 EA	1 EA
DOOR BELL SYSTEM	1	LS	1 LS	1 LS

Figure 19.1

Means Forms

QUANTITY SHEET

PROJECT SAMPLE PROJECT | SHEET NO. 16-2/2

ESTIMATE NO. 1

LOCATION | ARCHITECT HOME PLANNERS | DATE

TAKE OFF BY WJD | EXTENSIONS BY: WJD | CHECKED BY: KF

DESCRIPTION	NO.	DIMENSIONS				UNIT		UNIT		UNIT		UNIT
CABLE TV OUTLETS	5	EA			5	EA					5	EA
TELEPHONE OUTLETS	5	EA			5	EA					5	EA
METER SOCKET												
MAST & WEATHERHEAD	1	EA			1	EA					1	EA
LOAD CENTER												
(150 AMP)	1	EA			1	EA					1	EA
ELECTRICAL PERMIT	1	EA			1	EA					1	EA

Figure 19.1 (continued)

APPENDIX

APPENDIX A

Appendix A includes commonly used symbols and their definitions. **Graphic Symbols** are frequently used on building plans to illustrate elements such as gas and water service lines, window types, or existing work. **Material Indication Symbols** are used to represent various building materials, such as brick, ceramic tile, and windows. The chart shows how materials appear on plans, elevations, and sections. **Landscape Symbols and Graphics** illustrate elements in the finished landscape, such as stone walls, trees, rock, and water. Finally, **Trade Specific Symbols** are included for the electrical, HVAC, and plumbing trades.

Graphic Symbols

SYMBOL	DEFINITION
— — — — — — — 123	EXISTING CONTOUR
———————— 124	NEW CONTOUR
———— G ————	GAS SERVICE LINE
———— W ————	WATER SERVICE LINE
———— SS ————	SANITARY SEWER LINE
———— D ————	DRAIN LINE
———— E ————	ELECTRIC SERVICE
———— UE ————	UNDERGROUND ELECTRIC SERVICE
———— OE ————	OVERHEAD ELECTRIC SERVICE
— — — — — — — — — —	EXISTING WORK (GENERAL)
——— - ——— ——— - ———	CONTRACT LIMIT LINE
+ 12.34	SPOT GRADE
○— SMH #	SEWER MANHOLE #
○— DMH #	DRAIN MANHOLE #
□— CB #	CATCH BASIN #
—▷◁—	VALVE
⊕	BORING OR TEST PIT
BM 134.36	BENCH MARK ELEV.
⌖	FIRE HYDRANT
o #	UTILITY POLE #
RIM = 34.56'	ELEV. OF MANHOLE RIM
INV. IN = 34.56'	ELEV. OF INVERT IN
INV. OUT = 34.56'	ELEV. OF INVERT OUT

SYMBOL	DEFINITION
2 / A-4	WALL SECTION NO. 2 CAN BE SEEN ON DRAWING NO. A-4.
3 / A-5	DETAIL SECTION NO. 3 CAN BE SEEN ON DRAWING NO. A-5.
AA / A-6	BUILDING SECTION A-A CAN BE SEEN ON DRAWING NO. A-6.
———	MAIN OBJECT LINE
- - - - - - - - -	HIDDEN OR INVISIBLE LINE
—— - —— - ——	INDICATES CENTERLINE
→ 3" ← 3'-4"	DIMENSION LINES
↓ ↓	EXTENSION LINES
℄	SYMBOL INDICATES CENTER LINE
/////////////	INDICATES WALL SURFACE
(N)	INDICATES NORTH DIRECTION

SYMBOL	DEFINITION
(COL) ——— - - ———	COLUMN LINE GRID
◇5 — OR ▷5 —	PARTITION TYPE
⬡A	WINDOW TYPE
(05)	DOOR NUMBER
⌷05⌷	ROOM NUMBER
(10'-0")	CEILING HEIGHT
△2	REVISION MARKER
	BREAK IN A CONTINUOUS LINE
(3) ⟶	REFER TO NOTE #3
100'-0"	ELEVATION MARKER
1, 4, A-5, 2, 3	INTERIOR ELEVATIONS 1, 2, 3, & 4 CAN BE SEEN ON DRWG. A-5. DIRECTION OF TRIANGLE INDICATES ELEVATION.

Material Indication Symbols

MATERIAL INDICATION SYMBOLS			
MATERIAL	PLAN	ELEVATION	SECTION
EARTH	NONE	NONE	
POROUS FILL	NONE	NONE	
CONCRETE			SAME AS PLAN
BRICK			SAME AS PLAN
CONCRETE BLOCK			SAME AS PLAN
STONE	CUT STONE RUBBLE	CUT STONE RUBBLE	SAME AS PLAN
STEEL	OR OR	NONE	
FRAMING WOOD	WALL OR PARTITION	SIDING FINISH WOOD	BLOCKING FRAMING
PLYWOOD	INDICATED BY NOTE	INDICATED BY NOTE	
SHEET METAL FLASHING	INDICATED BY NOTE		SHAPED HEAVY LINE

MATERIAL INDICATION SYMBOLS

MATERIAL	PLAN	ELEVATION	SECTION
BATT INSULATION		NONE	SAME AS PLAN
RIGID INSULATION		NONE	SAME AS PLAN
GLASS			SMALL SCALE LARGE SCALE
GYPSUM WALLBOARD			SAME AS PLAN
ACOUSTICAL TILE		NONE	
CERAMIC WALL TILE			SAME AS PLAN
FLOOR TILE		NONE	

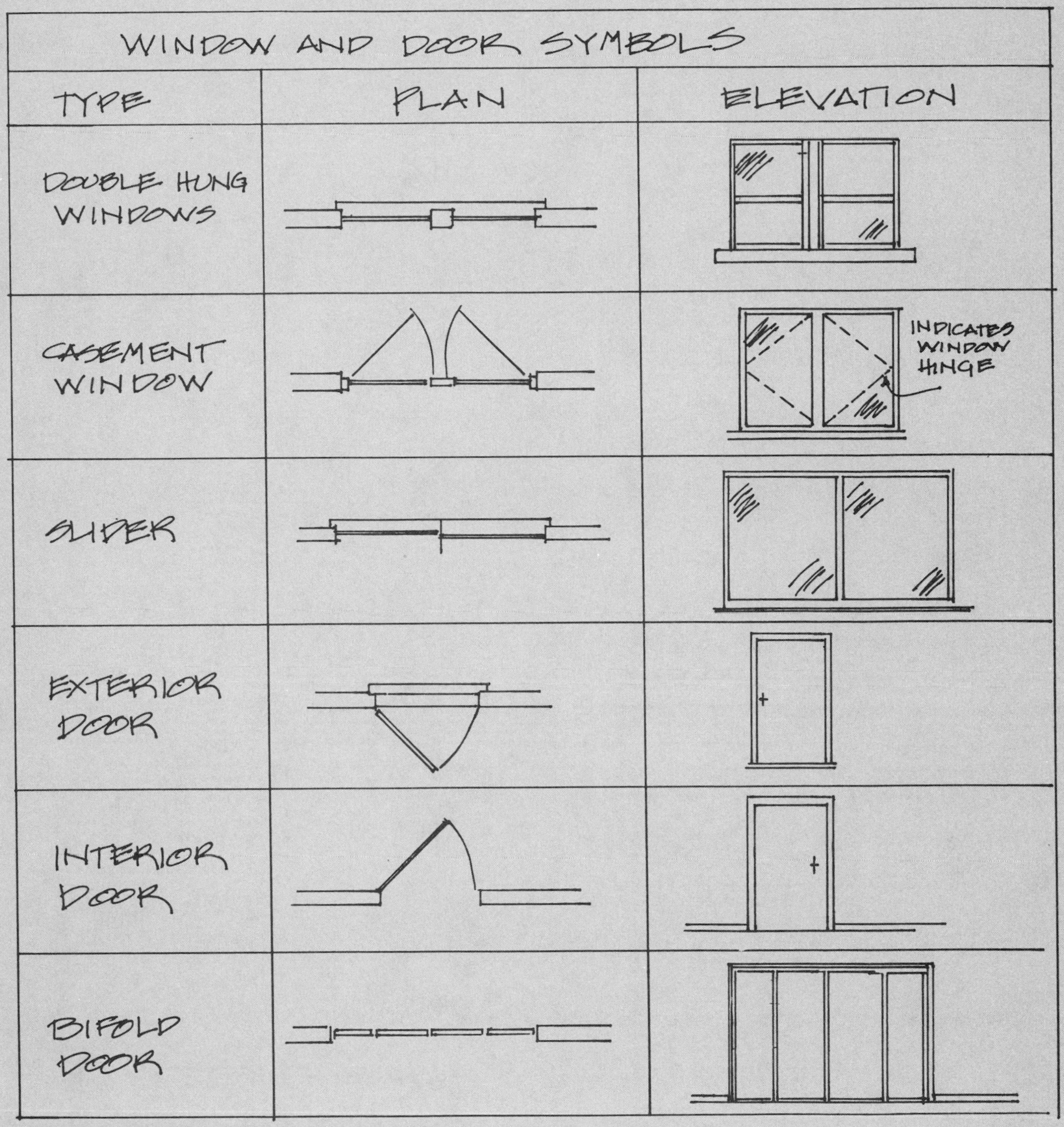

WINDOW AND DOOR SYMBOLS		
TYPE	PLAN	ELEVATION
DOUBLE HUNG WINDOWS		
CASEMENT WINDOW		INDICATES WINDOW HINGE
SLIDER		
EXTERIOR DOOR		
INTERIOR DOOR		
BIFOLD DOOR		

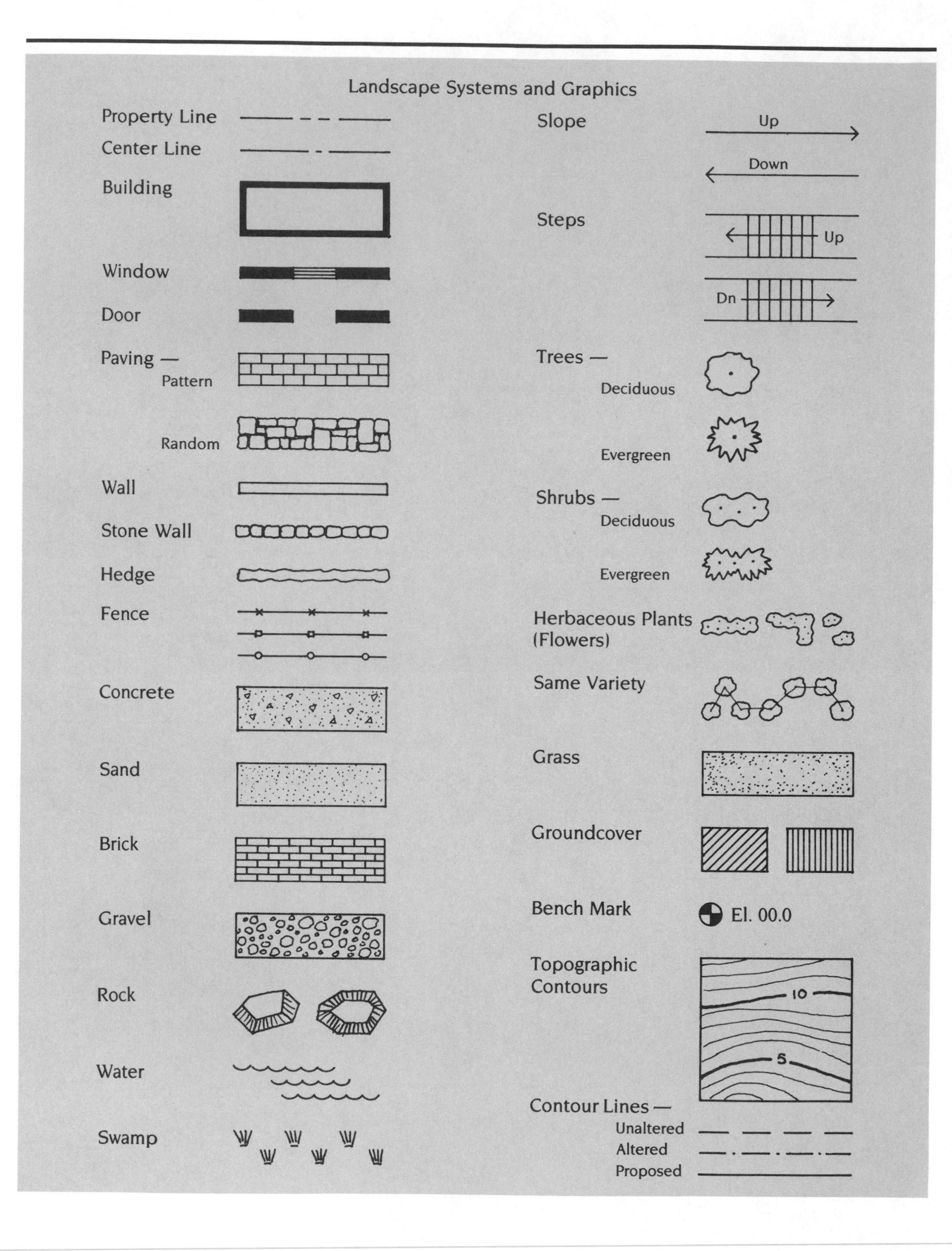
Landscape Systems and Graphics
Property Line
Center Line
Building
Window
Door
Paving —
Pattern
Random
Wall
Stone Wall
Hedge
Fence
Concrete
Sand
Brick
Gravel
Rock
Water
Swamp
Slope
Up
Down
Steps
Up
Dn
Trees —
Deciduous
Evergreen
Shrubs —
Deciduous
Evergreen
Herbaceous Plants
(Flowers)
Same Variety
Grass
Groundcover
Bench Mark
El. 00.0
Topographic
Contours
10
5
Contour Lines —
Unaltered
Altered
Proposed

Trade Specific Symbols

Lighting Outlets

- Ceiling Surface Incandescent Fixture
- Wall Surface Incandescent Fixture
- Ceiling Recess Incandescent Fixture
- Wall Recess Incandescent Fixture
- Standard Designation for All Lighting Fixtures – A = Fixture Type, 3 = Circuit Number, b = Switch Control
- Ceiling Blanked Outlet
- Wall Blanked Outlet
- Ceiling Electrical Outlet
- Wall Electrical Outlet
- Ceiling Junction Box
- Wall Junction Box
- Ceiling Lamp Holder with Pull Switch
- Wall Lamp Holder with Pull Switch
- Ceiling Outlet Controlled by Low Voltage Switching when Relay is Installed in Outlet Box
- Wall Outlet – Same as Above
- Outlet Box with Extension Ring
- Exit Sign with Arrow as Indicated
- Surface Fluorescent Fixture
- Pendant Fluorescent Fixture
- Recessed Fluorescent Fixture
- Wall Surface Fluorescent Fixture
- Channel Mounted Fluorescent Fixture
- Surface or Pendant Continous Row Fluorescent Fixtures
- Recessed Continuous Row Fluorescent Fixtures
- Incandescent Fixture on Emergency Circuit
- Fluorescent Fixture on Emergency Circuit

Receptacle Outlets

- Single Receptacle Outlet
- Duplex Receptacle Outlet
- Duplex Receptacle Outlet "X" Indicates above Counter Max. Height = 42″ or above Counter
- Weatherproof Receptacle Outlet
- Triplex Receptacle Outlet
- Quadruplex Receptacle Outlet
- Duplex Receptacle Outlet – Split Wired
- Triplex Receptacle Outlet – Split Wired
- Single Special Purpose Receptacle Outlet
- Duplex Special Purpose Receptacle Outlet
- Range Outlet
- Special Purpose Connection – Dishwasher
- Explosion-proof Receptacle Outlet Max. Height = 36″ to ℄
- Multi-outlet Assembly
- Clock Hanger Receptacle
- Fan Hanger Receptacle
- Floor Single Receptacle Outlet

Receptacle Outlets (Cont).

Floor Duplex Receptacle Outlet

Floor Special Purpose Outlet

Floor Telephone Outlet – Public

Floor Telephone Outlet – Private

Underfloor Duct and Junction Box for Triple, Double, or Single Duct System as Indicated by Number of Parallel Lines

Cellular Floor Header Duct

Switch Outlets

S	Single Pole Switch Max. Height = 42″ to ℄
S_2	Double Pole Switch
S_3	Three-way Switch
S_4	Four-way Switch
S_D	Automatic Door Switch
S_K	Key Operated Switch
S_P	Switch & Pilot Lamp
S_{CB}	Circuit Breaker
S_{WCB}	Weatherproof Circuit Breaker
S_{MC}	Momentary Contact Switch
S_{RC}	Remote Control Switch (Receiver)
S_{WP}	Weatherproof Switch
S_F	Fused Switch
S_L	Switch for Low Voltage Switching System
S_{LM}	Master Switch for Low Voltage Switching System
S_T	Time Switch
S_{TH}	Thermal Rated Motor Switch
S_{DM}	Incandescent Dimmer Swtich
S_{FDM}	Fluorescent Dimmer Switch
S	Switch & Single Receptacle
S	Switch & Double Receptacle
A S_A	Special Outlet Circuits

Institutional, Commercial, & Industrial System Outlets

Nurses Call System Devices – any Type

Paging System Devices – any Type

Fire Alarm System Devices – any Type

F — Fire Alarm Manual Station – Max. Height = 48″ to ℄

F — Fire Alarm Horn with Integral Warning Light

Fire Alarm Thermodetector, Fixed Temperature

S — Smoke Detector

Fire Alarm Thermodetector, Rate of Rise

F — Fire Alarm Master Box – Max. Height per Fire Department

H — Magnetic Door Holder

ANN — Fire Alarm Annunciator

Staff Register System – any Type

Electrical Clock System Devices – any Type

Public Telephone System Devices

Institutional, Commercial, & Industrial System Outlets (Cont).

Private Telephone System Devices

Watchman System Devices

Sound System, L = Speaker, V = Volume Control

Other Signal System Devices – CTV = Television Antenna, DP = Data Processing

SC Signal Central Station

Telephone Interconnection Box

PE Pneumatic/Electric Switch

EP Electric/Pneumatic Switch

GP Operating Room Grounding Plate

P 6 Patient Ground Point – 6 = Number of Jacks

Panelboards

Flush Mounted Panelboard & Cabinet

Surface Mounted Panelboard & Cabinet

Lighting Panel

Power Panel

Heating Panel

Controller (Starter)

Externally Operated Disconnect Switch

Busducts & Wireways

T T T Trolley Duct

B B B Busway (Service, Feeder, or Plug-in)

C C C Cable through Ladder or Channel

W W W Wireway

J Bus Duct Junction Box

Electrical Distribution or Lighting System, Aerial, Lightning Protection

Pole

Street Light & Bracket

Transformer

Primary Circuit

Secondary Circuit

Auxiliary System Circuits

Down Guy

Head Guy

Sidewalk Guy

Service Weather Head

Lightning Rod

—L— Lightning Protection System Conductor

Residential Signaling System Outlets

Push Botton

Buzzer

Bell

Bell and Buzzer Combination

Annunciator

Outside Telephone

Interconnecting Telephone

Telephone Switchboard

BT Bell Ringing Transformer

Residential Signaling System Outlets (Cont).

D (in box)	Electric Door Opener
M (in box)	Maid's Signal Plug
R (in box)	Radio Antenna Outlet
CH (in box)	Chime
TV (in box)	Television Antenna Outlet
T (in circle)	Thermostat

Underground Electrical Distribution or Lighting System

M (in box)	Manhole
H (in box)	Handhole
TM	Transformer – Manhole or Vault
TP	Transformer Pad
----	Underground Direct Burial Cable
	Underground Duct Line
	Street Light Standard Fed from Underground Circuit

Panel Circuits & Miscellaneous

----	Conduit Concealed in Floor or Walls
------	Wiring Exposed
	Home Run to Panelboard – Number of Arrows Indicate Number of Circuits
	Home Run to Panelboard – Two-Wire Circuit
	Home Run to Panelboard – Number of Slashes Indicate Number of Wires (When more than Two)
LS-L1,3,5	Home Run to Panelboard – 'LS' Indicates Panel Designation; LI, 3, 5, Indicates Circuit Breaker No.
—C—	Clock Circuit, Conduit and Wire
—E—	Emergency Conduit and Wiring
—T—	Telephone Conduit and Wiring
	Feeders
	Conduit Turned Up
	Conduit Turned Down
G (in circle)	Generator
M (in circle)	Motor
5 (in circle)	Motor – Numeral Indicates Horsepower
I (in circle)	Instrument (Specify)
T (in box)	Transformer
	Remote Start-Stop Push Button Station
	Remote Start-Stop Push Button Station w/Pilot Light
HTR	Electric Heater Wall Unit

Common Abbreviations on Drawings

EWC	Electric Water Cooler
EDH	Electric Duct Heater
AFF	Above Finished Floor
UH	Unit Heater
GFI	Ground Fault Interrupter
GFP	Ground Fault Protector
GFCB	Ground Fault Circuit Breaker
EC	Empty Conduit
WP	Weatherproof

Common Abbreviations on Drawings (Cont).

VP	Vaporproof	**ATS**	Automatic Transfer Switch
EXP	Explosion-proof	**IS**	Isolating Switch
AD	Auto Damper	**FATC**	Fire Alarm Terminal Cabinet
LP	Lighting Panel	**FA**	Fire Alarm
PP	Power Panel	**CAM**	Closed Circuit TV Camera
IP	Isolation Panel	**MON**	Closed Circuit TV Monitor
MC	Motor Controller	**MG**	Motor-Generator Set
MCC	Motor Control Center		

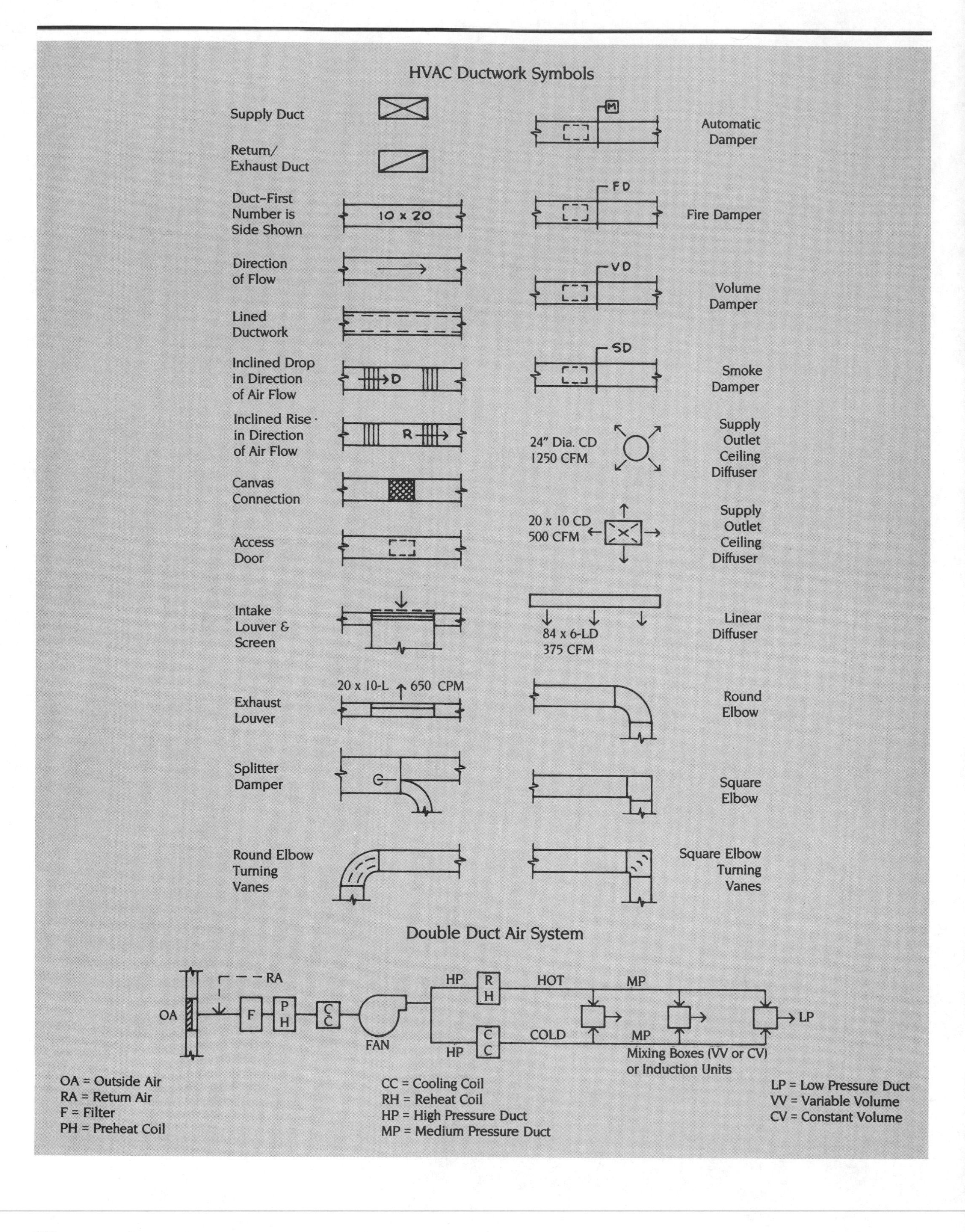
HVAC Ductwork Symbols
Supply Duct
Return/
Exhaust Duct
Duct-First
Number is
Side Shown
10 x 20
Direction
of Flow
Lined
Ductwork
Inclined Drop
in Direction
of Air Flow
D
Inclined Rise
in Direction
of Air Flow
R
Canvas
Connection
Access
Door
Intake
Louver &
Screen
Exhaust
Louver
20 x 10-L
650 CPM
Splitter
Damper
Round Elbow
Turning
Vanes
M
Automatic
Damper
FD
Fire Damper
VD
Volume
Damper
SD
Smoke
Damper
24" Dia. CD
1250 CFM
Supply
Outlet
Ceiling
Diffuser
20 x 10 CD
500 CFM
Supply
Outlet
Ceiling
Diffuser
84 x 6-LD
375 CFM
Linear
Diffuser
Round
Elbow
Square
Elbow
Square Elbow
Turning
Vanes
Double Duct Air System
RA
OA
F
PH
CC
FAN
HP
RH
HOT
MP
HP
CC
COLD
MP
LP
Mixing Boxes (VV or CV)
or Induction Units
OA = Outside Air
RA = Return Air
F = Filter
PH = Preheat Coil
CC = Cooling Coil
RH = Reheat Coil
HP = High Pressure Duct
MP = Medium Pressure Duct
LP = Low Pressure Duct
VV = Variable Volume
CV = Constant Volume

HVAC (Cont.)

Description	Symbol	Symbol	Description
Hot Water Heating Supply	—— HWS ——	—— FOG ——	Fuel Oil Gauge Line
Hot Water Heating Return	—— HWR ——	—o— PD —o—	Pump Discharge
Chilled Water Supply	—— CHWS ——	– – – – – – –	Low Pressure Condensate Return
Chilled Water Return	—— CHWR ——	—— LPS ——	Low Pressure Steam
Drain Line	—— D ——	—— MPS ——	Medium Pressure Steam
City Water	—— CW ——	—— HPS ——	High Pressure Steam
Fuel Oil Supply	—— FOS ——	—— BD ——	Boiler Blow-Down
Fuel Oil Return	—— FOR ——		
Fuel Oil Vent	—— FOV ——		

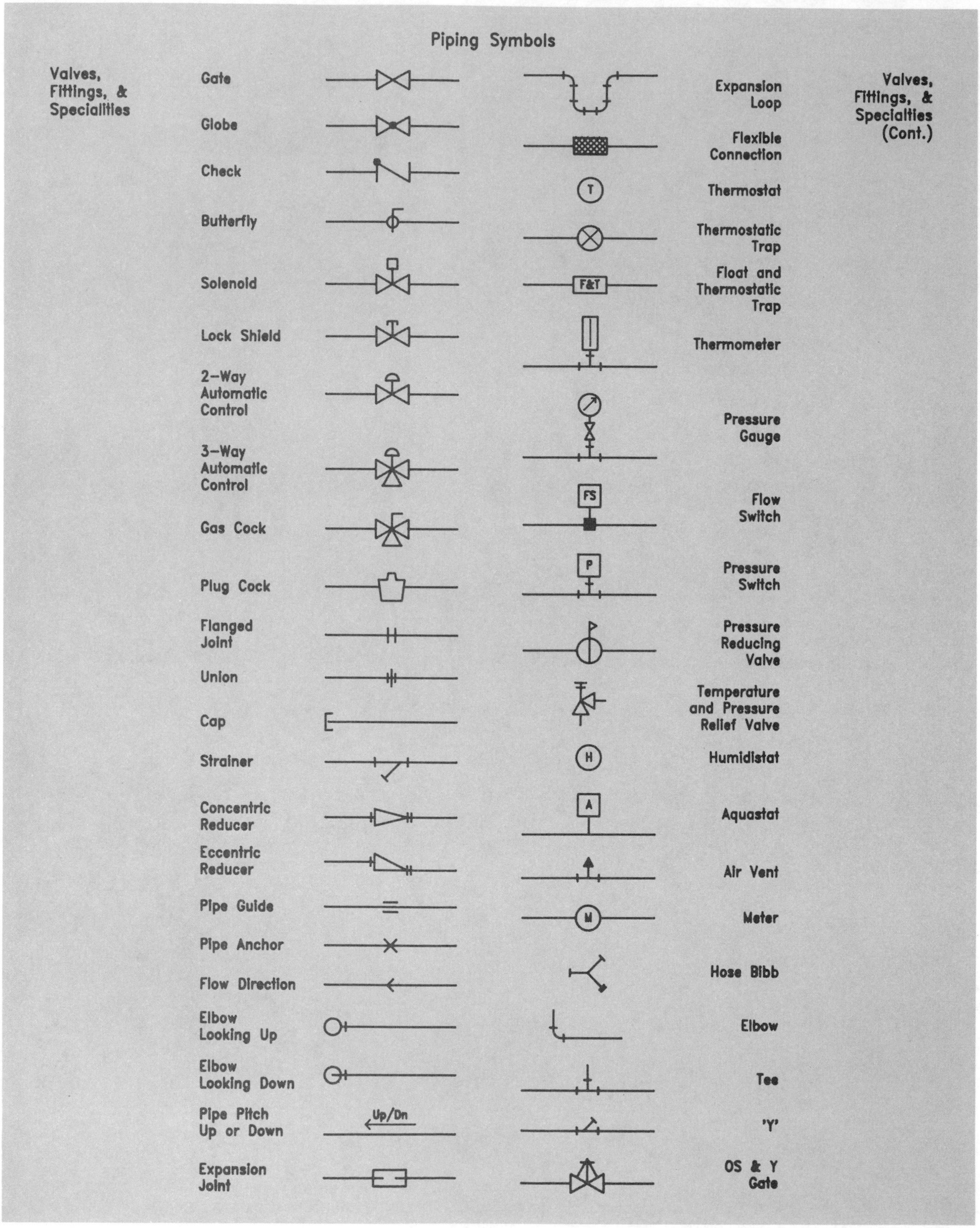

Piping Symbols

Valves, Fittings, & Specialities	Valves, Fittings, & Specialities (Cont.)
Gate	Expansion Loop
Globe	Flexible Connection
Check	Thermostat
Butterfly	Thermostatic Trap
Solenoid	Float and Thermostatic Trap
Lock Shield	Thermometer
2-Way Automatic Control	Pressure Gauge
3-Way Automatic Control	Flow Switch
Gas Cock	Pressure Switch
Plug Cock	Pressure Reducing Valve
Flanged Joint	Temperature and Pressure Relief Valve
Union	Humidistat
Cap	Aquastat
Strainer	Air Vent
Concentric Reducer	Meter
Eccentric Reducer	Hose Bibb
Pipe Guide	Elbow
Pipe Anchor	Tee
Flow Direction	'Y'
Elbow Looking Up	OS & Y Gate
Elbow Looking Down	
Pipe Pitch Up or Down (Up/Dn)	
Expansion Joint	

Symbol labels: T, F&T, FS, P, H, A, M

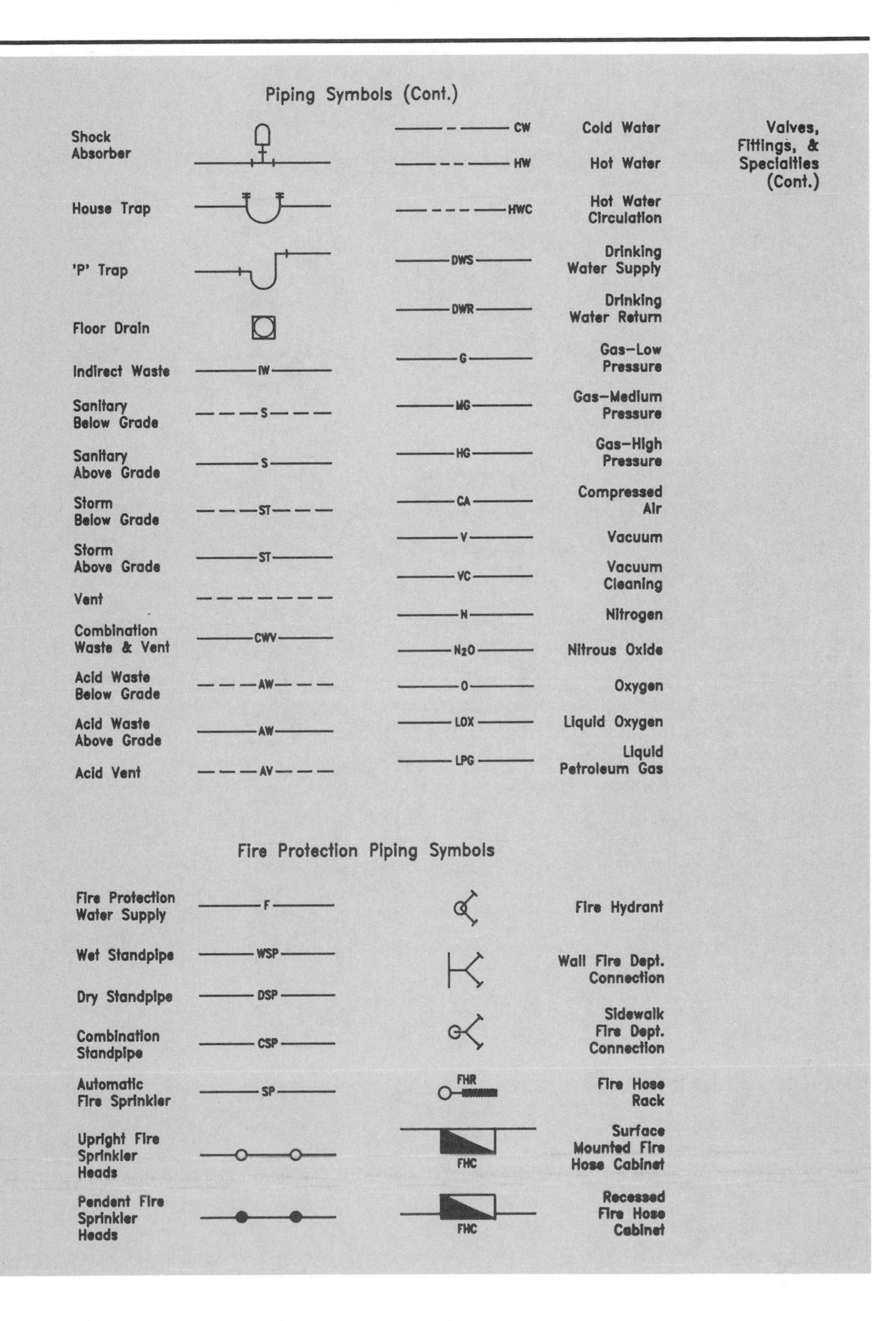
Piping Symbols (Cont.)
Valves, Fittings, & Specialties (Cont.)
Shock Absorber
House Trap
'P' Trap
Floor Drain
Indirect Waste
IW
Sanitary Below Grade
S
Sanitary Above Grade
S
Storm Below Grade
ST
Storm Above Grade
ST
Vent
Combination Waste & Vent
CWV
Acid Waste Below Grade
AW
Acid Waste Above Grade
AW
Acid Vent
AV
CW
Cold Water
HW
Hot Water
HWC
Hot Water Circulation
DWS
Drinking Water Supply
DWR
Drinking Water Return
G
Gas-Low Pressure
MG
Gas-Medium Pressure
HG
Gas-High Pressure
CA
Compressed Air
V
Vacuum
VC
Vacuum Cleaning
N
Nitrogen
N2O
Nitrous Oxide
O
Oxygen
LOX
Liquid Oxygen
LPG
Liquid Petroleum Gas
Valves, Fittings, & Specialties (Cont.)
Fire Protection Piping Symbols
Fire Protection Water Supply
F
Wet Standpipe
WSP
Dry Standpipe
DSP
Combination Standpipe
CSP
Automatic Fire Sprinkler
SP
Upright Fire Sprinkler Heads
Pendent Fire Sprinkler Heads
Fire Hydrant
Wall Fire Dept. Connection
Sidewalk Fire Dept. Connection
FHR
Fire Hose Rack
FHC
Surface Mounted Fire Hose Cabinet
FHC
Recessed Fire Hose Cabinet

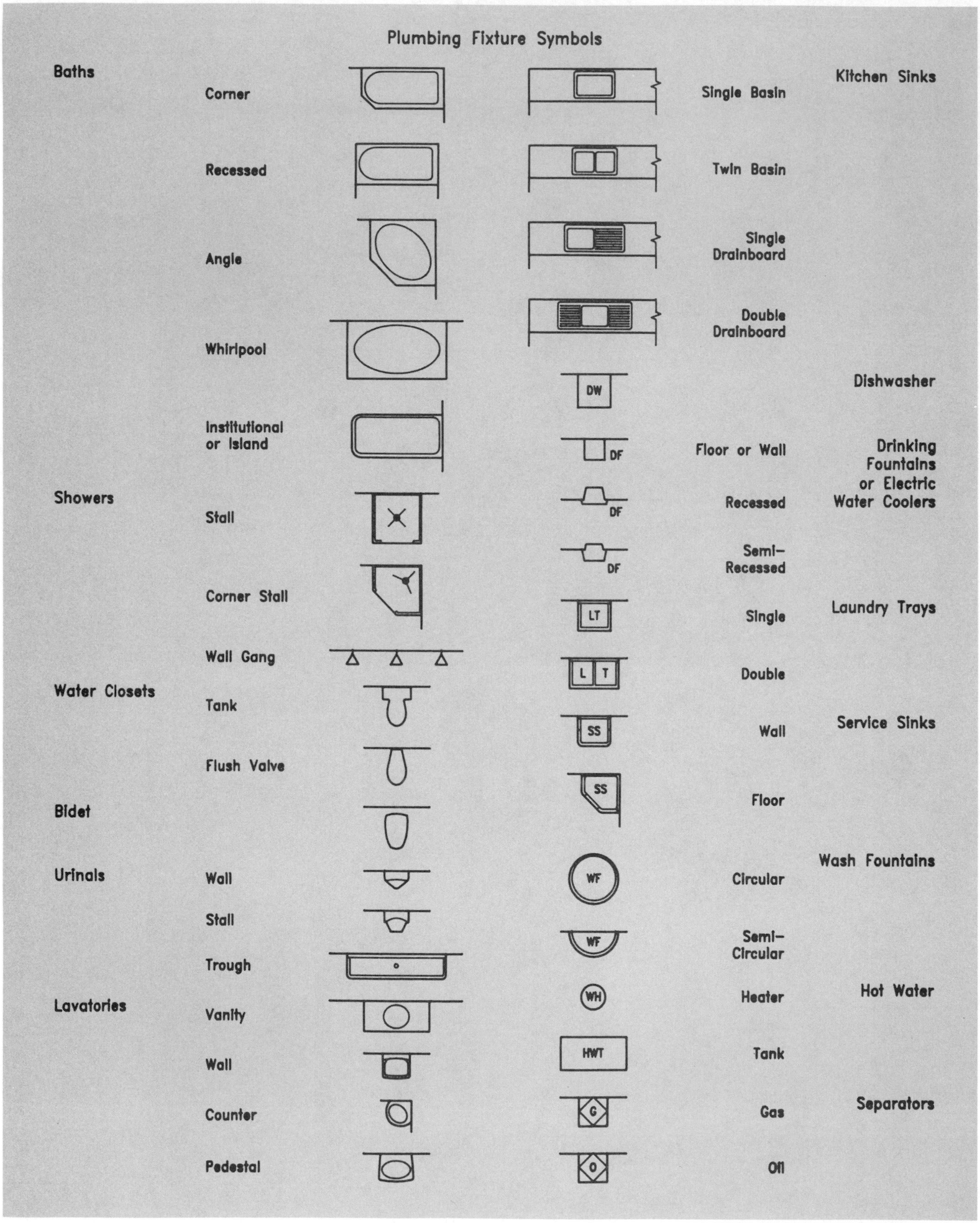

Plumbing Fixture Symbols

Baths	Corner
	Recessed
	Angle
	Whirlpool
	Institutional or Island
Showers	Stall
	Corner Stall
	Wall Gang
Water Closets	Tank
	Flush Valve
Bidet	
Urinals	Wall
	Stall
	Trough
Lavatories	Vanity
	Wall
	Counter
	Pedestal

Fixture	Type	Symbol text
Kitchen Sinks	Single Basin	
	Twin Basin	
	Single Drainboard	
	Double Drainboard	
Dishwasher		DW
Drinking Fountains or Electric Water Coolers	Floor or Wall	DF
	Recessed	DF
	Semi-Recessed	DF
Laundry Trays	Single	LT
	Double	L T
Service Sinks	Wall	SS
	Floor	SS
Wash Fountains	Circular	WF
	Semi-Circular	WF
Hot Water	Heater	WH
	Tank	HWT
Separators	Gas	G
	Oil	O

APPENDIX B

Appendix B includes a set of blueprints provided by Home Planners, Inc. The cover sheet provides a rendering of how the house will look once it is built. The foundation plan illustrates the complete foundation layout, including support walls and excavated and unexcavated areas. The floor plans show sections of the foundation, interior and exterior walls, floors, stairways, and roof details. Interior elevations show the design and placement of kitchen and bathroom cabinets, laundry areas, fireplaces, and other built-ins. Exterior elevations show the front, rear, and side views of the house.

Home Planners, Inc. is the leading supplier of residential home plans for consumers and builders in the United States and Canada.

EXPANDED DESIGN

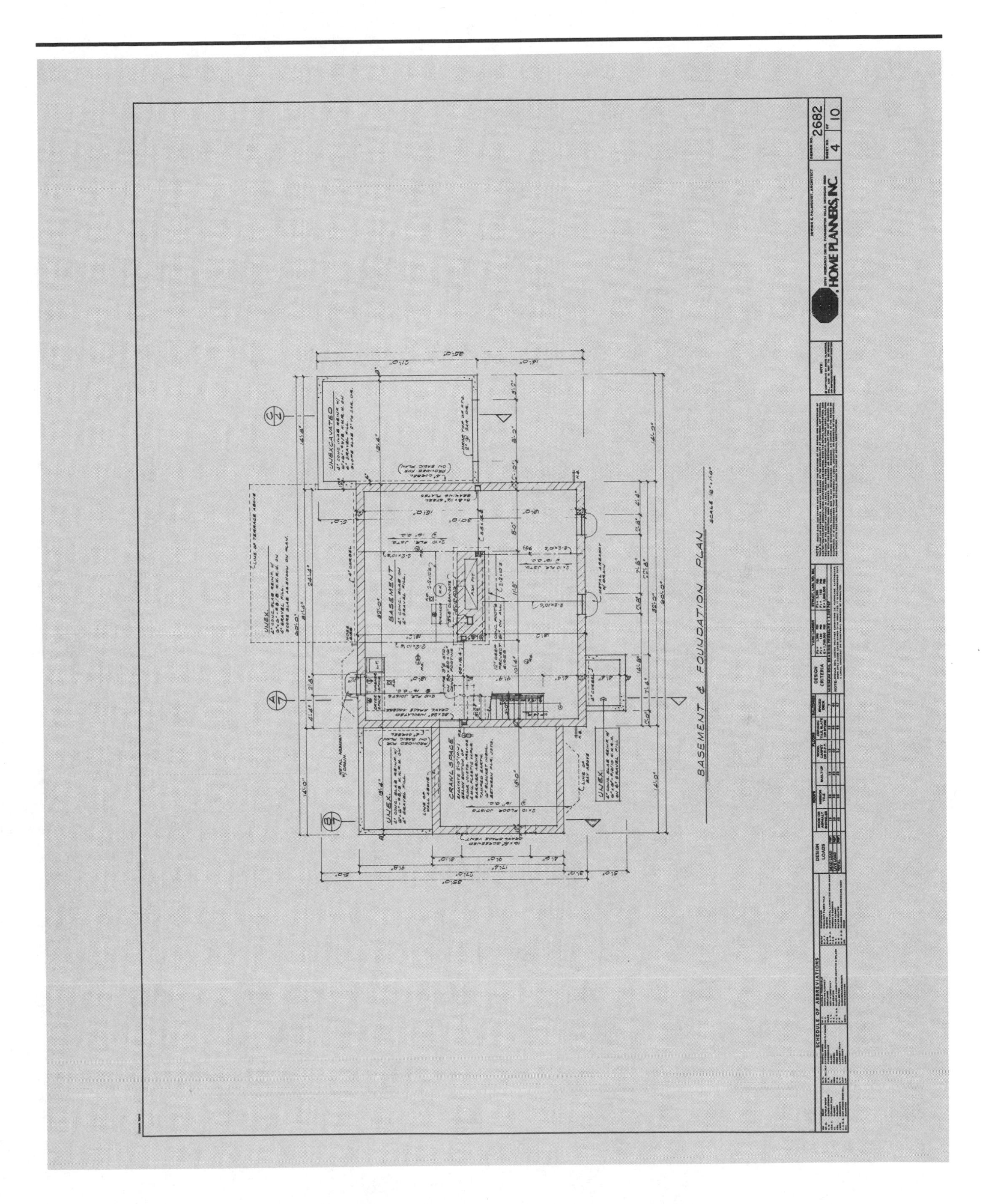
BASEMENT & FOUNDATION PLAN
BASEMENT
CRAWL SPACE
UNEXCAVATED
UNEX.
ASH PIT
SCHEDULE OF ABBREVIATIONS
DESIGN LOADS
DESIGN CRITERIA
HOME PLANNERS, INC.
2682
4
10

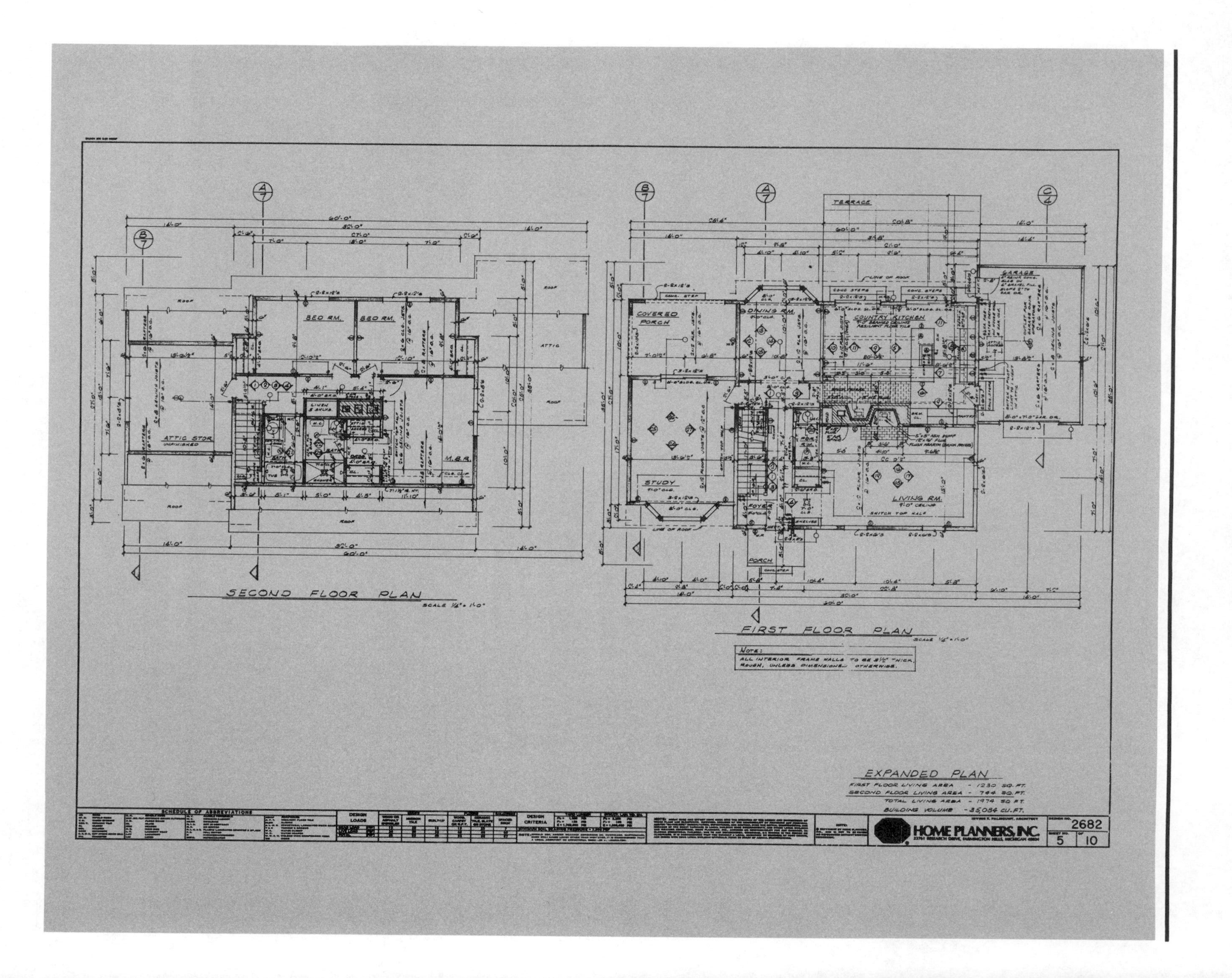
SECOND FLOOR PLAN
SCALE 1/4" = 1'-0"
FIRST FLOOR PLAN
SCALE 1/4" = 1'-0"
NOTE:
ALL INTERIOR FRAME WALLS TO BE 3 1/2" THICK, ROUGH, UNLESS DIMENSIONED OTHERWISE.
EXPANDED PLAN
FIRST FLOOR LIVING AREA - 1230 SQ. FT.
SECOND FLOOR LIVING AREA - 744 SQ. FT.
TOTAL LIVING AREA - 1974 SQ. FT.
BUILDING VOLUME - 35,084 CU. FT.
HOME PLANNERS, INC.
2682
5
10
SCHEDULE OF ABBREVIATIONS
DESIGN LOADS
DESIGN CRITERIA
TERRACE
GARAGE
COUNTRY KITCHEN
DINING RM.
COVERED PORCH
STUDY
FOYER
PORCH
LIVING RM.
BED RM.
BED RM.
M.B.R.
ATTIC STOR.
ATTIC
ROOF

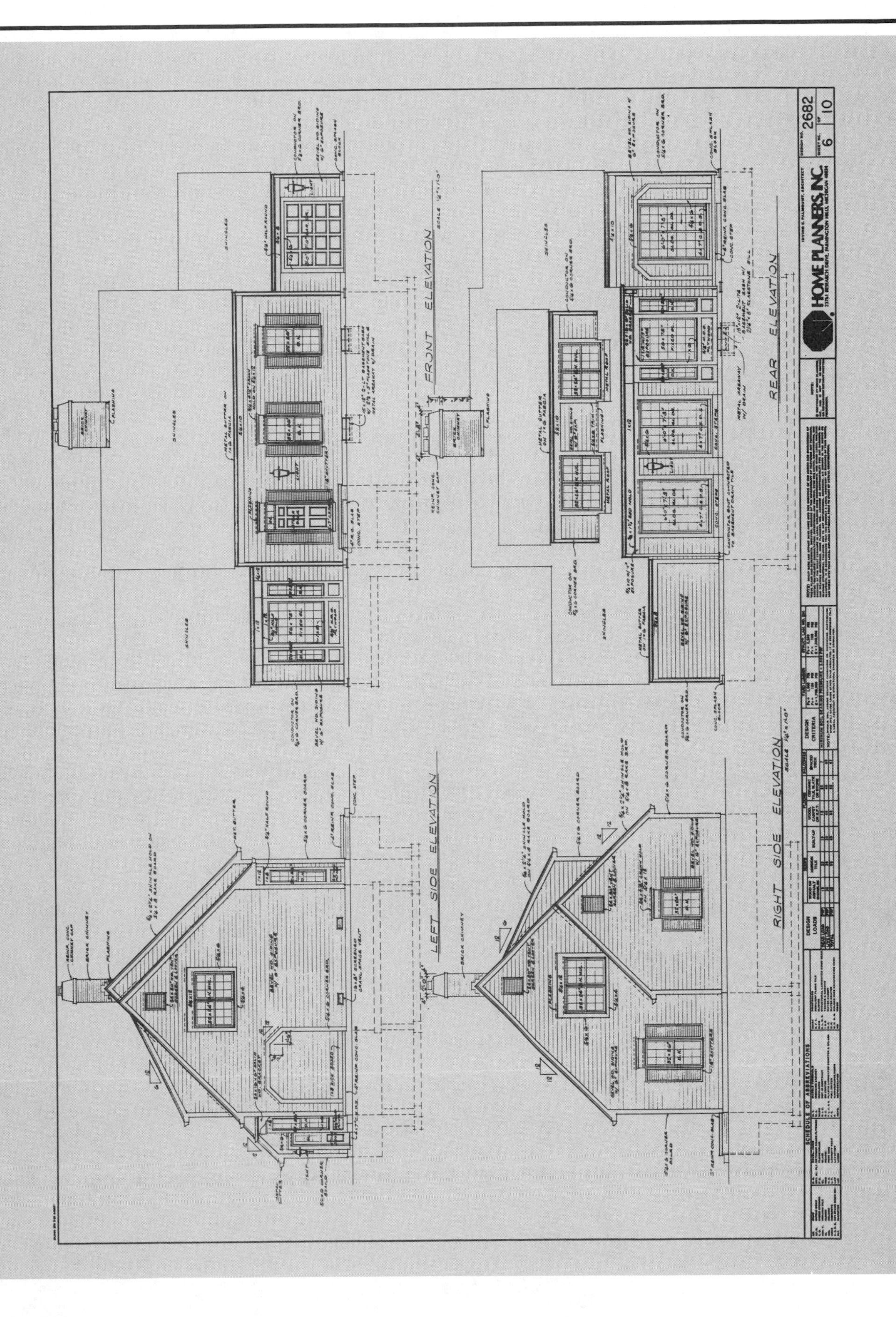
FRONT ELEVATION
REAR ELEVATION
LEFT SIDE ELEVATION
RIGHT SIDE ELEVATION
SCHEDULE OF ABBREVIATIONS
HOME PLANNERS, INC.
2682
6
10

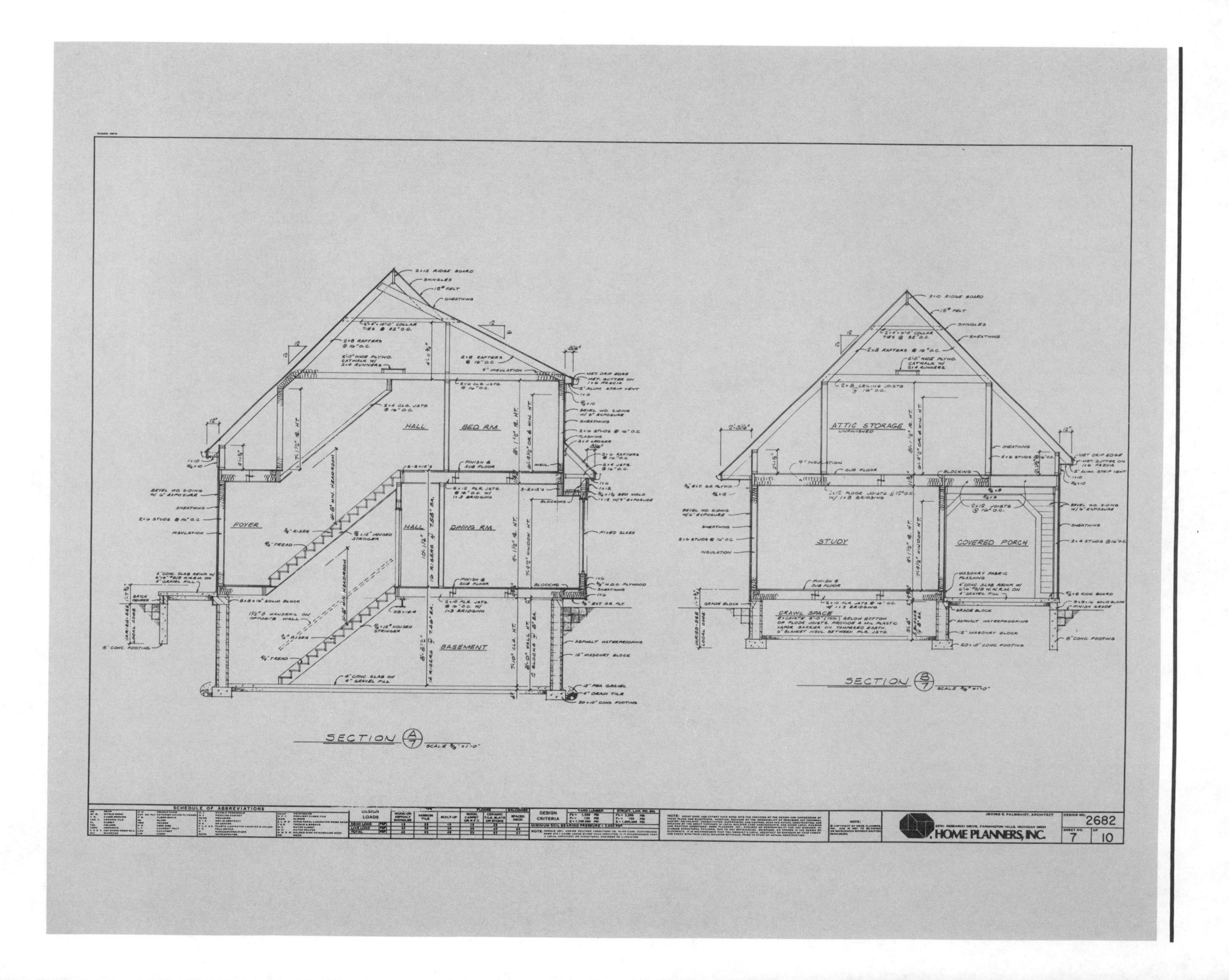
SECTION A/7
SECTION B/7
ATTIC STORAGE
UNFINISHED
STUDY
COVERED PORCH
CRAWL SPACE
HALL
BED RM.
HALL
DINING RM.
FOYER
BASEMENT
SCHEDULE OF ABBREVIATIONS
DESIGN LOADS
DESIGN CRITERIA
IRVING E. PALMQUIST, ARCHITECT
HOME PLANNERS, INC.
DESIGN NO. 2682
SHEET NO. 7 OF 10

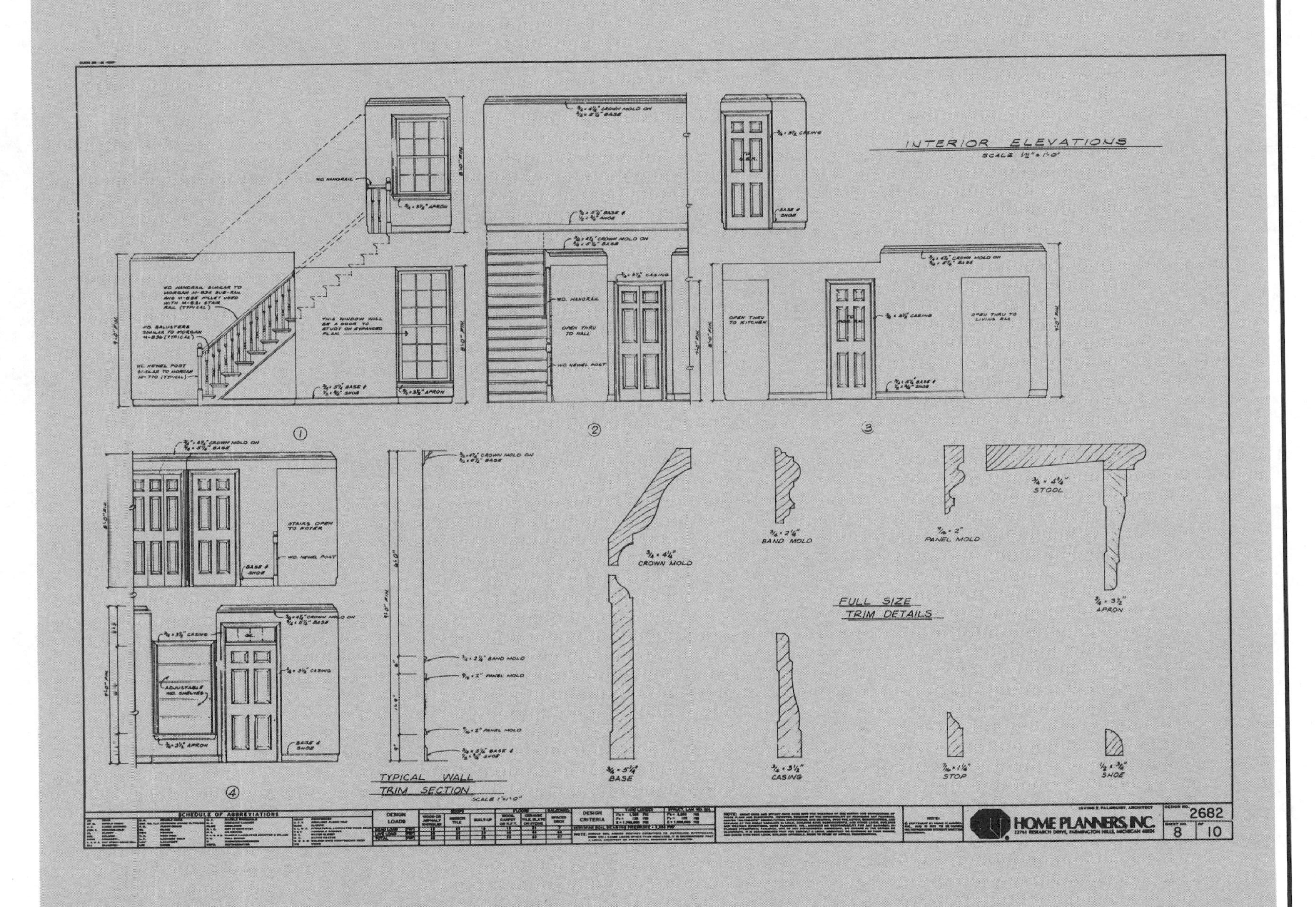
INTERIOR ELEVATIONS
SCALE 1/2" = 1'-0"
FULL SIZE TRIM DETAILS
TYPICAL WALL TRIM SECTION
3/4 × 4 1/4" CROWN MOLD
3/4 × 2 1/4" BAND MOLD
9/16 × 2" PANEL MOLD
3/4 × 4 3/4" STOOL
3/4 × 3 1/2" APRON
3/4 × 5 1/4" BASE
3/4 × 3 1/2" CASING
7/16 × 1 1/4" STOP
1/2 × 3/4" SHOE
SCHEDULE OF ABBREVIATIONS
HOME PLANNERS, INC.
2682
8
10

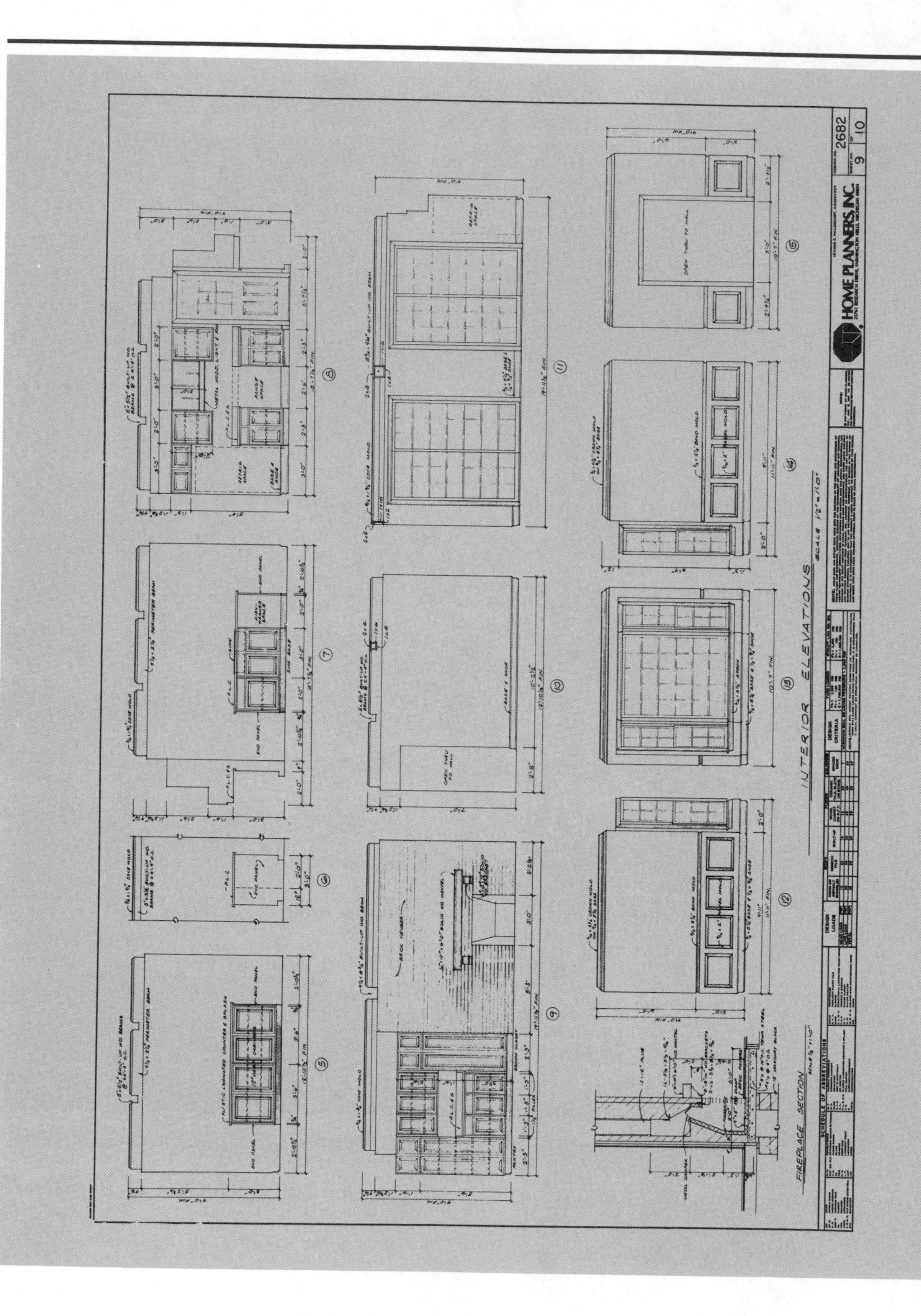
INTERIOR ELEVATIONS
FIREPLACE SECTION
SCHEDULE OF ABBREVIATIONS
DESIGN LOADS
DESIGN CRITERIA
HOME PLANNERS, INC.
2682
9
10

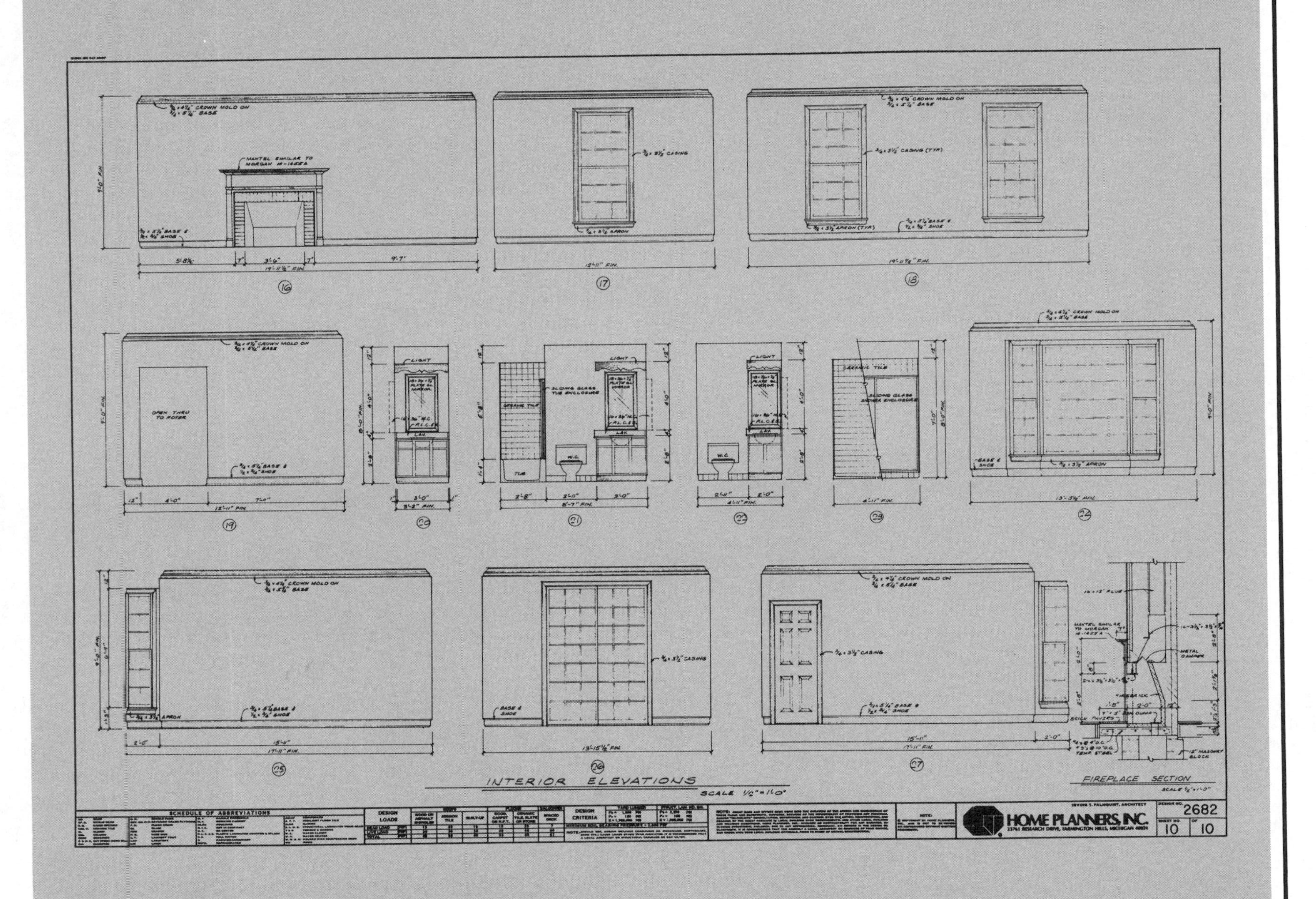
INTERIOR ELEVATIONS
SCALE 1/2"=1'-0"
FIREPLACE SECTION
SCHEDULE OF ABBREVIATIONS
DESIGN LOADS
DESIGN CRITERIA
HOME PLANNERS, INC.
23761 RESEARCH DRIVE, FARMINGTON HILLS, MICHIGAN 48024
2682
10
10

ABBREVIATIONS & BIBLIOGRAPHY

ABBREVIATIONS

AB	Anchor bolt
AC	Alternating current
A/C	Air-conditioning
ACC	Access
ACPL	Acoustical plaster
ACT	Acoustical ceiling tile
AD	Area drain
ADD	Addendum
ADH	Adhesive
ADJ	Adjacent
ADJT	Adjustable
A/E	Architect/Engineer
AFF	Above finished floor
AGG	Aggregate
ALUM	Aluminum
ALT	Alternate
ANC	Anchor
ANOD	Anodized
APPROX	Approximate
APT	Apartment
ARCH	Architect
ASB	Asbestos
ASPH	Asphalt
BD	Board
BEL	Below
BET	Between
BIT	Bituminous
CONC	Bituminous concrete
BLDG	Building
BLK	Block
BLKG	Blocking
BM	Beam or benchmark
BOF	Bottom of footing
BOTT	Bottom
BPL	Bearing plate
BRG	Bearing
BRK	Brick
BRZ	Bronze
BS	Both sides
BSMT	Basement
BUR	Built-up roofing
BVL	Beveled
BW	Both ways
CAB	Cabinet
CB	Catch basin
CEM	Cement
CER	Ceramic
CF	Cubic foot
CH	Ceiling height
CI	Cast iron
CIP	Cast in place
CIR	Circle
CIRC	Circumference
CJT	Control joint
CLG	Ceiling
HGT	Ceiling height
CLL	Contract limit line
CLO	Closet
CLR	Clear
CMU	Concrete masonry unit
CNTR	Counter
CO	Clean-out
COL	Column
COMB	Combination
COMP	Composition
CONC	Concrete
CONST	Construction
CONT	Continuous
CONTR	Contractor
CPT	Carpet
CRS	Course
CSMT	Casement
CT	Ceramic tile
CU FT	Cubic foot
CU YD, CY	Cubic yard
CW	Cold water
D	Drain
DA	Double acting
DC	Direct current
DEM	Demolish
DEP	Depressed

DET	Detail
DF	Drinking fountain
DH	Double hung
DIA	Diameter
DIAG	Diagonal
DIM	Dimension
DIV	Division
DL	Dead load
DMT	Demountable
DN	Down
DO	Ditto
DP	Dampproofing
DPR	Damper
DR	Door or dining room
DS	Downspout
DT	Drain tile
DTL	Detail or dentil
DW	Dishwasher or dumbwaiter
DWG	Drawing
DWR	Drawer
E	East
EA	Each
EF	Each face
EL	Elevation
ELEC	Electric
ELEV	Elevation or elevator
EMER	Emergency
ENCL	Enclosure
ENT	Entrance
EP	Electric panel
EQ	Equal
EQUIP	Equipment
ESC	Escalator
EWC	Electric water cooler
EXC	Excavate
EXG	Existing
EXH	Exhaust
EXP JT	Expansion joint
EXT	Exterior
FA	Fire alarm
FAS	Fasten
FB	Face brick
FBD	Fiber board
FBO	Furnished by others
FBRK	Fire brick
FD	Floor drain
FE	Fire extinguisher
FEC	Fire extinguisher cabinet
FIN FL	Finish floor
FFL	Finish floor line
FIN	Finish
FJT	Flush joint
FL	Flashing
FL CO	Floor clean-out
FLG	Flooring or flashing
FLR	Floor
FLUOR	Fluorescent
FLX	Flexible
FN	Fence
FND	Foundation
FOB	Freight on board
FOC	Face of concrete
FOF	Face of finish
FOM	Face of masonry
FOS	Face of stud
FP	Fireproof
FPL	Fireplace
FRT	Fire retardant
FS	Full size
FTG	Footing
FUR	Furred
FUT	Future
G	Gas
GA	Gauge
GAL	Gallon
GALV	Galvanized
GB	Grab bar
GC	General contractor
GI	Galvanized iron
GL	Glass or glazing
GL BK	Glass block
GRBM	Grade beam
GRD	Grade
GRN	Granite
GRT	Grout
GVL	Gravel
GYP	Gypsum
HA	Hot air
HB	Hose bib
HBD	Hardboard
HC	Hollow core
HCP	Handicap
HCWC	Handicap watercloset
HD	Heavy duty
HDR	Header
HDW	Hardware
HES	High early strength concrete
HGT	Height
HH	Handhole
HM	Hollow metal
HPT	High point
HOR	Horizontal
HT	Height
HTG	Heating
HVAC	Heating, ventilating, and air conditioning
HW	Hot water
HWD	Hardwood
HWH	Hot water heater
HWT	Hot water tank
I	Iron
ID	Inside diameter
IN	Inches
INCL	Included
INSUL	Insulation
INT	Interior
INV	Invert

IPS	Iron pipe size
JC	Janitor's closet
JF	Joint filler
JMB	Jamb
JST	Joist
JT	Joint
KD	Knocked-down
KIT	Kitchen
KO	Knock-out
KPL	Kick plate
KW	Kilowatt
L	Length
LAB	Laboratory
LAD	Ladder
LAM	Laminated
LAV	Lavatory
LBL	Label
LF	Linear foot
LH	Left hand
LIN	Linen closet
LINO	Linoleum
LL	Live load
LPT	Low point
LT	Light
LTL	Lintel
LVL	Laminated veneer lumber
LVR	Louver
LW	Lightweight
M	Meter
MAS	Masonry
MAX	Maximum
MC	Medicine cabinet
MDO	Medium density overlay
MECH	Mechanical
MED	Medium
MFR	Manufacturer
MIN	Minimum
MIR	Mirror
MISC	Miscellaneous
MKWK	Millwork
MLDG	Moulding
MM	Millimeter
MMB	Membrane
MO	Masonry opening
MOR	Mortar
MRB	Marble
MRD	Metal roof deck
MTD	Mounted
MTL	Metal
MTL FR	Metal Frame
MULL	Mullion
N	North
N/A	Not applicable
NIC	Not in contract
NO	Number
NOM	Nominal
NRC	Noise reduction coefficient
NTS	Not to scale
OA	Overall
OC	On center
OD	Outside diameter
OH	Overhead
OJ	Open web joist
OPG	Opening
OPH	Opposite hand
OPP	Opposite
PAR	Parallel
PBD	Particle board
PCC	Precast concrete
PCF	Pounds per cubic foot
PERF	Perforated
PL	Plate
PLAM	Plastic laminate
PLF	Pounds per linear foot
PNT	Painted
PT	Pressure treated
PTD	Painted or paper towel dispenser
PTN	Partition
PRM	Perimeter
PSF	Pounds per square foot
PSI	Pounds per square inch
PVC	Polyvinyl choride
PVMT	Pavement
PWD	Plywood
QT	Quarry tile
R	Riser
R	Radius
RA	Return air
RB	Rubber base
RBT	Rubber tile
RCP	Reinforced concrete pipe
RD	Roof drain
REINF	Reinforced
REF	Reference
REFR	Refrigerator
REG	Register
REM	Remove
REQ'D	Required
RES	Resilient
RET	Return
REV	Revised
RFG	Roofing
RFL	Reflected
RH	Right hand
RM	Room
RO	Rough opening
RVS	Reverse
RWC	Rain water conductor
S	South
SC	Solid core
SCH	Schedule
SD	Storm drain
SF	Square foot

SEC	Section	TSL	Top of slab
SHO	Shoring	TOS	Top of steel
SHT	Sheet	TOW	Top of wall
SHTHG	Sheathing	TYP	Typical
SIM	Similar	UC	Undercut
SKL	Skylight	UNEXC	Unexcavated
SNT	Sealant	UNO	Unless noted otherwise
SPEC	Specifications	UNF	Unfinished
SPK	Speaker	V	Volts
SQ	Square	VAT	Vinyl asbestos tile
SS	Sanitary sewer	VB	Vapor barrier
SST, S/S	Stainless steel	VCT	Vinyl composition tile
ST	Steel	VERT	Vertical
STA	Station	VIF	Verify in field
STC	Sound transmission coefficient	VG	Vertical grain
STD	Standard	VT	Vinyl tile
STO, STOR	Storage	VPB	Veneer plywood
STRUCT	Structural	VNR	Veneer
SUS	Suspended	W	West
SYM	Symmetrical	WC	Water closet
SYN	Synthetic	WD	Wood
SYP	Southern yellow pine	WG	Wire glass
SYS	System	WH	Wall hung
T	Tread	WI	Wrought iron
TB	Towel bar	W/	With
TEL	Telephone	W/O	Without
T&G	Tongue and groove	WP'G	Waterproofing
THK	Thick	WSCT	Wainscoting
THR	Threshold	WTW	Wall to wall
TOL	Tolerance	WWF	Welded wire fabric
TPD	Toilet paper dispenser	WWM	Welded wire mesh
TR	Transom		

BIBLIOGRAPHY

Basics for Builders: Framing and Rough Carpentry. By Scot Simpson. R.S. Means Company, Inc., 1991.

"Construction Documents Fundamentals and Formats Module" of the *Manual of Practice.* Construction Specifications Institute (Alexandria, VA: Construction Specifications Institute, 1992), p. FF/030.8.

MasterFormat. Construction Specifications Institute and Construction Specifications Canada. (Alexandria, VA: Construction Specifications Institute, 1988), pp. 10-12.

Means Assemblies Cost Data 1994. R.S. Means Company, Inc., 1994.

Means Electrical Estimating: Standards and Procedures. R.S. Means Company, Inc., 1986.

Means Estimating Handbook. R.S. Means Company, Inc., 1990.

Means Forms for Building Construction Professionals. R.S. Means Company, Inc., 1986.

Means Forms for Contractors. R.S. Means Company, Inc., 1990.

Means Illustrated Construction Dictionary, Condensed Edition. R.S. Means Company, Inc., 1991.

Means Illustrated Construction Dictionary, New Unabridged Edition. R.S. Means Company, Inc., 1991.

Means Interior Estimating. R.S. Means Company, Inc., 1987.

Means Landscape Estimating. Second Edition. By Sylvia H. Fee. R.S. Means Company, Inc., 1991.

Means Mechanical Estimating: Standards and Procedures. Second Edition. R.S. Means Company, Inc., 1992.

Means Plumbing Estimating. By Joseph J. Galeno and Sheldon T. Greene. R.S. Means Company, Inc., 1991.

Means Scheduling Manual. Third Edition. By F. William Horsley. R.S. Means Company, Inc., 1991.

Means Site Work & Landscape Cost Data 1994. R.S. Means Company, Inc., 1994.

Means Structural Steel Estimating. By S. Paul Bunea, PhD, PE. R.S. Means Company, Inc. 1987.

Roofing: Design Criteria, Options, Selection. By R. D. Herbert, III. R.S. Means Company, Inc., 1989.

INDEX

INDEX

D

N

P

R